Using Information Technology

Technology

A Practical Introduction to Computers & Communications

Brief Version

Second Edition

Using Information Technology

A Practical Introduction to Computers & Communications

Brief Version

Second Edition

Stacey C. Sawyer

Brian K. Williams

Sarah E. Hutchinson

IRWIN

Chicago • Bogotá • Boston • Buenos Aires • Caracas
London • Madrid • Mexico City • Sydney • Toronto

Publisher: Tom Casson
Sponsoring editor: Garrett Glanz
Editorial assistant: Carrie Berkshire
Marketing manager: Michelle Hudson
Development: Burrston House, Ltd., Burr Ridge, IL
Project supervisor: Gladys True
Production supervisor: Bob Lange
Project editor, layout, & production management: Stacey C. Sawyer,
    Sawyer & Williams, Incline Village, NV
Copy editor: Anita Wagner, Takoma Park, MD
Multimedia coordination: Lew Gossage, Bill Bayer, and Brian Nacik
Design coordination: Laurie Entringer, Matthew Baldwin
Designer: Ellen Pettengell
Cover Illustration: © Dick Palulian/SIS
Cover Design: Matthew Baldwin
Photo research: Monica Suder, San Francisco
Art & composition: GTS Graphics, Commerce, CA
Prepress buyer: Charlene R. Perez
Permissions: David Sweet, San Francisco
Compositor: GTS Graphics
Typeface: 10/12 Trump Mediaeval
Printer: Quebecor Printing/Dubuque

ISBN 0-256-20980-4

Library of Congress Cataloging-in-Publication Data

Sawyer, Stacey C.
    Using information technology: a practical introduction to computers & communications / Stacey C. Sawyer, Brian K. Williams, Sarah E. Hutchinson. — Brief 2nd ed.
       p.   cm.
    ISBN   0-256-20980-4
    Includes index.
    1. Computers.   2. Telecommunications systems.   3. Information technology.
I. Williams, Brian K., 1938-   II. Hutchinson, Sarah E. III. Title
QA76.5.S2193   1997
004 —dc20   96–31006

Photo and other credits are listed in the back of the book.

Printed in the United States of America
2 3 4 5 6 7 8 9 0 QD 3 2 1 0 9 8 7

Brief Contents

Preface to the Instructor

The Promises of This Book

USING INFORMATION TECHNOLOGY: *A Practical Introduction to Computers & Communications—Brief Version*, SECOND EDITION, is intended for use as a concepts textbook to accompany a one-semester or one-quarter introductory course on computers or microcomputers. The **key features** are as follows. We offer:

1. **Careful revision in response to extensive instructor feedback.**
2. **Brief coverage of computers and information systems.**
3. **Emphasis on unification of computer and communications systems.**
4. **Emphasis throughout on ethics.**
5. **Use of techniques for reinforcing student learning.**
6. **Up-to-the-minute material—in the book and on our Web site.**

We elaborate on these features next.

Key Feature #1: Careful Revision in Response to Extensive Instructor Feedback

We were delighted to learn from our publisher that the First Edition of USING INFORMATION TECHNOLOGY, in both versions, was apparently the most successful new text in the field in 1995, with over 300 schools adopting it. An important reason for this success, we feel, was all the valuable contributions of the reviewers.

Both the printed version of the First Edition and the manuscript and proofs of the SECOND EDITION underwent a highly disciplined and wide-ranging reviewing process. This process of expert appraisal drew on instructors who were both users and nonusers, who were from a variety of educational institutions, and who expressed their ideas in both written form and in focus groups.

We have sometimes been overwhelmed with the amount of information, but we have tried to respond to all consensus criticisms and countless individual suggestions. It is not an exaggeration to say that every page of the SECOND EDITION has been influenced by instructor feedback. The result, we think, is **a book reflecting the wishes of most instructors.** In particular, we have addressed the following matters:

- **Old Chapters 1 and 2 combined into one:** We combined old Chapters 1 and 2 into a single chapter, so that there would be less introductory material for the student to get through. Some instructors had found some of the old introductory material too technical for a first chapter and had expressed the wish to have the "overview" material moved closer to the beginning. The new chapter reflects their wishes.

- **Input and output material made two chapters:** The single chapter "Input & Output" became two chapters because of the amount of new material that has become available.

- **New chapter added on information systems and software development:** *New to this edition!* Some users and reviewers said they needed coverage of information systems, systems analysis and design, programming, and languages. Thus, we have added the short chapter "Systems: Development, Programming, & Languages." Other users may find no need for this chapter.

In addition to these major structural changes, we have made hundreds of line-by-line and word-by-word adjustments to conform with instructors' requests.

Key Feature #2: Brief Coverage of Computers & Information Systems

This text is a brief version of our more comprehensive book by the same title. The purpose of the *Brief Version* is to offer essential coverage of hardware, software, telecommunications, and societal issues related to computers and communications systems. In this way, we hope to offer a concepts book for instructors teaching abbreviated courses or those whose principal emphasis is on software tutorials and labs.

Chapters are organized according to the topic coverage of traditional introductory computer texts. Thus, most instructors can continue to follow their present course outlines.

NOTE: The text allows for **a good deal of instructor flexibility.** After Chapter 1, the remaining 9 chapters may be taught in any sequence, or selectively omitted, at the instructor's discretion. To make this possible, the authors have occasionally **repeated the definitions of key terms throughout the text** (also a part of the book's deliberate strategy of reinforcement).

The end-of-chapter essay appearing in the Experience Box is optional material but may be assigned if the instructor wishes.

Key Feature #3: Emphasis on Unification of Computers & Communications

The text emphasizes the technological merger of the computer, communications, consumer electronics, and media industries through the exchange of information in the digital format used by computers. This is the relatively new phenomenon known as **technological convergence.**

This theme covers much of the technology currently found under such phrases as *the Information Superhighway, the Multimedia Revolution,* and *the Digital Age: mobile computing, the Internet, Web search tools, online services, workgroup computing, the virtual office, video compression, PC/TVs, information appliances,* "intelligent agents," and so on.

The theme of convergence is given in-depth treatment in five chapters—the introduction, systems software, communications, storage and databases, and challenges and promises (Chapters 1, 3, 7, 8, 10)—and is also brought out in examples throughout other chapters.

Key Feature #4: Emphasis Throughout on Ethics

New to this edition! Many texts discuss ethics only once, usually in one of the final chapters. We believe this topic is too important to be treated last or lightly. Thus, **we cover ethical matters in numerous places** throughout the book, as indicated by the special sign shown here in the margin. For example, the all-important question of what kind of software can be legally copied is discussed in Chapter 2 ("Applications Software"), an appropriate place for students just starting software labs. Other ethical matters discussed are the manipulation of truth through digitizing of photographs, intellectual property rights, netiquette, censorship, privacy, and computer crime.

A list of pages of ethics coverage appears on the inside cover. Instructors wishing to teach all ethical matters as a single unit may refer to this list.

Key Feature #5: Reinforcement for Learning

Having individually or together written over a dozen successful textbooks and scores of labs, the authors are vitally concerned with reinforcing students in acquiring knowledge and developing critical thinking. Accordingly, we offer the following **to provide learning reinforcement:**

- **Interesting writing:** Studies have found that textbooks **written in an imaginative style** significantly improve students' ability to retain information. Thus, the authors have employed a number of journalistic devices—such as the colorful fact and the apt direct quote—to make the material as interesting as possible. We also use real anecdotes and examples rather than fictionalized ones.

- **Key terms and definitions in boldface:** Each **key term AND its definition is printed in boldface** within the text, in order to help readers avoid any confusion about which terms are important and what they actually mean.

- **Learning objectives to aid students:** ***New to this edition!*** Lists of learning objectives at the start of chapters are common in textbooks—and most students simply skip them. Because we believe learning objectives are excellent instruments for reinforcement, we have crafted ours to make them more helpful to students. We do this by **tying the numbered learning objectives to the end-of-chapter summary.** That is, we have numbered the objectives. Then, in the summary at the end of the chapter, we have given corresponding numbers to the terms and concepts that relate to the particular objectives.

   For example, in Chapter 2, *Learning Objective 2* is "After reading this chapter, you should be able to: 2. Discuss the ethics of copying software." Terms and concepts appearing in the end-of-chapter summary that relate to this objective—such as "copyright," "freeware," and "intellectual property"—are identified with the notation *LO 2.*

- **"Preview & Review" presents abstracts of each section for learning reinforcement:** Each main section heading throughout the book is followed by **an abstract or précis entitled Preview & Review.** This enables the student to get a *preview* of the material before reading it and then to *review* it afterward, for maximum learning reinforcement.

- **Innovative chapter summaries for learning reinforcement:** ***New to this edition!*** The chapter summary is especially innovative—and especially helpful to students. In fact, research through student focus groups has shown that this format was clearly first among five different choices of summary formats. Each concept is discussed under **two columns, headed "What It Is/What It Does" and "Why It's Important."**

Each concept or term is also given a cross-reference page number that refers the reader to the main discussion within the chapter.

In addition, as mentioned, the term or concept is also given a number (such as *LO 1, LO 2,* and so on) corresponding to the appropriate learning objective at the beginning of the chapter.

- **Cross-referencing system for key terms and concepts:** *New to this edition!* Wherever important key terms and concepts appear throughout the text that students might need to remind themselves about, we have added **"check the cross reference"** information, as in: (✓ p. 111). In student focus groups during the last two years, this device was found to rank *first* out of 20-plus study/learning aids.

- **Material in "bite-size" portions:** Major ideas are presented **bite-size form,** with generous use of advance organizers, bulleted lists, and new paragraphing when a new idea is introduced.

- **Short sentences:** Most sentences have been kept short, the majority not exceeding **22–25 words** in length.

- **Innovative use of art:** Artwork in the book is designed principally to be **didactic.** There are no unnecessary space-filling photo "galleries," for instance. To support learning concepts, photographs are often coupled with *additional* information—an elaboration of the discussion in the text, some how-to advice, an interesting quotation, or a piece of line art.

- **End-of-chapter exercises:** For practice purposes, students will benefit from *several exercises* at the end of each chapter: **short-answer questions, fill-in-the-blank questions, multiple-choice questions,** and **true-false questions.** Answers to selected exercises appear in the back of the book.

  In addition, we present several **projects/critical-thinking questions,** generally of a practical nature, to help students absorb the material.

- **Internet exercises:** *New to this edition!* In keeping with the practical and communications orientation of the book, we present **exercises on the use of the Internet** at the end of every chapter.

  Internet connection and software requirements: In general, the exercises assume an Internet setup is readily available to most college students. In some chapters, we assume students connect to the Internet using a command-line Unix interface. In others, we assume students use the Navigator Netscape 2.0 or 2.01 Web browser. If these are not compatible with your setup, please check out Irwin's UIT Web site *(http://www.irwin.com/cit/uit)* for information regarding the publisher's additional Internet offerings.

Key Feature #6: Up-to-the-Minute Material—in the Text & on the Irwin Web Site

New to this edition! The number of technological developments that have occurred since we wrote the First Edition has been awesome, and every day seems to bring reports of something new and important. As we write this, our 1996 publication date is only three months away. However, because our publisher has allowed us to do several steps concurrently (writing, reviewing, editing, production), readers will find several score 1996 references in the notes to the book. As evidence for our being current, our text includes coverage of the following material:

Advanced TV. AT&T Internet offer. Cable modems. DVD-ROMs. E-money. Internet PC. Intranets. Java. Microsoft Exchange. Microsoft Internet Explorer. PC/TVs. Telecommunications Act of 1996. 3-D displays. 3-D sound. V chip. VRML. Web indexes. Windows 95 (latest information). Yahoo. Zip drives. . . . And more.

Still, we recognize that a Gutenberg-era lag exists between our last-minute scribbling and the book's publication date. And of course we also realize that fast-moving events will unquestionably overtake some of the facts in this book by the time it is the student's hands. Accordingly, after publication we are periodically offering instructors updated material and other interaction on the Irwin UIT Web Site (**http://www.irwin.com/cit/uit**).

Complete Course Solutions: Supplements That Work—Four Distinctive Offerings

It's not important how many supplements a textbook has but whether they are truly useful, accurate, and of high quality. Irwin presents **four distinctive kinds of supplement offerings** to complement the text:

1. **Application software tutorials**
2. **Interactive software**
3. **Classroom presentation software**
4. **Instructor support materials, including software support program**

We elaborate on these below.

Supplement Offering #1: Application Software Tutorials—Two Types

Our publisher, Richard D. Irwin, offers two different series of application software tutorials, or lab manuals, which present two different hands-on approaches to learning software. An Irwin sales representative can explain the specific software covered in each series.

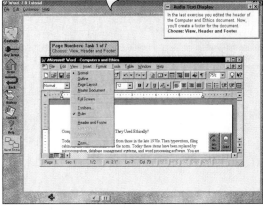

New for fall of '96: Advantage Interactive Software Tutorials for Windows '95.

- **The Irwin Advantage Series for Computer Education:** Written by Sarah E. Hutchinson and Glen J. Coulthard, the *Irwin Advantage Series for Computer Education* covers the complete Microsoft Office Professional with your choice of either one comprehensive spiral-bound package or individual editions featuring full-color layouts and large screen captures.

    Manuals are available for Microsoft *Windows 3.1* and *Windows 95*, as well as for Microsoft *Word, Excel, Access,* and *PowerPoint,* both for Windows 3.1 and for Windows 95. A manual called *Integrated Microsoft Office* is also available for Windows 3.1 and Windows 95.

    Each tutorial leads students through step-by-step instructions not only for the most common methods of executing commands but also for alternative methods. Each lesson begins with a case scenario and concludes with case problems, showing the real-world application of the software. Quick Reference guides summarizing important functions and shortcuts appear throughout. Annotated Toolbar screen shots provide easy and quick reference. Boxes introduce unusual functions that will enhance the user's productivity. Hands-on exercises and short-answer questions allow students to practice their skills.

• **The Irwin Effective Series:**   Written specifically for the first-time computer user, by Fritz J. Erickson and John A. Vonk, the **Irwin Effective Series** is based on the premise that success breeds confidence and confident students learn more effectively. Exercises embedded within each lesson allow students to experience success before moving on to a more advanced topic. The "why" as well as the "how" is always carefully explained. Each lesson features several applications projects and a comprehensive problem for student solution.

Manuals are available for Microsoft *Windows 3.1* and *Windows 95,* as well as for Microsoft *Word, Excel, Access,* and *Works* for both Windows 3.1 and Windows 95 and for *PowerPoint* for Windows 95.

Important Note—Custom Publishing: The contents of these products can be tailored to meet your course needs through **custom publishing.** Titles or specific lessons from several titles in these series can be combined. *Irwin will happily send you an examination copy of the custom-published text you want so you can see exactly what your students will get.* Ask your Irwin sales representative for details.

Supplement Offering #2: Interactive Software— Two Types

Irwin offers two types of interactive software to accompany the text—*Info Tech, Version 2.0* and *Internet: A Knowledge Odyssey.*

• **Info Tech, Version 2.0:**   Developed by Irwin New Media and Tony Baxter, **Info Tech Version 2.0** is a revised, updated, and expanded version of interactive multimedia software provided with the first edition of this text. The CD-ROM gives students several self-paced learning modules, on topics ranging from applications software to networks. Combining text, illustrations, and animation, the Info Tech interactive tool may also be used by instructors in a lecture setting.

Info Tech 2.0 includes coverage of:

Data Into Information	*Data Representation*
Application Software	*Networks*
User Interfaces	*The Internet*
Processors	*Querying a Database*
Secondary Storage	*Client/Server*
Peripheral Devices	*Encryption/Decryption*
Backing Up Data	*Security* . . . and more.
Multimedia	

Each module provides three levels of learning: (1) The *introduction level* provides text and animated enhancement of computer concepts. (2) the *exploratory level* allows the user to experiment with various scenarios and see the immediate results. (3) The *practice level* poses cases and problems for which the user must provide solutions based on information learned in the first two levels.

System requirements: (a) IBM PC or compatible with at least 2 MB of RAM running Windows 3.1 or Windows 95, or (b) Macintosh with at least 2 MB of RAM running System 6.01 or later; CD-ROM drive.

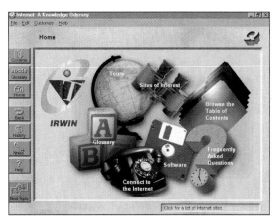

- **Internet: A Knowledge Odyssey:** A multimedia CD-ROM developed by MindQ Publishing, Inc., **Internet: A Knowledge Odyssey,** *Business Edition,* explains the history and workings of the Internet through 14 self-paced, interactive guided tours, 400 hyperlinked glossary terms, 90 minutes of video clips, and optional audio narration. Available to students for less than $10 (when packaged with other Irwin products), the CD-ROM also allows students to connect directly to specific sites on the Internet, automatically launching their Internet applications and inserting the correct URL address. Details and an evaluation copy are available from an Irwin representative.

    System requirements: IBM PC or compatible 486SX or higher with at least 4 MB of RAM running Windows 3.1 or Windows 95, SVGA graphics card (256 colors), and CD-ROM drive.

Supplement Offering #3: Classroom Presentation Software—Two Options

To help instructors enhance their lecture presentations, Irwin makes available software in two options, in a PowerPoint version and an Astound version. Both segment the course into seven topical modules: *introductory topics, hardware, software, communications, systems development, database and information management,* and *additional topics.*

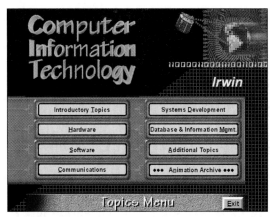

- **Irwin CIT Classroom Presentation Software—Astound option:** Developed by the Quality & Excellence Institute and Linda Behrens, the *Astound version* of classroom presentation software is a CD-ROM that allows instructors to make true multimedia lecture-enhancement presentations. A flexible menu-driven tool, the Astound version offers a complete lecture outline and a navigation interface, topical menus, animations, audio, and video clips. An Irwin sales representative can provide a demonstration of this tool.

    System requirements: IBM PC or compatible running Windows 3.1 or higher with a '486 processor (66 MHz or higher recommended) and at least 8 MB of RAM, SVGA graphics capability, and at least a dual-speed CD-ROM drive.

- **Irwin CIT Classroom Presentation Software—PowerPoint option:** Also from the same developers, the **PowerPoint version** of this classroom presentation software is a graphics-intensive set of lecture slides that helps instructors explain topics that may otherwise be difficult to present. The PowerPoint version, available on several 3.5-inch diskettes, offers lecture outline material with various graphics, backgrounds, and transitions.

System requirements: IBM PC or compatible with at least 2 MB of RAM running Windows 3.1. An LCD panel is needed if the images are to be shown to a large audience.

Supplement Offering #4: Instructor Support Materials

We offer the instructor the following other kinds of supplements and support to complement the text:

- **Instructor's Resource Guide:** This complete guide, prepared by Linda Behrens, supports instruction in any course environment. The **Instructor's Resource Guide** includes: *a student questionnaire, course planning and evaluation grid, suggestions for writing course objectives, suggested pace and coverage for courses of various lengths, suggestions for using the exercises in various class structures,* and *projects for small and large classes.*

  For each chapter, the IRG provides an overview, chapter outline, lecture notes, notes regarding the boxes (README boxes) from the text, solutions, and suggestions and additional information to enhance your course.

- **Color transparencies:** There are **150 full-color transparency acetates** available to the instructor. Transparencies have been specially *upsized*—enlarged and enhanced for clear projection.

- **Test bank:** The test bank, prepared by Margaret Batchelor, contains **2000 different questions,** which are directly referenced to the text. Specifically, it contains *true/false, multiple-choice,* and *fill-in questions,* categorized by difficulty and by type; *short-essay questions; sample midterm exam; sample final exam;* and *answers to all questions.*

- **Computerized testing software:** Called *Computest,* Irwin's popular computerized testing software is a **user-friendly, menu-driven, microcomputer-based test-generating system** that is free to qualified adopters. Containing all the questions from the test bank described above, Computest's Version 4 allows instructors to customize test sheets, entering their own questions and generating review sheets and answer keys.

  Available for DOS, Windows, and Macintosh formats, Computest has advanced printing features that allow instructors to print all types of graphics; Windows and Macintosh versions use easily remembered icons. All versions support over 250 dot-matrix and laser printers.

  System requirements: (a) IBM PC or compatible with at least 2 MB of RAM running Windows 3.1 or (b) Macintosh with at least 2 MB of RAM running System 6.01 or later; CD-ROM drive or 3.5-inch diskette drives.

- **Videos:** A broad selection from **21 new video segments** of the acclaimed PBS television series, *Computer Chronicles,* is available. Each video is approximately 30 minutes long. The videos cover topics ranging from computers and politics, to CD-ROM, to visual programming languages, to the Internet.

- **Instructor's data disks:** Instructor's data disks are avaliable for instructors whose students are using the tutorials for software education in the Irwin Advantage Series for Computer Education and Irwin Effective Series. These are **diskettes containing files** used in the DOS-, Windows 3.X, and Windows 95–based software labs. Specifically, the diskettes contain the letters and memos that the student will use in the word processing labs, sample budgets and other files that the student will retrieve and modify in the spreadsheet labs, and the data and reports that the student will work with in the database labs.

- **Phone, fax, and e-mail instructor support services:** Richard D. Irwin's College New Media Department offers **telephone- or computer-linked support services to instructors** in matters related to Irwin software, such as Computest and data disks used for the student tutorials in the Irwin Advantage Series or Effective Series. Software support analysts are available to help solve technical questions not covered in the documentation for any Irwin software product.

  Three kinds of support are offered: (1) toll-free telephone numbers, available 9:00 A.M. to 5 P.M. Central, Monday through Friday (except holidays); (2) support-on-demand FAX-BACK service, available 24 hours a day, seven days a week; (3) e-mail, accessible 24 hours a day. Directions for getting this support appear in the *Instructor's Resource Guide.*

- **Irwin Web site:** It's appropriate that a text with a strong communications focus also find a way to employ the new communications technologies available. Accordingly, a text-specific Irwin UIT **Web Site (http://www.irwin.com/cit/uit)** has been developed as a place to go for periodic updates of text material, relevant links, downloads of supplements, an instructors' forum for sharing information with colleagues, and other value-added features.

Acknowledgments

Three names are on the front of this book, but a great many others are powerful contributors to its development.

First among the staff of Richard D. Irwin is our sponsoring editor, Garrett Glanz, our lifeline, who did a fantastic job of supporting us and of coordinating the many talented people whose efforts on development and supplementary materials help strengthen our own. Garrett, you've been sensational. We're also grateful to Garrett's predecessor, Paul Ducham, for his initial spadework on this edition. It should go without saying that we owe a lot to our publisher, Tom Casson, who was not only the midwife on the first edition but has been one of our great fans on the second. Tom, it was good to feel your presence here.

We also appreciate the cheerfulness and efficiency of other people in Tom's and Garrett's group—the good and great Michelle Hudson, Carrie Berkshire, Linda Eiermann, and Sharon Pass. Mike Beamer did yeoman work on advertising pieces for us. Irwin's top management—the very supportive John Black, as well as David Littlehale, Jerry Saykes, and Jeff Sund—actively backed our revision, and we are extremely grateful to them. Many others at Irwin have also closely assisted us, and we would like to single out Laurie Entringer, Matt Baldwin, and Gladys True, who worked actively with us. In addition, we owe Bill Bayer, Lew Gossage, Charlie Hess, Bob Lange, Merrily Mazza, Keith McPherson, Evelyn Mosley-Harris, Brian Nacik, and Charlene Perez our special thanks. We appreciate having you all in our corner.

We would like to thank every single one of the Irwin sales reps, who did such an outstanding job on our behalf in promoting the first edition. We would particularly like to thank regional managers Jimmy Bartlett and Bob Bryan and district managers Barbara Anson, Bunny Barr, Greg Bowman, Frank Chihowski, Jr., Chad Douglas, Denise Mariani, and Jerry Swanson, plus manager for in-house sales Jean Geracie. By name we need to single out Irwin reps Rob Brown, Gary Rodgers, Steve Edmonson, Bill Firth, Liz Lindboe-Mulcahy, Julie Daniels, Stewart Mattson, Kitty Cavanaugh, Carol Preston, and Tony Noel, all of whom were honored at the national sales meeting for their spectacular efforts. Many thanks!

Outside of Irwin we were fortunate to find ourselves in a community of first-rate publishing professionals. We are ecstatic fans of the editorial development company of Burrston House, Ltd., and their active participants on this book, the highly experienced Glenn and Meg Turner and the very hard-working and always supportive Cathy Crow.

Two-thirds of the author team would like once again to sing the praises of the third, Stacey Sawyer, who not only co-wrote this book but also massaged in all the reviewers' comments, picked the photos, conceptualized and laid out much of the art, remade the pages, and directed the production of the entire enterprise under truly frantic deadlines—an exhausting business. Stacey, believe us when we say this book couldn't have happened were it not for you. Thanks for everything.

Stacey worked with freelance designer Ellen Pettengell, who bore with us through all the picky changes that we raised in order to try to get close to what we wanted. Ellen, we appreciate your efforts and patience.

Again, we can't say enough for Monica Suder, our photo researcher. Monica once again had to work with extremely tight deadlines and, miracle of miracles, came up with all the new photos that, we hope, make this book didactic and interesting. Photographer Frank Bevans fleshed out what was missing by supplying us with some outstanding photos.

Anita Wagner, an old publishing colleague and top-notch copy editor, performed her usual careful scrutiny of the manuscript. Standing behind Anita were our able proofreaders Kathy Pope and Linda Smith, who saved us from ourselves by removing inconsistencies and other potential embarrassments. Lois Oster, one of the finer talents in her line of work, did her customary unbeatable job of indexing. David Sweet did his always highly competent job of obtaining permissions.

GTS Graphics turned in their usual top-drawer performance in handling prepress production. We especially want to thank Elliott and Bennett Derman, Gloria Fontana, and their dynamite production coordination team of Donna Machado and Angie Armendarez. We also want to express our delight with the illustration program they did for us, with the creative art planning by Charlene Locke and renderings by Yvonne Welch.

Finally, the authors are grateful to a number of people for their superb work on the ancillary materials. They include Anthony Baxter, who helped develop Info Tech 2.0; MindQ Publishing, which developed Internet: A Knowledge Odyssey, Business Edition; The Quality & Excellence Institute and Linda Behrens, who developed both the PowerPoint and Astound versions of the Irwin CIT Classroom Presentation Software; Linda Behrens, again, who prepared the Instructor's Resource Guide; and Margaret Batchelor, who created the Test Bank.

Acknowledgment of Focus Group Participants, Survey Respondents, & Reviewers

We are grateful to the following people for their participation in focus groups, response to surveys, or reviews on manuscript drafts or page proofs of all or part of the book. We cannot overstate their importance and contributions in helping us to make this the most market-driven book possible.

Focus Group Participants

Patrick Callan
Concordia University

Joe Chambers
Triton College

Hiram Crawford
Olive Harvey College

Edouard Desautels
University of Wisconsin–Madison

William Dorin
Indiana University–Northwest

Bonita Ellis
Wright City College

Charles Geigner
Illinois State University

Julie Giles
DeVry Institute of Technology

Dwight Graham
Prairie State College

Stan Honacki
Moraine Valley Community College

Tom Hrubec
Waubonsee Community College

Alan Iliff
North Park College

Julie Jordahl
Rock Valley College

John Longstreet
Harold Washington College

Pattie Riden
Western Illinois University

Behrooz Saghafi
Chicago State University

Naj Shaik
Heartland Community College

Survey Respondents

Nancy Alderdice
Murray State University

Margaret Allison
University of Texas-Pan American

Angela Amin
Great Lakes Junior College

Connie Aragon
Seattle Central Community College

Gigi Beaton
Tyler Junior College

William C. Brough
University of Texas–Pan American

Jeff Butterfield
University of Idaho

Helen Corrigan-McFadyen
Massuchusetts Bay Community College

James Frost
Idaho State Universtiy

Candace Gerrod
Red Rocks Community College

Julie Heine
Southern Oregon State College

Jerry Humphrey
Tulsa Junior College

Jan Karasz
Cameron University

Alan Maples
Cedar Valley College

Norman Muller
Greenfield Community College

Paul Murphy
Massachusetts Bay Community College

Sonia Nayle
Los Angeles City College

Janet Olpert
Cameron University

Pat Ormond
Utah Valley State College

Marie Planchard
Massachusetts Bay Community College

Fernando Rivera
University of Puerto Rico–Mayaguez Campus

Naj Shaik
Heartland Community College

Jack Shorter
Texas A&M University

Randy Stolze
Marist College

Ron Wallace
Blue Mountain Community College

Steve Wedwick
Heartland Community College

Reviewers

Nancy Alderdice
Murray State University

Sharon Anderson
Western Iowa Tech Community College

Bonnie Bailey
Morehead State University

David Brent Bandy
University of Wisconsin–Oshkosh

Robert Barrett
Indiana University Purdue University at Fort Wayne

Anthony Baxter
University of Kentucky

Virginia Bender
William Rainey Harper College

Warren Boe
University of Iowa

Randall Bower
Iowa State University

Phyllis Broughton
Pitt Community College

J. Wesley Cain
City University, Bellevue

Judy Cameron
Spokane Community College

Kris Chandler
Pikes Peak Community College

William Chandler
University of Southern Colorado

John Chenoweth
East Tennessee State University

Ashraful Chowdhury
Dekalb College

Erline Cocke
Northwest Mississippi Community College

Robert Coleman
Pima County Community College

Glen Coulthard
Okanagan University

Robert Crandall
Denver Business School

John Durham
Fort Hays State University

John Enomoto
East Los Angeles College

Ray Fanselau
American River College

Eleanor Flanigan
Montclair State University

Ken Frizane
Oakton Community College

James Frost
Idaho State University

Jill Gebelt
Salt Lake Community College

Charles Geigner
Illinois State University

Frank Gillespie
University of Georgia

Myron Goldberg
Pace University

Sallyann Hanson
Mercer County Community College

Albert Harris
Appalachian State University

Jan Harris
Lewis & Clark Community College

Michael Hasset
Fort Hays State University

Martin Hochhauser
Dutchess Community College

James D. Holland
Okaloosa-Waltoon Community College

Wayne Horn
Pensacola Junior College

Christopher Hundhausen
University of Oregon

Jim Johnson
Valencia Community College

Jorene Kirkland
Amarillo College

Victor Lafrenz
Mohawk Valley Community College

Stephen Leach
Florida State University

Chang-Yang Lin
Eastern Kentucky University

Paul Lou
Diablo Valley College

Deborah Ludford
Glendale Community College

Peter MacGregor
Estrella Mountain Community College

Donna Madsen
Kirkwood Community College

Kenneth E. Martin
University of North Florida

Curtis Meadow
University of Maine

Marty Murray
Portland Community College

Charles Nelson
Rock Valley College

Wanda Nolden
Delgado Community College

E. Gladys Norman
Linn-Benton Community College

John Panzica
Community College of Rhode Island

Rajesh Parekh
Iowa State University

Merrill Parker
*Chattanooga State Technical
Community College*

Leonard Presby
William Patterson State College

Delores Pusins
Hillsborough Community College

Eugene Rathswohl
Universtiy of San Diego

Jerry Reed
Valencia Community College

John Rezac
Johnson County Community College

Jane Ritter
University of Oregon

Stan Ross
Newbury College

Al Schroeder
Richland College

Earl Schweppe
University of Kansas

Tom Seymour
Minot State University

Elaine Shillito
Clark State Community College

Denis Titchenell
Los Angeles City College

Jim Vogel
Sanford Brown College

Dale Walikainen
Christopher Newport University

Reneva Walker
Valencia Community College

Patricia Lynn Wermers
North Shore Community College

Edward Winter
Salem State College

Floyd Winters
Manatee Community College

Eileen Zisk
Community College of Rhode Island

Write to Us

We welcome your response to this book, for we are truly trying to make it as useful as possible. Write to us in care of Garrett Glanz, Editor, Richard D. Irwin, 1333 Burr Ridge Parkway, Burr Ridge, IL 60521 or via his e-mail: *citmail@irwin.com* (or directly to the authors at *76570.1533@ compuserve.com*).

Stacey C. Sawyer
Brian K. Williams
Sarah E. Hutchinson

Contents

Chapter 6

Output: The Many Uses of Computers & Communications 161

Chapter 8

Communications: Starting Along the Information Highway   231

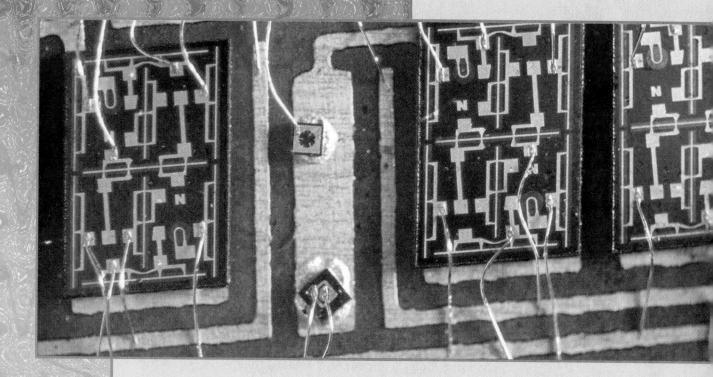

CHAPTER 1

The Digital Age
Overview of the Revolution in Computers & Communications

Concepts You Should Know

After reading this chapter, you should be able to:

1. Define the terms *information technology*, *technological convergence*, *computer*, and *communications*

2. Briefly define *analog* and *digital*

3. Identify the six major elements of a computer-and-communications system

4. Describe the difference between an information technology professional and an end-user

5. Define *data* and *information*

6. Briefly explain the five operations of a computer-and-communications system: input, processing, output, storage, and communications, as well as the corresponding categories of hardware

7. Explain the difference between applications software and systems software

8. Identify the five major categories of computers

9. Discuss the important trends in computer technology

10. Define the term *Information Superhighway*

We should all be concerned about the future," said engineer and inventor Charles Kettering, "because we will have to spend the rest of our lives there."

This book is about your future, a future rapidly becoming present. It is about a revolution that will make—indeed, is making now—profound changes in your life. The revolution has many names: The Digital Age. The Information Age. The Age of Convergence. The Interactive Revolution. The Multimedia Revolution. The Information Superhighway—or "Infobahn" or I-way or Dataway. Whatever it's called, the revolution is happening in all parts of society and in all parts of the world, and its consequences will reverberate throughout our lifetimes.

The technological systems and industries that this revolution is bringing forth may seem awesomely complex. However, the concept on which they are based is as simple as the flick of a light switch: *on* and *off*. Let us begin to see how this works.

From the Analog to the Digital Age: The "New Story" of Computers & Communications

Preview & Review: Information technology is technology that merges computers and high-speed communications links. The fusion of computer and communications technologies is producing "technological convergence."

Computers are based on digital, binary (two-state) signals. However, most phenomena in the world are analog, representing continuously variable quantities. Some formerly analog devices are now taking digital form. To transmit a computer's digital signals over an analog telephone line requires a modem. To digitize analog sound or images, as in recording a live performance for a CD, requires sampling and averaging.

The essence of all revolution, stated philosopher Hannah Arendt, is the start of a *new story* in human experience. For us, the new story may be said to have begun in 1991. In that year, according to one report, "companies for the first time spent more on computing and communications gear . . . than on industrial, mining, farm, and construction machines." It adds: "Info tech is now as vital . . . as the air we breathe."[1]

"Info tech"—information technology—is what this book is about. **Information technology is technology that merges computing with high-speed communications links carrying data, sound, and video.**[2] The arrival of information technology is having powerful consequences, the most notable being the gradual fusion of several important industries in a phenomenon that has been called *technological convergence.*

What Is "Technological Convergence"?

Technological convergence, **also known as** *digital convergence,* **is the technological merger of several industries through various devices that exchange information in the electronic, or digital, format used by computers. The industries are computers, communications, consumer electronics, entertainment, and mass media.**

Technological convergence has tremendous significance. It means that, from a common electronic base, information can be communicated in all the ways we are accustomed to receiving it. These include the familiar media of newspapers, photographs, films, recordings, radio, and television. However, it can also be communicated through newer technology—satellite, fiber-optic cable, cellular phone, fax machine, or compact disk, for example. More important, as time goes on, *the same information may be exchanged among many kinds of equipment, using the language of computers.* Understanding this shift from single, isolated technologies to a unified digital technology means understanding the effects of this convergence on your life—such as:

* The increased need for continuous learning
* Adapting to less well-defined jobs as an "information worker"
* The stepped-up pace of change
* Exposure to relatively unregulated technical and social information from other cultures via global networks.

Is this consolidation of technologies an overnight phenomenon? Actually, it has been developing over several years, as we explain next. (■ *See Panel 1.1.*)

The Fusion of Computer & Communications Technologies

Technological convergence is derived from a combination of two recent technologies—*computer* and *communications.*

* **Computer technology:** Is there anyone reading this book who has not seen a computer by now? Nevertheless, let's define what it is. **A *computer***

PANEL 1.1

Fusion of computer and communications technology

Today's new information environment came about gradually from two separate streams of technological development. (*Continued on pages 4–8.*)

Computer Technology

3000 BC	200 BC
Abacus, used for arithmetic calculations, developed in Orient	Chinese artisans develop an entire mechanical orchestra

Communications Technology

35,000 BC	4000 BC	3000 BC	1800 BC	600 BC	1453 AD
Language probably existed	Sumerian writing on clay tablets	Early Egyptian hieroglyphics	Phoenician alphabet	Book printing in China	First book printed in Europe

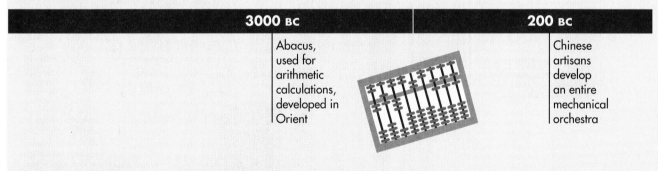

is a programmable, multiuse machine that accepts data—raw facts and figures—and processes, or manipulates, it into information we can use, such as summaries or totals. Its purpose is to speed up problem solving and increase productivity.

If you've actually touched a computer it's probably been a personal computer, such as the widely advertised desktop or portable models from Apple, IBM, Compaq, or Packard Bell. However, many other machines, such as microwave ovens and portable phones, use miniature electronic processing devices (microprocessors, or microcontrollers) similar to those that control personal computers.

- **Communications technology:** Unquestionably you've been using communications technology for years. *Communications, or telecommunications, technology* **consists of electromagnetic devices and systems for communicating over long distances.** The principal examples are telephone, radio, television, and cable.

Before the 1950s, computer technology and communications technology developed independently, like rails in a railroad track that never merge. Since then, however, they have gradually fused together, producing a new information environment.

Why have the worlds of computers and of telecommunications been so long in coming together? The answer is this: *computers are digital, but most of the world is analog.* Let us explain what this means.

PANEL 1.1
Continued

Computer Technology

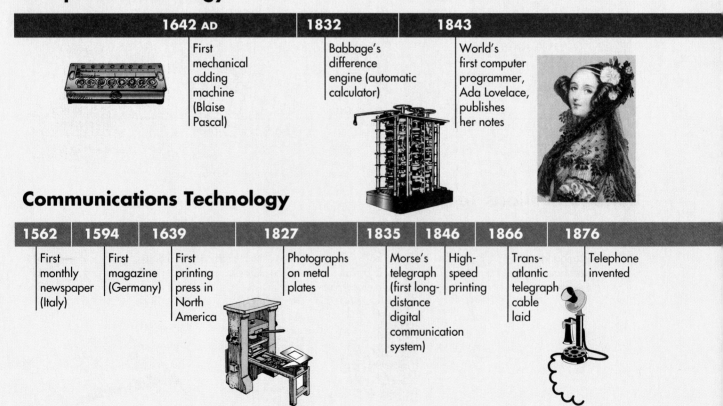

	1642 AD	1832	1843
	First mechanical adding machine (Blaise Pascal)	Babbage's difference engine (automatic calculator)	World's first computer programmer, Ada Lovelace, publishes her notes

Communications Technology

1562	1594	1639	1827	1835	1846	1866	1876
First monthly newspaper (Italy)	First magazine (Germany)	First printing press in North America	Photographs on metal plates	Morse's telegraph (first long-distance digital communication system)	High-speed printing	Trans-atlantic telegraph cable laid	Telephone invented

The Digital Basis of Computers

Computers may seem like incredibly complicated devices, but their underlying principle is simple. When you open up a personal computer, what you see is mainly electronic circuitry. And what is the most basic statement that can be made about electricity? It is simply this: it can be either *turned on* or *turned off*, or switched between *high voltage* and *low voltage.*

With a two-state on/off, high/low, open/closed, present/absent, positive/negative, yes/no arrangement, one state can represent a 1 digit, the other a 0 digit. People are most comfortable with the *decimal system,* which has ten digits (0, 1, 2, 3, 4, 5, 6, 7, 8, 9). Because computers are based on on/off or other two-state conditions, they use the **binary system, which consists of only two digits—0 and 1.**

The word *digit* simply means numeral. The word *digital* is derived from "digit," referring to the fingers people used to count with. Today, however, ***digital* is almost synonymous with "computer-based." More specifically, it refers to communications signals or information represented in a discrete (individually distinct) form—usually in a binary or two-state way.**

In the binary system, **each 0 or 1 is called a** *bit*—**short for** *b*inary dig*it*. In turn, bits can be grouped in various combinations to represent characters of data—numbers, letters, punctuation marks, and so on. For example, the letter *H* could correspond to the electronic signal 01001000 (that is, off-on-off-off-on-off-off-off). In computing, **a group of eight bits is called a** *byte.*

***Digital data*, then, consists of data in discrete, discontinuous form—usually 0s and 1s.** This is the method of data representation by which computers process and store data and communicate with each other.

Computers are digital [handwritten annotation]

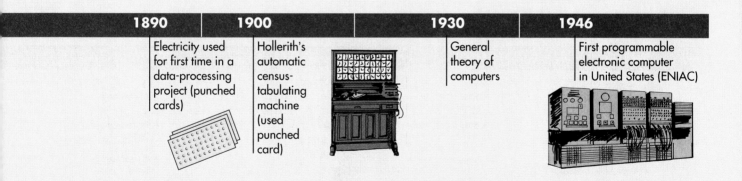

1890	1900		1930	1946
Electricity used for first time in a data-processing project (punched cards)	Hollerith's automatic census-tabulating machine (used punched card)		General theory of computers	First programmable electronic computer in United States (ENIAC)

1888	1894	1895	1912	1915	1928	1939	1946	1947
Radio waves identified	Edison makes a movie	Marconi develops radio; motion-picture camera invented	Motion pictures become a big business	AT&T long-distance service reaches San Francisco	First TV demonstrated; first sound movie	Commercial TV broad-casting	Color TV demon-strated	Transistor invented

The Analog Basis of Life

Most phenomena of the world are *analog,* **having continuously variable values.** Sound, light, temperature, and pressure values, for instance, can fall anywhere on a continuum or range. The highs, lows, and in-between states have historically been represented with analog devices rather than in digital form. Examples of analog devices are humidity recorder, thermometer, and pressure sensor, which can measure continuous fluctuations. Thus, *analog data* **is transmitted in a continuous form that closely resembles the information it represents.** The electrical signals on a telephone line are analog-data representations of the original voices. Telephone, radio, television, and cable-TV have traditionally transmitted analog data.

The differences between analog and digital transmission are apparent when you look at a drawing of a wavy analog signal, such as a voice message appearing on a standard telephone line, and an on/off digital signal. **In general, to transmit your computer's digital signals over telephone lines, you need to use a *modem* to translate them into analog signals.** (■ *See Panel 1.2.*)

The modem provides a means for computers to communicate with one another while the old-fashioned copper-wire telephone network—an analog system that was built to transmit the human voice—still exists. Our concern, however, goes far beyond telephone transmission. How can the analog realities of the world be expressed in digital form? How can light, sounds, colors, temperatures, and other dynamic values be represented so that they can be manipulated by a computer? Let us consider this.

PANEL 1.1

Continued

Computer Technology

1952	1964	1970	1971	1977
UNIVAC computer correctly predicts election of Eisenhower as U.S. President	IBM introduces 360 line of computers	Microprocessor chips come into use; floppy disk introduced for storing data	First pocket calculator	Apple II computer (first personal computer sold in assembled form)

Communications Technology

1950	1952	1957	1961	1968	1975	1976	1977
Cable TV	Direct-distance dialing (no need to go through operator); transistor radio introduced	First satellite launched (Russia)	Push-button telephones	Portable video recorders; video cassettes	Flat-screen TV	First wide-scale marketing of TV computer games (Atari)	First inter-active cable-TV

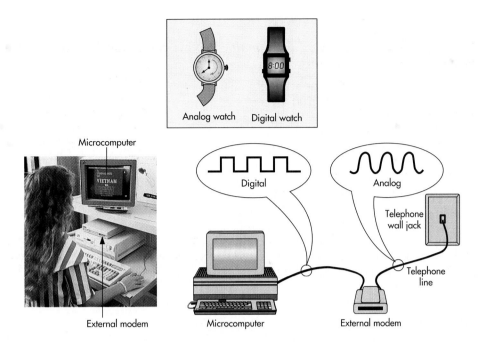

PANEL 1.2

Analog versus digital signals, and the modem

(*Top*) On an analog watch, the hands move continuously around the watch face; on a digital watch, the display changes once each minute. (*Bottom*) Note the wavy line for an analog signal and the on/off (or high-voltage/low voltage) line for a digital signal. The modem shown here is outside the computer. Today most modems are inside the computer and not visible.

Converting Reality to Digital Form: Sampling & Averaging

Suppose you are using an analog tape recorder to record a singer during a performance. The analog process will produce a near duplicate of the sounds. This will include distortions, such as buzzings and clicks. It will also include aberrations introduced by the recording process itself.

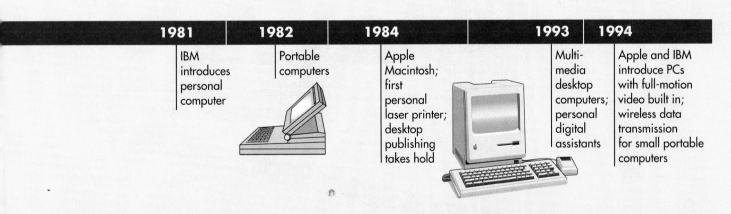

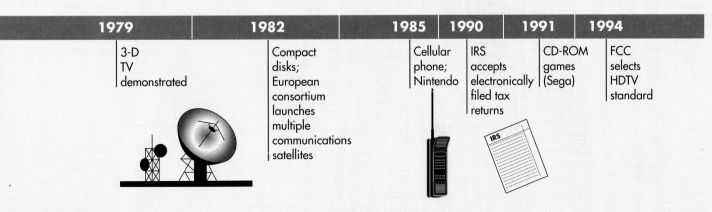

The digital recording process is different. The way in which music is captured for music CDs (compact disks) does not provide a duplicate of a musical performance. Rather, the digital process uses *sampling and averaging* to record the sounds and produce a copy that is virtually exact and free from distortion and noise. Sounds are sampled by computer-based equipment at regular intervals—nearly 44,100 times a second. They are then converted to numbers and averaged. Similarly, for visual material, values such as brightness or color can also be sampled and then averaged. The same is true of other aspects of real-life experience, such as pressure, temperature, and motion.

Are we being cheated out of experiencing "reality" by allowing computers to do sampling and averaging of sounds, images, and so on? Actually, people willingly made this compromise years ago, before computers were invented. Movies, for instance, carve up reality into 24 frames a second. Television frames are drawn at 30 lines per second. These processes happen so quickly that our eyes and brains easily jump the visual gaps. Digital processing of analog experience represents just one more degree of compromise.

Let us now look at how a computer-and-communications system works.

The Six Elements of a Computer-&-Communications System

Preview & Review: A computer-and-communications system has six elements: (1) people, (2) procedures, (3) data/information, (4) hardware, (5) software, and (6) communications.

PANEL 1.1
Continued

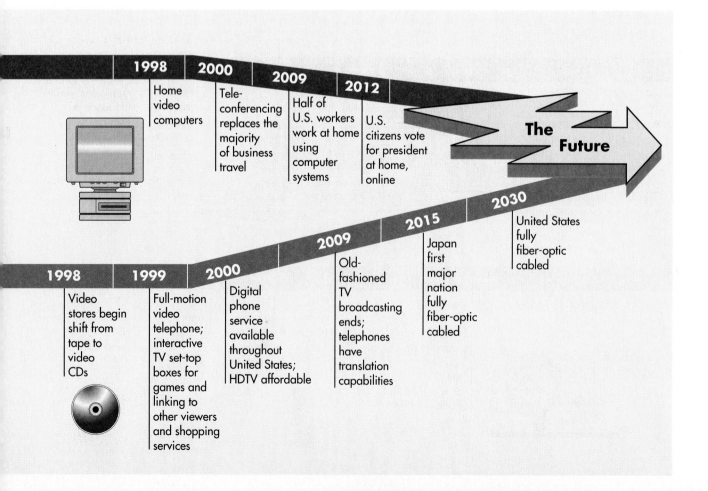

1998	2000	2009	2012
Home video computers	Tele-conferencing replaces the majority of business travel	Half of U.S. workers work at home using computer systems	U.S. citizens vote for president at home, online

The Future

1998	1999	2000	2009	2015	2030
Video stores begin shift from tape to video CDs	Full-motion video telephone; interactive TV set-top boxes for games and linking to other viewers and shopping services	Digital phone service available throughout United States; HDTV affordable	Old-fashioned TV broadcasting ends; telephones have translation capabilities	Japan first major nation fully fiber-optic cabled	United States fully fiber-optic cabled

A *system* is a group of related components and operations that interact to perform a task. A system can be many things: registration day at your college, the 52 bones in the foot, a weather storm front, the monarchy of Great Britain. Here we are concerned with a technological kind of system. **A computer-and-communications system is made up of six elements: (1) people, (2) procedures, (3) data/information, (4) hardware, (5) software, and (6) communications. (■ *See Panel 1.3.*)** We briefly describe these elements in the next six sections and elaborate on them in subsequent chapters.

ppdhsc

System Element 1: People

Preview & Review: People are the most important part of, and the beneficiaries of, a computer-and-communications system.

There are two types of people using information technology—professionals and end-users.

People can analyze, develop, and improve computer systems. They can also complicate the operation of a system.

Although we will not say a lot about them here, people are the most important part of a computer-and-communications system. People of all levels and skills, from novices to programmers, are the users and operators of the system. The whole point of the system, of course, is to benefit people.

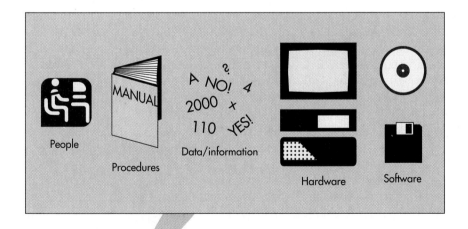

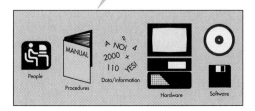

PANEL 1.3

A computer-and-communications system

The six elements of the system include people, procedures, data/information, hardware, software, and communications.

Two Types of Users: Professionals & End-Users

Two types of people use information technology—*professionals* and *"end-users."*

- **Professionals:** An *information technology professional* is a person who has had formal education in the technical aspects of using a computer-and-communications system. For example, a *computer programmer* creates the programs (software) that process the data in a computer system.

- **End-users:** An "end-user" is a person probably much like yourself. An *end-user,* or simply a *user,* is someone without much technical knowledge of information technology who uses computers for entertainment, education, or work-related tasks. The user is not a technology expert but knows enough about it to use it for his or her own purposes, such as for career advancement. Lawyers, for example, may know how to use a computer to search a huge data bank of information for legal decisions relevant to their cases.

Both computer professionals and end-users can work to improve computer systems by analyzing old systems and suggesting or developing new ones. However, people can also be a complicating factor in a computer system.

People as a Complicating Factor

When experts speak of the "unintended effects of technology," what they are usually referring to are the unexpected things people do with it. People can complicate the workings of a system in three ways:[3]

- **Faulty assessment of information needs:** Humans often are not good at assessing their own information needs. Thus, for example, many users will acquire a computer-and-communications system that either is not sophisticated enough or is far more complex than they need. If all you need is a personal computer on which to type research papers, for instance, you don't need to spend $10,000 on a state-of-the-art system. An outdated system bought used for $500 or less may do just fine. In addition, *people don't always know what information is needed to make a decision.* In other words, they don't know what they *need* to know.

- **Human emotions affect performance:** Of course, human emotions can also affect the performance of a system. For example, one frustrating experience with a computer is enough to make some people abandon the whole system. Hammering on the keyboard or bashing the display screen is certainly not going to advance the learning experience. Also, many people are afraid of computers. However, this feeling is common and diminishes with experience.

- **Human perceptions may be too slow:** Humans act on their perceptions, which in modern information environments are often too slow to keep up with the equipment. You can be so overwhelmed by information overload, for example, that decision making may be just as faulty as if you had too little information.

In summary, although people are the supposed beneficiaries of a computer-and-communications system, they can be the most complicating factor in it.

System Element 2: Procedures

Preview & Review: Procedures are steps for accomplishing a result. Some procedures may be expressed in manuals or documentation. Documentation is also available online.

***Procedures* are descriptions of how things are done, steps for accomplishing a result.** Sometimes procedures are unstated, the result of tradition or common practice. You may find this out when you join a club or are a guest in someone's house for the first time. Sometimes procedures are laid out in great detail in manuals, as is true, say, of tax laws.

When you use a bank ATM—a form of computer system—the procedures for making a withdrawal or a deposit are given in on-screen messages. In other computer systems, procedures are spelled out in manuals. **Manuals, called *documentation*, contain instructions, rules, or guidelines to follow when using hardware or software.** When you buy a microcomputer or a software package, it comes with documentation, or procedures. Nowadays, in fact, many such procedures come not only in a book or pamphlet but also on a computer disk, which presents directions on your display screen. Many companies also offer documentation online, through a phone line connection.

System Element 3: Data/Information

Preview & Review: The distinction is made between raw data, which is unprocessed, and information, which is processed data. Units of measurement of data/information capacity include kilobytes, megabytes, gigabytes, and terabytes.

Though used loosely all the time, the word *data* has some precise and distinct meanings.

"Raw Data" Versus Information

Data can be considered the raw material—whether in paper, electronic, or other form—that is processed by the computer. In other words, ***data* consists of the raw facts and figures that are processed into information.**

***Information* is summarized data or otherwise manipulated data that is useful for decision making.** Thus, the raw data of employees' hours worked and wage rates is processed by a computer into the information of paychecks and payrolls. Some characteristics of useful information are that it is *relevant, timely, accurate, concise,* and *complete.*

Actually, in ordinary usage the words *data* and *information* are often used synonymously. After all, one person's information may be another person's data. The "information" of paychecks and payrolls may become the "data" that goes into someone's yearly financial projections or tax returns.

Units of Measurement for Capacity: From Bytes to Terabytes

A common concern of computer users is "How much data can this gadget hold?" The gadget might be a diskette, a hard disk, or a computer's main memory. The question is a crucial one. If you have too much data, the computer may not be able to handle it. Or if a software package takes up too much storage space, it cannot be run on a particular computer.

We mentioned that computers deal with "on" and "off" (or high-voltage and low-voltage) electrical states, which are represented in the hardware in terms of 0s and 1s, called *bits*. Bits are combined in groups of eight, called *bytes*, to hold the equivalent of a character. A *character* is a single letter, number, or special symbol (such as a punctuation mark or dollar sign). Examples of characters are A, 1, and ?.

A computer system's data/information storage capacity is represented by bytes, kilobytes, megabytes, gigabytes, and terabytes:

- **Kilobyte:** A *kilobyte* (abbreviated *K* or *KB*) is equivalent to approximately 1000 bytes (or characters). More precisely, 1 kilobyte is 1024 (2^{10}) bytes, but the figure is commonly rounded off. Kilobytes are a common unit of measure for the data-holding (memory) capacity of personal computers.

- **Megabyte:** A *megabyte* (abbreviated *M* or *MB*) is about 1 million bytes. Some personal computers can run programs requiring 16 or less megabytes, or about 16 million bytes, of memory.

- **Gigabyte:** A *gigabyte* (*G* or *GB*) is about 1 billion bytes. Pronounced "*gig-a-bite*" (not "*jig*-a-bite"), this unit of measure is used not only with "big iron" computers (mainframes and supercomputers) but also with newer personal computers.

- **Terabyte:** A *terabyte* (*T* or *TB*) is about 1 trillion bytes.

System Element 4: Hardware

Preview & Review: The basic operations of computing consist of (1) input, (2) processing, (3) output, and (4) storage. Communications (5) adds an extension capability to each phase.

Hardware devices are categorized according to which of these five operations they perform. (1) Input hardware includes the keyboard, mouse, and scanner. (2) Processing and memory hardware consists of the CPU (the processor) and main memory. (3) Output hardware includes the display screen, printer, and sound devices. (4) Secondary storage hardware stores data on diskette, hard disk, magnetic tape devices, and optical-disk. (5) Communications hardware includes modems.

As we said earlier, a *system* is a group of related components and operations that interact to perform a task. Once you know how the pieces of the system fit together, you can then make better judgments about any one of them. And you can make knowledgeable decisions about buying and operating a computer system.

The Basic Operations of Computing

How does a computer system process data into information? It goes through four operations: *(1) input, (2) processing, (3) output,* and *(4) storage.* (■ *See Panel 1.4.*)

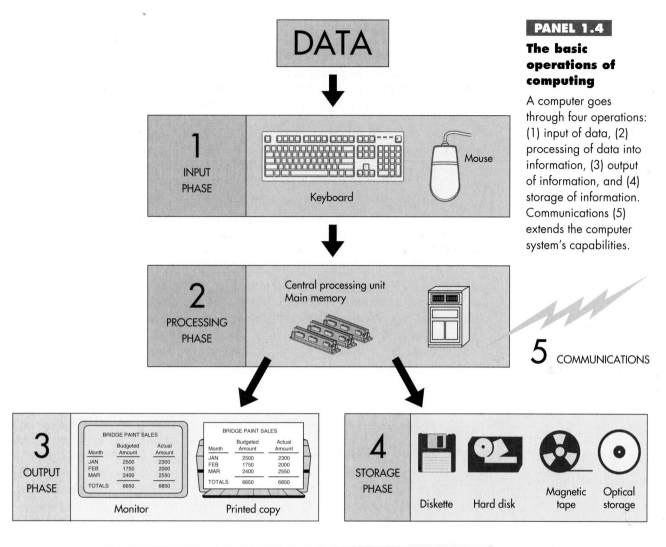

PANEL 1.4

The basic operations of computing

A computer goes through four operations: (1) input of data, (2) processing of data into information, (3) output of information, and (4) storage of information. Communications (5) extends the computer system's capabilities.

1. *Input operation:* In the ***input operation*, data is entered or otherwise captured electronically and is converted to a form that can be processed by the computer.** The means for "capturing" data (the raw, unsorted facts) is input hardware, such as a keyboard.
2. *Processing operation:* In the ***processing operation*, the data is manipulated to process or transform it into information** (such as summaries or totals). For example, numbers may be added or subtracted.
3. *Output operation:* In the ***output operation*, the information, which has been processed from the data, is produced in a form usable by people.** Examples of output are printed text, sound, and charts and graphs displayed on a computer screen.
4. *Secondary storage operation:* In the ***storage operation*, data, information, and programs are stored in computer-processable form.** Diskettes are examples of materials used for storage.

Often these four operations occur so quickly that they seem to be happening simultaneously.

Where does communications fit in here? In the four operations of computing, communications offers an *extension* capability. Data may be input from afar, processed in a remote area, output in several different locations, and stored in yet other places. And information can be transmitted to other computers. All this is done through a wired or wireless connection to the computer.

Hardware Categories

Hardware is what most people think of when they picture computers. *Hardware* **consists of all the machinery and equipment** in a computer system. The hardware includes, among other devices, the keyboard, the screen, the printer, and the computer or processing device itself.

As computing and telecommunications have drawn together, people have begun to refer loosely to *any* machinery or equipment having to do with either one as "hardware." This is the case whether the equipment is a "smart box," such as a cable-TV set-top controller, or (sometimes) the connecting cables, transmitters, or other communications devices.

In general, computer hardware is categorized according to which of the five computer operations it performs:

- Input
- Output
- Communications
- Processing and memory
- Secondary storage

External devices that are connected to the main computer cabinet are referred to as "peripheral devices." **A *peripheral device* is any piece of hardware that is connected to a computer.** Examples are the keyboard, mouse, monitor, and printer.

We describe hardware in detail elsewhere (Chapters 4–8), but the following offers a quick overview.

Input Hardware

Input hardware **consists of devices that allow people to put data into the computer in a form that the computer can use.** For example, input may be by means of a *keyboard, mouse,* or *scanner.* The keyboard is the most obvious. The mouse is a pointing device attached to many microcomputers. An example of a scanner is the grocery-store bar-code scanner. (These and other input devices are discussed in Chapter 5.)

- **Keyboard: A *keyboard* includes the standard typewriter keys plus a number of specialized keys.** The standard keys are used mostly to enter words and numbers. Examples of specialized keys are the *function keys,* labeled *F1, F2,* and so on. These special keys are used to enter commands.
- **Mouse: A *mouse* is a device that can be rolled about on a desktop to direct a pointer on the computer's display screen.** The pointer is a symbol, usually an arrow, on the computer screen that is used to select items from lists (menus) or to position the cursor. **The *cursor* is the symbol on the screen that shows where data may be entered next,** such as text in a word processing program.
- **Scanners: *Scanners* translate images of text, drawings, and photos into digital form.** The images can then be processed by a computer, displayed on a monitor, stored on a storage device, or communicated to another computer.

Processing & Memory Hardware

The brains of the computer are the *processing* and *main memory* devices, housed in the computer's system unit. **The *system unit,* or *system cabinet,* houses the electronic circuitry, called the *CPU,* which does the actual processing and the main memory, which supports processing.** (■ *See Panel 1.5.*) (These are discussed in detail in Chapter 4.)

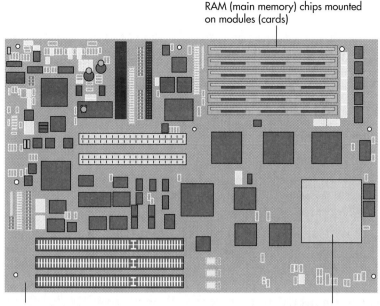

RAM (main memory) chips mounted
on modules (cards)

PANEL 1.5

The system unit

Motherboard (*top*) fits inside
system cabinet (*bottom*).

Motherboard
(system board)

Microprocessor chip
(with CPU)

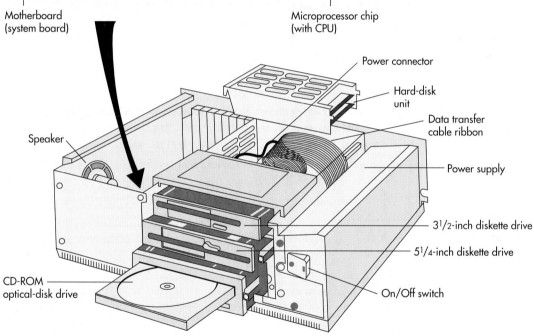

Power connector

Hard-disk
unit

Data transfer
cable ribbon

Speaker

Power supply

3¹/2-inch diskette drive

5¹/4-inch diskette drive

CD-ROM
optical-disk drive

On/Off switch

- CPU—the processor: **The *CPU*, for *Central Processing Unit*, is the
  processor, or computing part of the computer. It controls and manipulates
  data to produce information.** In a personal computer the CPU is usually a
  single fingernail-size "chip" called a *microprocessor,* with electrical circuits printed on it. This microprocessor and other components necessary
  to make it work are mounted on a main circuit board called a *motherboard* or *system board.*
- Memory—working storage: *Memory*—also known as *main memory,
  RAM,* or *primary storage*—is working storage. **Memory is the computer's
  "work space," where data and programs for immediate processing are held.**
  Computer memory is contained on memory chips mounted on the motherboard. Memory capacity is important because it determines how much
  data can be processed at once and how big and complex a program may
  be used to process the data.

[handwritten notes in margins]

internal

works

↗ the CPU on the motherboard

temporary
volatile

R E A D M E

Practical Matters: Common Measurements Used in Computers & Communications

When salespeople or friends rattle on about how fast a computer's "clock speed" is or how many "dpi" the printer uses, how will you know what they're talking about? The following is a quick guide to some common measurement terms. You can use this box as a reference as you work through the book.

bps (bits per second) Bps is often used to measure modem speeds. It is a measure of the actual number of bits that are transferred per second. (A bit is a 0 or 1, the smallest unit of information used in computing.)

capacity—from bits to terabytes *Capacity* refers to how much data/information a storage device will hold. Capacity is represented by bits, bytes, kilobytes, megabytes, gigabytes, and terabytes.

- *Bit:* Short for *binary digit;* a 0 or 1, which the computer hardware represents as an "on" or "off" (or high-voltage or low-voltage) electrical state.

- *Byte:* Usually a group of eight bits.

- *Kilobyte (K or KB):* About 1000 (1024 or 2^{10}) bytes. Bytes and their multiples are common units of measure for memory or storage capacity of personal computers.

- *Megabyte (M or MB):* About 1 million (specifically 1,048,576) bytes.

- *Gigabyte (G or GB):* Pronounced *"gig-a-bite"* (not *"jig-a-bite"*); about 1 billion (1,073,741,824) bytes.

- *Terabyte (T or TB):* About 1 trillion (specifically 1,009,511,627,776) bytes.

clock speed—Hz, kHz, and MHz *Clock speed* refers to how fast a computer processes. (The CPU, or central processing unit, is circuitry that controls the interpretation and execution of instructions.) The *CPU clock* uses a quartz crystal to generate a steady stream of pulses to the CPU to regulate the system's internal speed. The clock measures speed in hertz, kilohertz, and megahertz as frequency of electrical vibrations (cycles) per second.

- *Hertz (Hz):* A single clock cycle per second

- *Kilohertz (kHz):* 1000 cycles per second

- *Megahertz (MHz):* 1 million cycles per second

The clock speed of a microprocessor (a CPU on a single chip) in a microcomputer is measured in megahertz. For example, for newer microprocessors, speeds range from 25 MHz to 166 MHz or more.

dot pitch Measurement used to describe the clarity of the image on a display screen or of a printer's output.

- *Pixels:* For display screens, dot pitch is expressed in millimeters as the distance between individual dots, or *pixels (picture elements).* The smaller the dot pitch, the clearer the image. Typically, display screens vary from .28 to .51 millimeters.

- *dpi:* For printers, dot pitch is expressed in *dpi,* the number of dots that a printer can print in a linear inch.

dpi *Dots per inch;* dpi is a measurement used to describe the image clarity of printers and scanners. If you look closely, you will see that a printed image is made up of individual dots. The higher the number of dots per linear inch, the clearer the image. A 300 dpi printer prints 300 × 300, or 90,000, dots in 1 square inch. A 400 dpi printer produces 160,000 dots, a 500 dpi printer produces 250,000 dots. Common printer and scanner measurements range from 300 to 600 dpi and up.

fractions of a second In increasing order of rapidity:

- *Milliseconds (ms):* Thousandths of a second; measures the amount of time the computer takes to access information from a hard disk.

- *Microseconds (μs):* Millionths of a second; measures instruction execution.

- *Nanoseconds (ns):* Billionths of a second; measures the speed at which information travels through circuits, as for memory chips (70–60 ns).

- *Picoseconds (ps):* Trillionths of a second; measures transistor switching.

inch Measurement used to describe size of diskettes and of monitors.

- *Diskette sizes:* Diskettes come in two principal sizes: 3½ (or 3.5) inches, now the common standard; and 5¼ (or 5.25) inches, an older size, now less common.

- *Monitor sizes:* Like the screens of television sets, computer monitors, or display screens, are measured diagonally from one corner to the other. Common sizes are 14–17 inches or larger.

MIPS *million instructions per second;* MIPS are a measurement of the execution speed of a large computer.

word The standard unit of information natural to a particular system. The unit varies depending on the computer. For a computer with a microprocessor that processes 16 bits at a time, a word would be 16 bits.

Despite its name, memory does not remember. That is, once the power is turned off, all the data and programs within memory simply vanish. This is why data/information must also be stored in relatively permanent form on disks and tapes, which are called *secondary storage* to distinguish them from main memory's *primary storage.*

Output Hardware

Output hardware **consists of devices that translate information processed by the computer into a form that humans can understand.** We are now so exposed to products output by some sort of computer that we don't consider them unusual. Examples are grocery receipts, bank statements, and grade reports. More recent forms are digital recordings and even digital radio.

As a personal computer user, you will be dealing with three principal types of output hardware—*screens, printers,* and *sound output devices.* (These and other output devices are discussed in detail in Chapter 6.)

- Screen: **The** *screen* **is the display area of a computer.** A desktop computer or video terminal (such as those listing flight information in airports) will use a *monitor,* **a high-resolution screen.** The monitor is often called a *CRT,* **for** *Cathode-Ray Tube,* **the familiar TV-style picture tube.**

- Printer: **A** *printer* **is a device that converts computer output into printed images.** Printers are of many types, some noisy, some quiet, some able to print carbon copies, some not.

- Sound: Many computers emit chirps and beeps. Some go beyond those noises and contain sound processors and speakers that can play digital music or human-like speech. High-fidelity stereo sound is becoming more important as computer and communications technologies continue to merge.

Secondary Storage Hardware

Main memory (primary storage) is *internal* storage. It works with the CPU chip on the motherboard to hold data and programs for immediate processing. Secondary storage, by contrast, is *external* storage. It is not on the motherboard (although it may still be inside the system cabinet). **Secondary storage consists of devices that store data and programs permanently on disk or tape.**

nonvolatile

You may hear people use the term "storage media." *Media* refers to the material that stores data, such as diskette or magnetic tape. For microcomputers the principal storage media are the *diskette (floppy disk), hard disk, magnetic tape,* and *CD-ROM.* (■ *See Panel 1.6, next page.*) (These and other secondary storage devices are discussed in detail in Chapter 7.)

- Diskette: **A** *diskette,* **or** *floppy disk,* **is a removable round, flexible disk that stores data as magnetized spots.** The disk is contained in a plastic case or square paper envelope to prevent the disk surface from being touched by human hands.

Secondary storage for microcomputers

(*Left*) Examples of diskette, CD-ROM, and magnetic-tape drives (the hard-disk drive has no exterior opening). (*Right*) Inside of hard-disk drive. At a minimum, a personal computer will have a diskette drive.

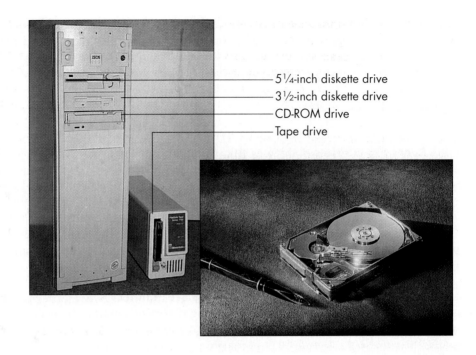

- 5¼-inch diskette drive
- 3½-inch diskette drive
- CD-ROM drive
- Tape drive

Diskettes

The most common size is 3½ inches (*top*); some older computers still use the 5¼-inch size (*bottom*).

Two sizes of diskettes are used for microcomputers. The older and larger size is *5¼ inches* in diameter. The smaller size, now by far the most common, is *3½ inches*. (■ *See Panel 1.7.*) The smaller disk, which can fit in a shirt pocket, has a compact and rigid case and actually does not feel "floppy" at all.

To use a diskette, you need a diskette drive. **A *diskette drive* is a device that holds and spins the diskette inside its case; it "reads" data from and "writes" data to the disk.** The words *read* and *write* are used a great deal in computing.

Read **means that the data represented in magnetized spots on the disk (or tape) is converted to electronic signals and transmitted to the memory in the computer.**

Write **means that the electronic information processed by the computer is recorded onto disk (or tape).**

- **Hard disk:** Diskettes are made out of tape-like material, which is what makes them "floppy." They are also removable. By contrast, **a *hard disk* is a disk made out of metal and covered with a magnetic recording surface. It also holds data represented by the presence (1) and absence (0) of magnetized spots.**

  Hard-disk drives *read* and *write* data in much the same way that diskette drives do. However, there are three significant differences. First, hard-disk drives can handle thousands of times more data than diskettes do. Second, hard-disk drives are usually built into the system cabinet, in which case they are not removable. Third, hard disks read and write data faster than diskettes do.

- **Magnetic tape:** Moviemakers used to love to represent computers with banks of spinning reels of magnetic tape. Indeed, with early computers, "mag tape" was the principal method of secondary storage.

  The magnetic tape used for computers is made from the same material as that used for audiotape and videotape. That is, **magnetic tape is made of flexible plastic coated on one side with a magnetic material; again, data is represented by the presence and absence of magnetized spots.** Because of its drawbacks (described in Chapter 7), nowadays tape is used mainly to provide low-cost duplicate storage, especially for microcomputers. A tape that is a duplicate or copy of another form of storage is referred to as a *backup.*

Because hard disks sometimes fail ("crash"), personal computer users who don't wish to do backup using a lot of diskettes will use magnetic tape instead.

• **Optical disk—CD-ROM:** If you have been using music CDs (compact disks), you are already familiar with optical disks. **An *optical disk* is a disk that is written and read by lasers. *CD-ROM,* which stands for Compact Disk—Read Only Memory, is one kind of optical-disk format that is used to hold text, graphics, and sound.** CD-ROMs can hold hundreds of times more data than diskettes, and can hold more data than many hard disks.

Communications Hardware

Computers can be "stand-alone" machines, meaning that they are not connected to anything else. Indeed, many students tote around portable personal computers on which they use word processing or other programs to help them with their work. Many people are quite happy using a computer that has no communications capabilities.

However, the *communications* component of the computer system vastly extends the range of a computer. Indeed, the range is so many orders of magnitude larger that comprehending it is difficult.

In general, computer communications is of two types: *wired connections,* such as telephone wire or cable, and *wireless connections,* such as via radio waves (covered in detail in Chapter 8).

The dominant communications media that have been developed during this century use analog transmission. Thus, for many years, the principal form of direct connection was via standard copper-wire telephone lines. Hundreds of these twisted-pair copper wires are bundled together in cables and strung on telephone poles or buried underground. As mentioned, a modem is communications hardware required to translate a computer's digital signals into analog form for transmission over telephone wires. Although copper wiring still exists in most places, it is gradually being supplanted by two other kinds of direct connections: coaxial cable and fiber-optic cable. Eventually, all transmission media will accommodate digital signals.

System Element 5: Software

Preview & Review: Software comprises the step-by-step instructions that tell the computer what to do. In general, software is divided into applications software and systems software.

Applications software, which may be customized or packaged, performs useful work on general-purpose tasks.

Systems software, which includes operating systems, enables the applications software to run on the computer.

Software, **or *programs,* consists of the step-by-step instructions that tell the computer how to perform a task.** In most instances, the words *software* and *program* are interchangeable. Although it may be contained on disks of some sort, software is invisible, being made up of electronic blips.

There are two major types of software:

- **Applications software:** This may be thought of as the kind of software that people use to perform a general-purpose task, such as word processing software used to prepare the text for a document.
- **Systems software:** This may be thought of as the underlying software that the computer uses to manage its own internal activities and run applications software.

Although you may not need a particular applications program, you must have systems software, or you will not even be able to "boot up" your computer (make it run).

Applications Software

Applications software is defined as software that can perform useful work on general-purpose tasks. Examples are programs that do word processing, desktop publishing, or payroll processing.

Applications software may be either *customized* or *packaged. Customized software* is software designed for a particular customer. This is the kind of software that you would hire a professional computer programmer—a software creator—to develop for you. Such software would perform a task that could not be done with standard off-the-shelf packaged software available from a computer store or mail-order house.

Packaged software, or a *software package,* is the kind of "off-the-shelf" program developed for sale to the general public. This is the principal kind that will be of interest to you. Examples of packaged software that you will most likely encounter are word processing and spreadsheet programs. (We discuss these in Chapter 2.)

Systems Software

As the user, you interact mostly with the applications software. **Systems software enables the applications software to interact with the computer and manages the computer's internal resources.**

Systems software consists of several programs, the most important of which is the operating system. **The *operating system* acts as the master control program that runs the computer.** It handles such activities as running and storing programs and storing and processing data. The purpose of the operating system is to allow applications to operate by standardizing access to shared resources such as disks and memory. Examples of operating systems are MS-DOS, Windows 95, OS/2 Warp, and the Macintosh operating system (MacOS). (We discuss these operating systems in detail in Chapter 3.)

▶ System Element 6: Communications

Preview & Review: "Communications" refers to the electronic transfer of data. The kind of data being communicated is rapidly changing from analog to digital.

Communications is defined as the electronic transfer of data from one place to another. Of all six elements in a computer-and-communications system, communications probably represents the most active frontier at this point.

We mentioned that, until now, most data being communicated has been analog data. However, as former analog methods of communication become digital, we will see a variety of suppliers, using wired or wireless connections, providing data in digital form: telephone companies, cable-TV services, news and information services, movie and television archives, interactive shopping channels, video catalogs, and more.

Developments in Computer Technology

Preview & Review: Computers are becoming increasingly smaller, more powerful, and less expensive.

Today the five types of computers are supercomputers, mainframe computers, minicomputers, microcomputers (both personal computers and workstations), and microcontrollers (embedded computers).

A human generation is not a very long time, about 30 years. During the short period of one and a half generations, computers have come from nowhere to transform society in unimaginable ways. One of the first computers, the outcome of military-related research, was delivered to the U.S. Army in 1946. ENIAC—short for *Electronic Numerical Integrator And Calculator*—weighed 30 tons, was 80 feet long and two stories high, and required 18,000 vacuum tubes. However, it could multiply a pair of numbers in the then-remarkable time of three-thousandths of a second. This was the first general-purpose, programmable electronic computer, the grandparent of today's lightweight handheld machines.

The Three Directions of Computer Development

Since the days of ENIAC, computers have developed in three directions:

- **Smaller size:** Everything has become smaller. ENIAC's old-fashioned radio-style vacuum tubes gave way to the smaller, faster, more reliable transistor. A *transistor* is a small device used as a gateway to transfer electrical signals along predetermined paths (circuits).
    The next step was the development of tiny integrated circuits. **Integrated circuits (ICs) are entire collections of electrical circuits or pathways etched on tiny squares of silicon** half the size of your thumbnail. *Silicon* is a natural element found in sand that is purified to form the base material for making computer processing devices.
- **More power:** In turn, miniaturization allowed computer makers to cram more power into their machines, providing faster processing speeds and more data storage capacity.
- **Less expense:** The miniaturized processor of a personal computer that sits on a desk performs the same sort of calculations once performed by a computer that filled an entire room. However, processor costs are only a fraction of what they were 15 years ago.

Five Kinds of Computers

Computers are categorized into five general types, based mainly on their processing speeds and their capacity to store data: *supercomputers, mainframe computers, minicomputers, microcomputers,* and *microcontrollers.* (■ *See Panel 1.8, next page.*)

- **Supercomputers:** *Supercomputers* **are high-capacity computers that cost millions of dollars, occupy special air-conditioned rooms, and are often used for research.** Among their uses are worldwide weather forecasting, oil exploration, aircraft design, and mathematical research.

- **Mainframe computers:** Less powerful than supercomputers, *mainframe computers* **are fast, large-capacity computers also occupying specially wired, air-conditioned rooms.** Mainframes are used by large organizations—banks, airlines, insurance companies, mail-order houses, universities, the Internal Revenue Service—to handle millions of transactions.

- **Minicomputers:** *Minicomputers,* **also called** *midrange computers,* **are generally refrigerator-size machines that are essentially scaled-down mainframes.** Because of their lesser processing speeds and data-storing capacities, they have been typically used by medium-sized companies for specific purposes, such as accounting. Minicomputers are being replaced by networks of microcomputers.

- **Microcomputers:** *Microcomputers* **are small computers that can fit on a desktop or in one's briefcase.** Microcomputers are of two types—*personal computers* and *workstations*—although the distinction is blurring rapidly.

    Personal computers (PCs) **are desktop or portable computers that can run easy-to-use programs, such as word processors or spreadsheets.** Whether desktop, laptop, notebook, or palmtop (in declining order of size),

[handwritten annotations: nano seconds (one billionth); measured in microseconds = one millionth of a second]

PANEL 1.8

The principal types of computers—and the microprocessor that powers them

(*Clockwise from top left*) A supercomputer, a mainframe computer, a minicomputer, and two kinds of microcomputers—a personal computer (PC) and a workstation—and a microcontroller. (*Photo lower right*) A microprocessor (this one is Intel's P6) the miniaturized circuitry that does the processing in computers. A PC may have only one of these, a supercomputer thousands.

personal computers are now found in most businesses. They are also found in about one-third of American homes.

Workstations are expensive, powerful desktop machines used mainly by engineers and scientists for sophisticated purposes. Providing many capabilities formerly found only in mainframes and minicomputers, workstations are used for such tasks as designing airplane fuselages or prescription drugs. Workstations are often connected to a larger computer system to facilitate the transfer of data and information.

- **Microcontrollers:** Also called *embedded computers, microcontrollers* are **installed in "smart" appliances like microwave ovens.**

Needless to say, as you progress from the microcontroller up to the supercomputer, cost, memory capacity, and speed all increase.

The Mighty Microprocessor

Computers by themselves are important. However, perhaps equally significant are the affiliated technologies made possible by the invention of the microprocessor. **A *microprocessor* is the miniaturized circuitry of the computer's processor—the part that manipulates data into information. The circuitry is etched on a sliver or "chip" of material, usually silicon.**

Microprocessors are the CPUs in personal computers. Equally important, microprocessors provide the "thinking" for most other new electronic devices, from CD players to music synthesizers to automobile fuel-injection systems. When you hear of all the things gadgetry is supposed to do for us, often you can credit the microprocessor.

Developments in Communications Technology

Preview & Review: Communications, or telecommunications, has had three important developments: better communications channels, better networks, and better sending and receiving devices.

Throughout the 1980s and early 1990s, telecommunications made great leaps forward. Three of the most important developments were:

- Better communications channels
- Better networks
- Better sending and receiving devices

Better Communications Channels

We mentioned that data may be sent by wired or wireless connections. The old copper-wire telephone connections have begun to yield to the more efficient coaxial cable and, more important, to fiber-optic cable, which can transmit vast quantities of information in both analog and digital form.

Even more interesting has been the expansion of wireless communication. Federal regulators have permitted existing types of wireless channels to be given over to new uses, as a result of which we now have many more kinds of two-way radio, cellular telephone, and paging devices than we had previously.

Better Networks

When you hear the word "network," you may think of a *broadcast network*, a group of radio or television broadcasting stations that cut costs by airing the same programs. Here, however, we are concerned with **communications networks, which connect one or more telephones or computers or associated devices.** The principal difference is that *broadcast networks transmit messages in only one direction, communications networks transmit in both directions.* Communications networks are crucial to technological convergence, for they allow information to be exchanged electronically.

A communications network may be large or small, public or private, wired or wireless or both. In addition, smaller networks may be connected to larger ones. For instance, a *local area network (LAN)* may be used to connect users located near one another, as in the same building. On some college campuses, for example, microcomputers in the rooms in residence halls are linked throughout the campus by a LAN. **A computer in a network shared by multiple users is called a *server*.**

Better Sending & Receiving Devices

Part of the excitement about telecommunications in the last decade or so has been the development of new devices for sending and receiving information. Two examples are the *cellular phone* and the *fax machine.*

- **Cellular phones:** *Cellular telephones* **use a system that divides a geographical service area into a grid of "cells." In each cell, low-powered, portable, wireless phones can be accessed and connected to the main (wire) telephone network.**

  The significance of the wireless, portable phone is not just that it allows people to make calls from their cars. Most important is its effect on worldwide communications. Countries with underdeveloped wired telephone systems, for instance, can use cellular phones as a fast way to install better communications. Such technology gives these nations a chance to join the world economy.

- **Fax machines:** *Fax* **stands for "facsimile," which means "a copy"; more specifically, *fax* stands for "facsimile transmission." A *fax machine* scans an image and sends a copy of it in the form of electronic signals over transmission lines to a receiving fax machine. The receiving machine re-creates the image on paper.** Fax messages may also be sent to and from microcomputers.

  Fax machines have been commonplace in offices and even many homes for some time, and new uses have been found for them. For example, some newspapers offer facsimile editions, which are transmitted daily to subscribers' fax machines. These editions look like the papers' regular editions, using the same type and headline styles, although they have no photographs. Toronto's *Globe & Mail* offers people who will be away from Canada a four-page fax that summarizes Canadian news. The *New York Times* sends a faxed edition, transmitted by satellite, to island resorts and to cruise ships in mid-ocean. (Fax machines and fax modems are covered in detail in Chapter 8.)

Computer & Communications Technology Combined: Connectivity & Interactivity

Preview & Review: Trends in information technology involve connectivity and interactivity.

Connectivity, or online information access, refers to connecting computers to one another by modem or network and communications lines. Connectivity, among other things provides the benefits of voice mail and e-mail, telecommuting, teleshopping, databases, and online services and networks.

Interactivity refers to the back-and-forth "dialog" between a user and a computer or communications device. Interactive devices include multimedia computers, personal digital assistants, and up-and-coming "smart boxes" and "Internet appliances."

Lee Taylor is what is known as a *lone eagle.* Once he was the manager of several technical writers for a California information services company. Then, taking a one-third pay cut, he moved with his wife to a tiny cabin near the ski-resort town of Telluride, Colorado. There he operates as a free-lance consultant for his old company, using phone, computer network, and fax machine to stay in touch.[4]

"Lone eagles" like Taylor constitute a growing number of professionals who, with information technology, can work almost anywhere they want, such as resort areas and backwoods towns. Although their income may be less, it is offset by such "quality of life" advantages as weekday skiing or reduced housing costs.

Taylor is one beneficiary of trends that will probably intensify as information technology continues to proliferate. These trends are:

* Connectivity
* Interactivity

Connectivity (Online Information Access)

As we discussed, small telecommunications networks may be connected to larger ones. This is called **connectivity, the ability to connect computers to one another by modem or network and communications lines to provide online information access.** It is this connectivity that is the foundation of the latest advances in the Digital Age.

The connectivity of telecommunications has made possible many kinds of activities. Although we cover these activities in more detail in Chapter 8, briefly they are as follows:

* **Voice mail and e-mail:** *Voice mail* acts like a telephone answering machine. Incoming voice messages are digitized and stored for your retrieval later. Retrieval is accomplished by dialing into your "mailbox" number from any telephone. You can get your own personal voice-mail setup by paying a monthly fee to a telephone company, such as AT&T.

    An alternative system is e-mail. **E-mail, or *electronic mail,* is a software-controlled system that links computers by wired or wireless connections. It allows users, through their keyboards, to post messages and to read responses on their computer screens.** Whether the network is a company's small local area network or a worldwide network, e-mail allows users to send messages anywhere on the system.

- **Telecommuting:** In standard commuting, one takes transportation (car, bus, train) from home to work and back. In *telecommuting*, one works at home and communicates with ("commutes to") the office by computer and communications technology. Already more than 9 million people—not including business owners or independent contractors—telecommute at least part of the time. By 1998, according to Link Resources, there will be 13 million telecommuters.[5] (■ *See Panel 1.9.*)

- **Teleshopping:** Teleshopping is the computer version of cable-TV shop-at-home services. With *teleshopping*, microcomputer users dial into a telephone-linked computer-based shopping service that lists prices and descriptions of products, which may be ordered through the computer. You charge the purchase to your credit card. The teleshopping service sends the merchandise to you by mail or other delivery service.

- **Databases:** A database may be a large collection of data located within your own unconnected personal computer. Here, however, we are concerned with databases located elsewhere. These are libraries of information at the other end of a communications connection that are available to you through your microcomputer. **A *database* is a collection of electronically stored data. The data is integrated, or cross-referenced, so that different people can access it for different purposes.**

  For example, suppose an unfamiliar company offered you a job. To find out about your prospective employer, you could go online to gain access to some helpful databases. Examples are Business Database Plus, Magazine Database Plus, and TRW Business Profiles. You could then study the company's products, review financial data, identify major competitors, or learn about recent sales increases or layoffs. You might even get an idea of whether or not you would be happy with the "corporate culture."[6]

PANEL 1.9

Telecommuting

The number of employees who telecommute—use computers and wired or wireless technology to work from home—for at least part of their working day has greatly increased. Telluride, Colorado, has become a testing site for telecommuting and other new communications ideas.

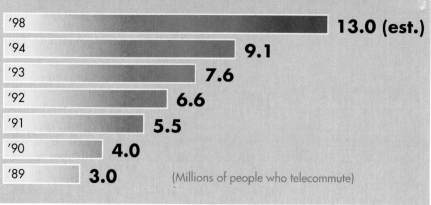

'98   13.0 (est.)
'94   9.1
'93   7.6
'92   6.6
'91   5.5
'90   4.0
'89   3.0   (Millions of people who telecommute)

- **Computer online services and networks and the Internet:** Established major commercial online services include America Online and CompuServe. **A *computer online service* is a commercial information service that, for a fee, makes various services available to subscribers through their telephone-linked microcomputers.**

  Among other things, consumers can research information in databases, go teleshopping, make airline reservations, or send messages via e-mail to others using the service.

  Through a computer online service you may also gain access to the greatest network of all, the Internet. **The *Internet* is an international network connecting approximately 36,000 smaller networks that link computers at academic, scientific, and commercial institutions.** An estimated 24 million people in the United States and Canada alone are already on the Internet—fully 11% of the North American population over age 16. The most well known part of the Internet is the World Wide Web, which stores information in multimedia form—sounds, photos, video, as well as text.

Interactivity: The Examples of Multimedia Computers, Personal Digital Assistants, & Futuristic "Smart Boxes" & "Internet Appliances"

The movie rolls on your TV/PC screen. The actors appear. Instead of passively watching the plot unfold, however, you are able to determine different plot developments by pressing keys on your keyboard. This is an example of interactivity. ***Interactivity* means that the user is able to make an immediate response to what is going on and modify the processes. That is, there is a dialog between the user and the computer or communications device.** Videogames, for example, are interactive. Interactivity allows users to be active rather than passive participants in the technological process.

Among the types of interactive devices are multimedia computers, personal digital assistants, and various kinds of up-and-coming "smart boxes" that work either with a TV or a PC.

- **Multimedia computers:** The word *multimedia*, one of the buzzwords of the '90s, has been variously defined. Essentially, however, **multimedia refers to technology that presents information in more than one medium, including text, graphics, animation, video, music, and voice.**

  Multimedia personal computers are powerful microcomputers that include sound and video capability, run CD-ROM disks, and allow users to play games or perform interactive tasks.

- **Personal digital assistants:** In 1988, handheld electronic organizers were introduced, consisting of tiny keypads and barely readable screens. They were unable to do much more than store phone numbers and daily "to do" lists.

  In 1993, electronic organizers began to be supplanted by personal digital assistants, such as Apple's Newton. *Personal digital assistants (PDAs)* are small pen-controlled, handheld computers that, in their most developed form, can do two-way wireless messaging. Instead of pecking at a tiny keyboard, you can use a special pen to write out commands on the computer screen. The newer generation of PDAs can be used not only to keep an appointment calendar and write memos but also to access the Internet and send and receive faxes and e-mail. With a PDA, then, you can immediately get information from some remote location—such as your microcomputer on your desk at home—and, if necessary, change it to update it.

- **Up-and-coming "smart boxes" and "Internet appliances":** Already envisioning a world of cross-breeding among televisions, telephones, and computers, enterprising manufacturers are experimenting with developing TV/PC set-top control boxes, or *"smart boxes,"* and *"Internet appliances."* With these futuristic devices, consumers presumably could listen to music CDs, watch movies, do computing, view multiple cable channels, and go online. Set-top boxes would provide two-way interactivity not only with videogames but also with online entertainment, news, and educational programs.

  Recently a *network computer,* or "hollow personal computer," has been developed, a machine intended to cost $500 or so that would be "hollowed out." Instead of having all the complex memory and storage capabilities built in, the network PC is designed to serve as an entry point to the online world, which is supposed to contain all the resources anyone would need.[7]

  Another gadget is the *cable modem,* which will allow cable-TV subscribers to connect their personal computers to various online computer services at speeds many times faster than traditional computer modems. This represents a way for cable operators to introduce voice, data, and video services on a large scale.[8]

  The converse of this is a kind of *"Internet TV" technology* known as Intercast, produced by chip maker Intel. Intercast lets specially equipped personal computers receive data from the Internet as well as television programming. Television networks could thus broadcast not only television shows but also additional data, such as geographical information about a country that is the subject of a news story, which computer owners could then look up.[9]

All these devices seem to be leading toward a kind of *"information appliance,"* as we describe next.

The "All-Purpose Machine": The Information Appliance That Will Change Your Future

Preview & Review: In the future, we may have an "information appliance," a device that combines telephone, television, VCR, and personal computer. This device would deliver digitized entertainment, communications, and information.

Computer pioneer John Von Neumann said that the computer should not be called the "computer" but rather the "all-purpose machine." After all, he pointed out, it is not just a gadget for doing calculations. The most striking thing about it is that it can be put to *any number of uses.*

More than ever, we are now seeing just how true that is.

The "Information Appliance": What Will It Be?

Recently, there has been enormous interest in what is perceived to be the coming Information Superhighway. This electronic delivery system would presumably direct a digitized stream of sound, video, text, and data to some sort of box, perhaps something called an *information appliance.* An information appliance would deliver digitized entertainment, communications, and information in a device that combines telephone, television, VCR, and personal computer. The vision inspired by this futuristic gadget has caused furious activity in the communications world. Telephone, cable, computer, consumer electronics, and entertainment companies have rushed to position themselves to take advantage of these developments.

So what, exactly, will the information appliance (a term coined by Apple Macintosh researcher Jef Raskin in 1978) turn out to be? Perhaps it could be the under-$500 network computer proposed by Oracle and others. Or it might be a variation on the TV set, like Gateway's Destination, a home-entertainment setup built around a personal computer and a 31-inch TV. Or it could be a new wrinkle on the videogame player, like the Pippin Atmark combination game player–Internet browser. Or it might be the grand fusion of the Internet and household appliances envisioned by Microsoft in its 1996 SIPC (Simply Interactive Personal Computer) standards. Whatever its final form, the information appliance will no doubt be adapted from a machine now present everywhere—the microcomputer.[10] (■ *See Panel 1.10, pp. 30–31*) Clearly, then, anyone who learns to use a microcomputer now is getting a head start on the revolution.

The Ethics of Information Technology

Ethical questions pervade all aspects of the use of information technology, as will be pointed out with a special symbol—**E**—throughout this book.

Every reader of this book at some point will have to wrestle with ethical issues related to computers-and-communications technology. *Ethics* is defined as a set of moral values or principles that govern the conduct of an individual or group. Indeed, ethical questions arise so often in connection with information technology that we have decided to earmark them wherever they appear in this book, starting in Chapter 2—with the special "E-for-ethics" symbol you see in the margin.

Onward: The Gateway to the Information Superhighway

The term *Information Superhighway* has roared into the nation's consciousness in recent times. Some say it promises what might be called a communications cornucopia—a *"communicopia"* of electronic interactive services. Others say it is surrounded "by more hype and inflated expectations than any technological proposal of recent memory."[11] What, in fact, is this electronic highway? Does it or will it really exist?

The *Information Superhighway* is a vision or a metaphor for a fusion of the two-way wired and wireless capabilities of telephones and networked computers with cable-TV's capacity to transmit hundreds of programs. The resulting interactive digitized traffic would include movies, TV shows, phone calls, databases, shopping services, and online services. This superhighway, it is hoped, would link all homes, schools, businesses, and governments.

At present, this electronic highway remains a vision, much as today's interstate highway system was a vision in the 1950s. It is as though we still had old-fashioned Highway 40s and Route 66s, along with networks of one-lane secondary and gravel backroads. These, of course, have largely been replaced by high-speed blacktop and eight-lane freeways. In 40 years, will the world be as changed by the electronic highway as North America has been by the interstate highways of the last four decades? It is the thesis of this book that it will be—and that we should prepare for it.

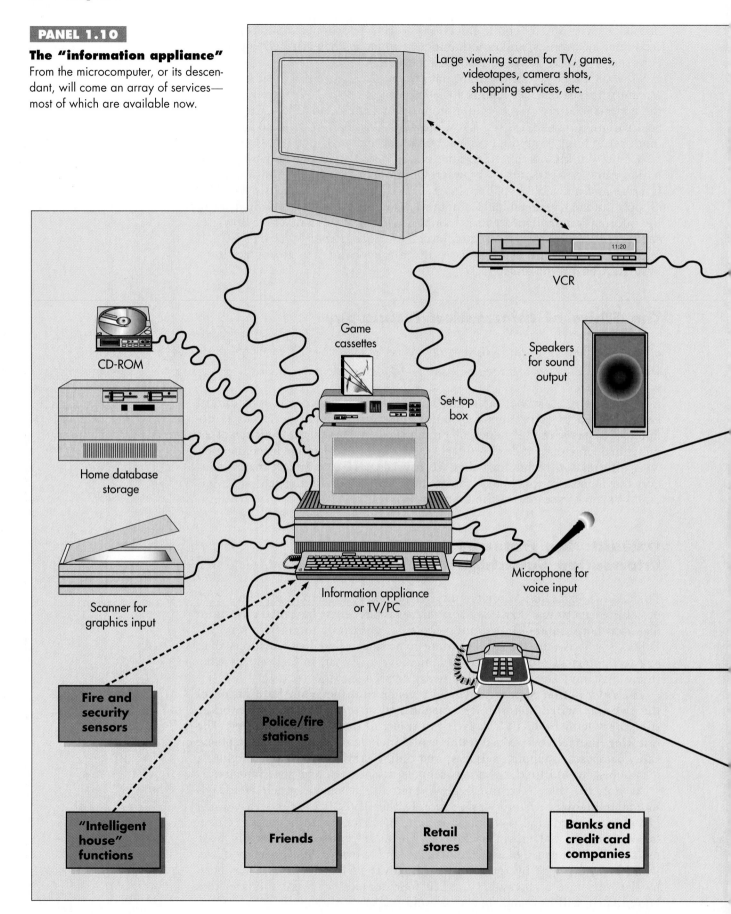

PANEL 1.10

The "information appliance"
From the microcomputer, or its descendant, will come an array of services—most of which are available now.

Large viewing screen for TV, games, videotapes, camera shots, shopping services, etc.

VCR

CD-ROM

Home database storage

Scanner for graphics input

Game cassettes

Set-top box

Speakers for sound output

Microphone for voice input

Information appliance or TV/PC

Fire and security sensors

Police/fire stations

"Intelligent house" functions

Friends

Retail stores

Banks and credit card companies

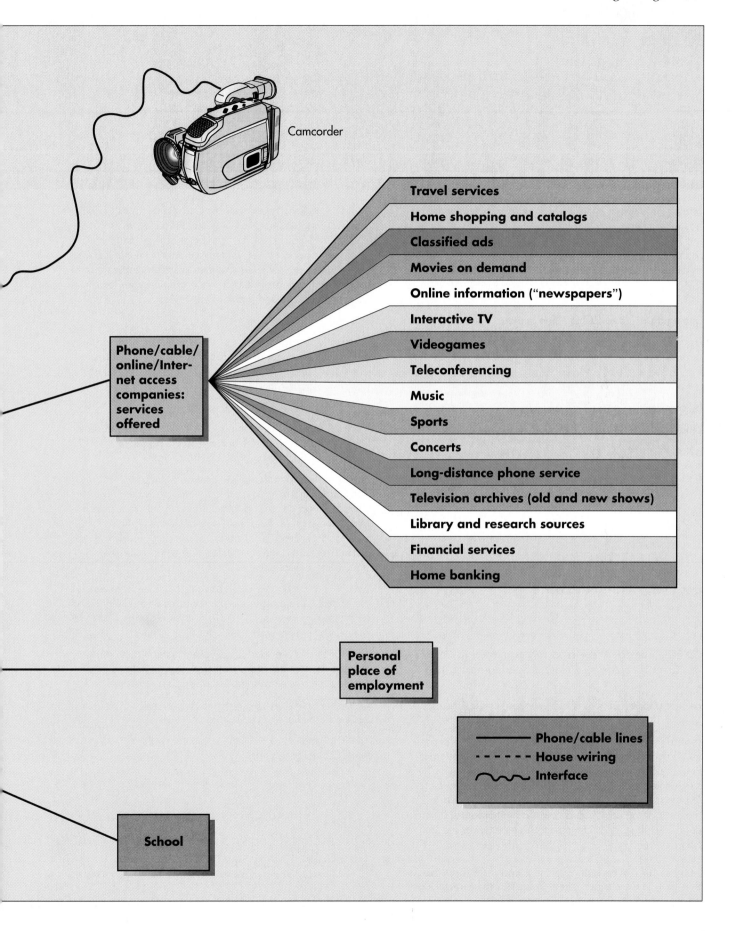

Camcorder

Phone/cable/ online/Inter- net access companies: services offered

- Travel services
- Home shopping and catalogs
- Classified ads
- Movies on demand
- Online information ("newspapers")
- Interactive TV
- Videogames
- Teleconferencing
- Music
- Sports
- Concerts
- Long-distance phone service
- Television archives (old and new shows)
- Library and research sources
- Financial services
- Home banking

Personal place of employment

School

——————— Phone/cable lines
- - - - - - House wiring
~~~~~~ Interface

SUMMARY

analog *(p. 6, LO 2*)* Refers to nondigital (noncomputer-based), continuously variable forms of data transmission, including voice and video. Most current telephone lines and radio, television, and cable-TV hookups are analog transmissions media. Analog is the opposite of digital.

You need to know about analog and digital forms of communication to understand what is required for you to connect your computer to other computer systems and information services. Computers cannot communicate over analog lines. A modem and communications software are usually required to connect a microcomputer user to other computer systems and information services.

analog data *(p. 6, LO 2)* See analog.

applications software *(p. 20, LO 7)* Software that can perform useful work on general-purpose tasks.

Applications software such as word processing, spreadsheet, database manager, graphics, and communications packages have become commonly used tools for increasing people's productivity.

binary digit *(p. 6, LO 2)* See bit.

binary system *(p. 5, LO 2)* Two-state system.

Computer systems use a binary system for data representation—two digits, 0 and 1, to refer to the presence or absence of electrical current or a pulse of light.

bit *(p. 5, LO 2)* Short for *binary digit,* which is either a 1 or a 0 within the binary system of data representation in computer systems.

The bit is the fundamental element of all data and information stored and manipulated in a computer system; 8 bits are combined according to a coding scheme to form a character, such as A. The bit is also the basic unit of measure for describing the capacity of computer hardware storage units and the processing speeds of some other hardware components.

byte *(p. 5, LO 2)* A group of 8 bits.

Bytes—such as 01101110 and 11101100—represent characters according to the particular coding scheme used in a computer system. Bytes are also used to describe computer storage hardware capacities.

CD-ROM (compact disk—read only memory) *(p. 19, LO 3)* Compact optical disk that holds text, graphics, and sound.

CD-ROM disks are used in computer systems to create fancy presentations, store multimedia presentations, and provide reference materials for research.

cellular phone *(p. 24, LO 9)* Mobile, wireless telephone.

Cellular phones further the availability of instant communication, no matter where you are.

central processing unit (CPU) *(p. 16, LO 3)* The processor; it controls and manipulates data to produce information. In a microcomputer the CPU is usually contained on a single integrated circuit or chip called a microprocessor. This chip and other components that make it work are mounted on a circuit board called a motherboard. In larger computers the CPU is contained on one or several circuit boards.

The CPU is the "brain" of the computer; without it, there would be no computers.

*Note to the reader: "LO" refers to Learning Objective; see the first page of the chapter. The number ties the summary term to the appropriate Learning Objective.

communications *(pp. 4, 20, LO 3)* The sixth element of a computer-and-communications system; the electronic transfer of data from one place to another.

Communications systems have helped to expand human communication beyond face-to-face meetings to electronic connections.

communications network *(p. 24, LO 9)* System of inter-connected computers, telephones, or other communications devices that can communicate with one another.

Communications networks allow users to share applications and data; without networks, information could not be electronically exchanged.

communications technology *(p. 4, LO 1)* Consists of electromagnetic devices and systems for communicating over long distances; also called *telecommunications*.

Communications technology enables computers and people to be connected in order to share information resources.

computer *(p. 3, LO 1)* Programmable, multiuse machine that accepts raw data—facts and figures—and processes (manipulates) it into useful information, such as summaries and totals.

Computers greatly speed up the process whereby people solve problems and accomplish many tasks and thus increase their productivity.

computer-and-communications system *(p. 9, LO 3)* System made up of six elements: people, procedures, data/information, hardware, software, and communications.

Users need to understand how the six elements of a computer-and-communications system relate to one another in order to make knowledgeable decisions about buying and using a computer system.

computer online service *(p. 27, LO 9)* Commercial information service that, for a fee, makes available to subscribers various services through their telephone-linked microcomputers.

Online services allow users to, among many other things, make airline reservations, research databases, check on the weather, send e-mail, shop, and bank—all through their keyboards.

connectivity *(p. 25, LO 9)* Ability to connect devices by telecommunications lines to other devices and sources of information.

Connectivity is the foundation of the latest advances in the Digital Age. It provides online access to countless types of information and services.

CRT (cathode-ray tube) *(p. 17, LO 3)* Familiar TV-style picture tube; vacuum tube used as a computer display screen on desktop computers.

CRT display screens provide one of the principal types of output.

cursor *(p. 16, LO 3)* The movable symbol on the screen that shows where data may be entered next.

All applications software packages use cursors to show users where their current "work location" is on the screen.

data *(p. 11, LO 3, 5)* Consists of the raw facts and figures that are processed into information; third element in a computer-and-communications system.

Users need data to create useful information.

database *(p. 26, LO 9)* Collection of integrated, or cross-referenced, electronically stored data that different people may access to use for different purposes.

Users with online connections to database services have enormous research resources at their disposal. In addition, businesses and organizations build databases to help them keep track of and manage their affairs.

digital *(p. 5, LO 2)* Term used synonymously with *computer*; refers to communications signals or information represented in a binary, or two-state, way—1s and 0s, on and off.

The whole concept of an information superhighway is based on the existence of communications in digital form, which allows computers to transmit voice, text, sound, graphics, color, and animation.

digital data *(p. 5, LO 2)* Data represented by discrete (individually distinct), discontinuous transmission bursts of power or light—0s and 1s.

Computers transmit data in digital form, as opposed to the analog form of data transmitted by regular telephone lines.

diskette (floppy disk) *(p. 17, LO 3)* Secondary storage medium; removable round, flexible disk that stores data as magnetized spots. The disk is contained in a plastic case or square paper envelope to prevent the disk from being touched by human hands. Most diskettes are 3½ inches in diameter.

Diskettes are used on all microcomputers.

diskette drive *(p. 18, LO 3)* Computer hardware device that holds, spins, reads from, and writes to magnetic or optical disks.

Users need diskette drives in order to use their disks. Diskette drives can be internal (built into the computer system cabinet) or external (connected to the computer by a cable).

What It Is / What It Does	Why It's Important

documentation *(p. 11, LO 3)* Also called *manuals;* set of instructions, rules, or guidelines (procedures) to follow when using hardware or software.

When users buy a microcomputer or software package, it comes with documentation, in booklet form and often also on disk. This documentation serves as reference material to help users learn how to use a product.

electronic mail *(p. 25, LO 9)* Also called *e-mail;* software-controlled system linking computers by wired or wireless connections that allow users, through their keyboards, to post messages and to read responses on their computer screens.

E-mail allows businesses and organizations to quickly and easily send messages to employees and outside people without having to use and distribute paper messages.

end-user *(p. 10, LO 3)* Also called *user;* a person without much technical knowledge of information technology who uses computers for entertainment, education, and/or work-related tasks.

End-users are the people for whom most computer-and-communications systems are created (by information technology professionals).

fax *(p. 24, LO 3)* Stands for *facsimile transmission;* the communication of text or graphic images between remote locations.

See fax machine.

fax machine *(p. 24, LO 9)* Device that scans an image and sends a copy of it as electronic signals over transmission lines to a receiving fax machine, which recreates the image on paper.

Fax availability has increased the pace at which business can be conducted and thus has improved productivity. Also, fax machines enable people in isolated areas to obtain copies of newspapers and printed information they may otherwise not be able to get.

gigabyte (G or GB) *(p. 12, LO 3)* Unit for measuring storage capacity; equals approximately 1 billion bytes.

This unit of measure is used with supercomputer, mainframe, and even with microcomputer secondary storage devices.

hard disk *(p. 18, LO 3)* Secondary storage medium; generally nonremovable disk made out of metal and covered with a magnetic recording surface. It holds data represented by the presence (1) and absence (0) of magnetized spots. Hard disks, which hold much more data than floppy disks do, are usually built into the computer's system cabinet.

Nearly all microcomputers now use hard disks as their principal secondary storage medium.

hardware *(p. 14, LO 3)* Fourth element in a computer-and-communications system; refers to all machinery and equipment in a computer system. Hardware is classified into five categories: input, processing and memory, output, secondary storage, and communications.

Hardware design determines the type of commands the computer system can follow. However, hardware runs under the control of software and is useless without it.

information *(p. 11, LO 3)* In general, refers to summarized data or otherwise manipulated data. Technically, data comprises raw facts and figures that are processed into information. However, information can also be raw data for the next person or job. Thus sometimes the terms are used interchangeably. Information/data is the third element in a computer-and-communications system.

The whole purpose of a computer (and communications) system is to produce (and transmit) usable information.

information superhighway *(p. 29, LO 10)* Vision or metaphor for a fusion of the two-way wired and wireless capabilities of telephones and networked computers with cable-TV's capacity to transmit hundreds of programs; the resulting interactive digitized traffic would include movies, TV shows, phone calls, databases, shopping services, and online services.

The information superhighway would fundamentally change the nature of communications and hence society, business, government, and personal life.

information technology *(p. 2, LO 1)* Technology that merges computing with high-speed communications links carrying data, sound, and video.

Information technology is bringing about the gradual fusion of several important industries in a phenomenon called *digital convergence* or *technological convergence.*

What It Is / What It Does	**Why It's Important**
information technology professional *(p. 10, LO 3)* Person who has had formal education in the technical aspects of using computer-and-communications systems.	Information technology professionals create and manage the software and systems that enable users (end-users) to accomplish many types of business, professional, and educational tasks and increase their productivity.
input hardware *(p. 14, LO 3)* Devices that allow people to put data into the computer in a form that the computer can use; that is, they perform *input operations*. Input may be by means of a keyboard, pointer, scanner, or voice-recognition device.	Useful information cannot be produced without input data.
input operation *(p. 13, LO 6)* The phase of information processing in which data is captured electronically and converted to a form that can be processed by the computer. The means for entering data is an input device such as a keyboard or scanner.	During this phase, the raw data for producing useful information is put into the computer system for processing.
integrated circuit (IC) *(p. 21, LO 8)* Collection of electrical circuits, or pathways, etched on tiny squares, or chips, of silicon half the size of a person's thumbnail.	The development of the IC enabled the manufacture of the small, powerful, and relatively inexpensive computers used today.
interactivity *(p. 27, LO 9)* Situation in which the user is able to make an immediate response to what is going on and modify processes; that is, there is a dialog between the user and the computer or communications device.	Interactive devices allow the user to actively participate in the ongoing processes instead of just reacting to them.
Internet *(p. 27, LO 9)* International network connecting approximately 36,000 smaller networks that link computers at academic, scientific, and commercial institutions.	The Internet makes possible the sharing of all types of information and services for millions of people all around the world.
keyboard *(p. 14, LO 3)* Input hardware device that uses standard typewriter keys plus a number of specialized keys to input data and issue commands.	Microcomputer users will probably use the keyboard more than any other input device.
kilobyte (K or KB) *(p. 12, LO 3)* Unit for measuring storage capacity; equals 1024 bytes (often rounded off to 1000 bytes).	The sizes of stored electronic files are often measured in kilobytes.
magnetic tape *(p. 18, LO 3)* Secondary storage medium made of flexible plastic coated on one side with magnetic material; data is represented by the presence and absence of magnetized spots.	Nowadays tape is used mainly to provide duplicate (backup) storage, especially for microcomputers.
mainframe computer *(p. 22, LO 8)* Second-largest type of computer available, after the supercomputer; occupies a specially wired, air-conditioned room and is capable of great processing speeds and data storage.	Mainframes are used by large organizations—such as banks, airlines, insurance companies, and colleges—for processing millions of transactions.
megabyte (M or MB) *(p. 12, LO 3)* Unit for measuring storage capacity; equals approximately 1 million bytes.	The storage capacities of most microcomputer hard disks are measured in megabytes. Users need to know how much data their hard disks can hold and how much space new software programs will take so that they do not run out of disk space.
memory *(p. 16, LO 3)* Also called *main memory, primary storage, RAM;* the computer's "work space," where data and programs for immediate processing are held. Memory is contained on chips mounted on the system board.	Memory size determines how much data can be processed at once and how big and complex a program may be used to process it. Memory is usually measured in megabytes.
microcomputer *(p. 22, LO 8)* Small computer that fits on a desktop; used either as a personal computer or a workstation.	Microcomputers are used in virtually every area of modern life. People going into business or professional life today are often required to have basic knowledge of the microcomputer.

What It Is / What It Does	Why It's Important
microcontroller *(p. 23, LO 8)* Also called an *embedded computer;* the smallest category of computer.	Microcontrollers are built into "smart" electronic devices as controlling agents.
microprocessor *(p. 23, LO 8)* Short for "microscopic processor," the miniaturized circuitry of the computer's processor—the part that manipulates data into information—which is etched on a sliver, or "chip," of material such as silicon.	Without the microprocessor, we would not have the microcomputer.
minicomputer *(p. 22, LO 8)* Third-largest type of computer; refrigerator-sized computer that is essentially a scaled-down mainframe; overlaps high-end microcomputers and low-end mainframe computers in price and performance.	Typically used by medium-sized companies for specific purposes, such as accounting.
modem *(p. 6, LO 2)* Communications hardware device for converting a computer's digital signals to analog signals—for transmission over copper telephone wires—and then back to digital signals.	Without modems, computers would not be able to transmit data over most existing telephone wiring.
mouse *(p. 14, LO 3)* Input hardware device that can be rolled about on a desktop to direct a pointer (cursor) on the computer's display screen.	With microcomputers, a mouse is needed to use most graphical user interface programs and to draw illustrations.
multimedia *(p. 27, LO 9)* Refers to technology that presents information in more than one medium, including text, graphics, animation, video, music, and voice.	Use of multimedia is becoming more common in business, the professions, and education as a means of improving the way information is communicated.
operating system *(p. 20, LO 7)* A component of systems software; it acts as the master control program that runs the computer.	The operating system sets the standards for the applications software programs. All programs must "talk to" the operating system.
optical disk *(p. 19, LO 3)* Disk that is written to and read by lasers.	Optical disks hold much more data than many types of magnetic disks.
output hardware *(p. 17, LO 3)* Consists of devices that translate information processed by the computer into a form that humans can understand; that is, they perform *output operations.* Common output devices are monitors (softcopy output) and printers (hardcopy output).	Without output devices, computer users would not be able to view or use their work.
output operation *(p. 13, LO 6)* The phase of data processing in which information, which has been processed from data, is produced in a form usable by people.	The output phase represents the productive aspect of computer-based information processing.
peripheral device *(p. 14, LO 3)* Any hardware device that is connected to a computer. Examples are keyboard, mouse, monitor, printer, and disk drives.	Most of a computer system's input and output functions are performed by peripheral devices.
personal computer (PC) *(p. 22, LO 8)* Desktop or portable (laptop, notebook, or palmtop) microcomputer that can run easy-to-use, personal-assistance programs such as word processing or spreadsheets.	The PC has enabled people to speed up many of their work and learning tasks, thus improving their productivity.
printer *(p. 17, LO 3)* Hardware device that converts computer output into printed images.	Printers provide one of the principal forms of computer output.
procedures *(p. 11, LO 3)* Descriptions of how things are done; steps for accomplishing a result.	Procedures are the second element in a computer-and-communications system. In the form of documentation, procedures help users learn to use hardware and software.
processing operation *(p. 13, LO 6)* The phase of data processing in which data is manipulated to process or transform it into information.	The processing phase represents the critical core of computer-based data processing. It enables people to solve problems quickly and improve their productivity.

read *(p. 18, LO 3)* Refers to the computer obtaining data from a secondary storage medium, such as disk or tape. It means that the data represented in the magnetized or laser-created spots on the disk (or tape) are converted to electronic signals and transmitted to the computer's memory (RAM).

Reading is an essential computer operation.

scanner *(p. 14, LO 3)* Input device that translates images of text, drawings, and photos into digital form.

Scanners simplify the input of complex data. The images can be processed by the computer, manipulated, displayed on a monitor, stored on a storage device, and/or communicated to another computer.

See CRT.

screen *(p. 17, LO 3)* Display area of a computer—also called a *monitor;* either CRT or flat-panel.

secondary storage *(p. 17, LO 3)* Refers to devices and media that store data and programs permanently—such as disks and disk drives, tape and tape drives. These devices perform *storage operations.* Storage capacity is measured in kilobytes, megabytes, gigabytes, and terabytes.

Without secondary storage media, users would not be able to save their work.

secondary storage operation *(p. 13, LO 6)* The phase of data processing in which data, information, or programs are stored in computer-processable form.

The storage phase enables people to save their work for later retrieval, manipulation, and output.

server *(p. 24, LO 9)* Computer shared by several users in a network.

Servers enable users to share data and applications.

software *(p. 19, LO 3)* Also called *programs;* step-by-step instructions that tell the computer hardware how to perform a task. Software represents the fifth element of a computer-and-communications system.

Without software, hardware would be useless.

supercomputer *(p. 22, LO 8)* Largest, fastest, and most expensive type of computer available, costing millions of dollars.

Supercomputers are used for research, weather forecasting, oil exploration, airplane building, complex mathematical operations, and movie special effects, for example.

system unit *(p. 14, LO 3)* Also called the *system cabinet;* housing that includes the electronic circuitry (CPU), which does the actual processing, and main memory, which supports processing.

The microcomputer was born when processing, memory, and power supply were made small enough to fit into a cabinet that would fit on a desktop.

systems software *(p. 20, LO 7)* Software that controls the computer and enables it to run applications software. Systems software, which includes the operating system, allows the computer to manage its internal resources.

Applications software cannot run without systems software.

technological convergence *(p. 2, LO 1)* Also called *digital convergence;* refers to the technological merger of several industries through various devices that exchange information in the electronic, or digital, format used by computers. The industries are computers, communications, consumer electronics, entertainment, and mass media.

From a common electronic base, the same information may be exchanged among many organizations and people.

terabyte (T or TB) *(p. 12, LO 3)* Unit for measuring storage capacity; equals approximately 1 trillion bytes.

The storage capacities of supercomputers are measured in terabytes.

workstation *(p. 23, LO 8)* Expensive, powerful desktop microcomputer used mainly by engineers and scientists for sophisticated purposes.

The development of high-performance workstations has speeded up and improved the previously laborious and time-consuming processes of design, testing, and manufacturing in many industries.

write *(p. 18, LO 3)* Refers to the electronic information processed by the computer being recorded magnetically or by laser onto disk (or tape).

Being able to electronically "write" to disk or tape enables users to save their work.

EXERCISES

(Selected answers appear at the back of the book.)

Short-Answer Questions

1. What does the term *digital* mean?
2. What is a modem used for?
3. What are a kilobyte, a megabyte, a gigabyte, and a terabyte?
4. Briefly describe the five categories of computer hardware.
5. What is the difference between systems and applications software?
6. What is the purpose of the microprocessor in a computer system?
7. To what does the term *connectivity* refer?
8. What do the terms *access* and *online* mean?
9. What is multimedia?
10. What is the Internet?

Fill-in-the-Blank Questions

1. A(n) _Computer_ is a programmable, multiuse machine that accepts data and processes it into information.
2. A(n) _comm. network_ allows two computers to communicate with each other over phone lines.
3. A person who doesn't have much technical knowledge of computers but who uses computers for entertainment, education, or work-related purposes is called a(n) _end-user_.
4. A(n) _megabyte_ is equal to approximately 1 million bytes.
5. Telephones send _analog_ signals, whereas computers send _digital_ signals.
6. The largest, most powerful type of computer is called a _supercomputer_.
7. _Storage_ _hardware_ consists of hardware devices that store data and programs permanently on disk or tape.
8. A(n) _____ _____ stores data on a metal disk and typically stores more than a removable _____.

9. In _telecommuting_, one works at home and communicates with the office via a computer and other technology.
10. _multimedia_ refers to the presentation of information in more than one medium, including text, graphics, animation, video, music, and voice.

Multiple-Choice Questions

1. In a(n) _output_, information is produced in a form usable by people.
   a. input operation
   b. processing operation
   c. output operation
   d. storage operation
   e. all of the above
2. _Data_ refers to the raw facts that are processed into _information_.
   a. Data, information
   b. Information, data
   c. Input, output
   d. Primary storage, secondary storage
   e. none of the above
3. All the machinery and equipment in a computer system is referred to as _hardware_.
   a. software
   b. hardware
   c. the system cabinet
   d. the central processing unit
   e. all of the above
4. Which of the following houses the central processing unit of a microcomputer?
   a. keyboard
   b. monitor
   c. system unit
   d. main memory
   e. none of the above
5. Which of the following stores data and programs for immediate processing?
   a. keyboard
   b. monitor
   c. system unit
   d. main memory
   e. none of the above

6. Which of the following types of software does the computer use to manage its internal resources?
   a. custom-written software
   b. packaged software
   c. applications software
   d. systems software
   e. all of the above

7. Which of the following best describes the direction of computer development over the past 50 years?
   a. smaller size
   b. more power
   c. less expense
   d. all of the above

8. The _____ controls and manipulates data to produce information.
   a. system unit
   b. monitor
   c. central processing unit
   d. CRT
   e. none of the above

9. A _____ is a small pen-controlled, handheld computer that can perform two-way messaging.
   a. smart box
   b. information appliance
   c. personal digital assistant
   d. minicomputer
   e. none of the above

True/False Questions

T F 1. A byte is made up of 8 bits.
T F 2. Most of today's phone lines can carry digital signals.
T F 3. In an input operation, data is transformed into information.
T F 4. A terabyte is bigger than a gigabyte.
T F 5. A piece of hardware that is connected to your computer is commonly referred to as a *peripheral device*.
T F 6. A mouse can translate images of drawings and photos into digital form.
T F 7. Diskettes are removable flexible disks that store data as magnetized spots.
T F 8. CD-ROMs typically store more data than diskettes.
T F 9. Computers are typically categorized into four general types: supercomputers, mainframe computers, minicomputers, and "smart boxes."

 T F 10. Three categories of hardware devices that provide interactivity are multimedia computers, personal digital assistants, and futuristic "smart boxes."

Projects/Critical-Thinking Questions

1. Determine what types of computers are being used where you work or go to school. Are microcomputers being used? Minicomputers? Mainframes? All types? What are they being used for? How are they connected, if at all?

2. Based on what you have learned so far, how do you think the five operations of a computer-and-communications system would be represented in your chosen profession? For example, how would data be input? Output? Under what circumstances? What kind of processing activities would take place? What kinds of communications activities?

3. Describe the computer you use at school, work, or home. Provide details about any input, processing and memory, output, secondary storage, and communications components that are part of your computer system. If you were to spend your own money to improve this computer system, what would you spend it on and why? Do you think you should buy a new computer system instead?

4. In an article for the *Harvard Business Review* (September/October 1992, p. 97), Peter Drucker predicted that, in the next 50 years, schools and universities will change more drastically than they have since they assumed their present form more than 300 years ago, when they reorganized themselves around the printed book. What will force these changes is, in part, new computer-and-communications technology and, in part, the demands of a knowledge-based society in which organized learning must become a life-long process for knowledge workers.

   How do you feel about the prospect of bookless reading and learning? What advantages and disadvantages can you see in using computers instead of books? And how do you feel about perhaps having to renew your fund of knowledge about your job or profession every 4 or 5 years?

5. Although more new information has been produced in the last 30 years than in the previous 5000, information is not knowledge. In our quest for knowledge in the Information Age, we are often overloaded with information that doesn't tell us what we want to know. Richard Wurman identified this problem in his book *Information Anxiety*. John Naisbett, in his books *Megatrends* and *Megatrends 2000*, said that

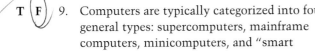

uncontrolled and unorganized information is no longer a resource in an information society. Instead, it becomes the enemy of the information worker.

Identify some of the problems of information overload in one or two departments in your school or place of employment—or in a local business, such as a real estate firm, health clinic, pharmacy, or accounting firm. What types of problems are people having? How are they trying to solve them? Are they rethinking their use of computer-related technologies?

(net) Using the Internet

Objective: *In this exercise we describe how to log onto the Internet and use some basic commands.*

Before you continue: *We assume that you have at least partial access to the Internet through your college, business, or commercial service provider and have a user ID and password. For this exercise, we assume your Internet connection uses: (1) a command-line interface, (2) VT100 terminal emulation, and (3) the Unix operating system. If any of these assumptions are incorrect, you may still be able to perform this Internet exercise; however, some of the commands and screen output may be different. If necessary, ask your instructor or system administrator for assistance.*

1. Access your Internet account using your user name and password. The procedure you use depends on your particular system. If necessary, ask your instructor or lab assistant.

   Sample procedure:

   TYPE: *your user name*
   PRESS: [Enter]
   TYPE: *your user password*
   PRESS: [Enter]

   Next, depending on your system, you may be prompted to choose a terminal emulation mode. If possible, choose VT100 terminal emulation. Otherwise, simply press [Enter] to bypass this option.

2. Guess what? You're on the Internet! At this point you are either presented with a command prompt such as $, %, or > and a blinking cursor, or a menu. If a menu appears, choose the option that exits you to command-line, or shell, mode.

3. The command prompt you see is referred to as the Unix prompt. As you work with the Internet, you will type commands after this prompt and then press [Enter] to execute the commands. Be forewarned that Unix is case sensitive. As a general rule, most Unix commands must be typed using lowercase letters, whereas file names can use a combination of both lower- and uppercase.

For example, to find out about yourself:

TYPE: who am i
PRESS: [Enter]

The output of this command shows your login name, the name of the terminal or line you are using, and the date and time you logged onto the Internet. Your screen may display information that is similar in format to the following:

```
$ who am i
sarahc     ttyp6     Jan 19 12:58     (router-1)
```

4. If you make an error when typing a command, Unix displays an error message followed by the Unix prompt. You can then try the command again.

   For example:

   TYPE: Who am I
   PRESS: [Enter]

   Because you didn't use all lowercase letters, a message similar to the following will appear:

   ```
   $ Who am I
   Who: not found
   ```

   (*Note:* Depending on the version of Unix that you're using on your computer, the error message you see may be slightly different.)

5. Your home directory, or working directory, is active when you log onto the Internet. To see the name of the working directory, you use the *pwd command* (print working directory):

   TYPE: pwd
   PRESS: [Enter]

   Your screen may display information that is similar in format to the following:

   ```
   $ pwd
   /users/home/sarahc
   ```

6. To list the files, if any, in the working directory, you use the *ls command* (list). The -l option provides additional information about any files stored in the working directory.

   TYPE: ls -l
   PRESS: [Enter]

7. Now that you know how to log onto the Internet and use some basic commands, let's disconnect from the Internet. You can exit by typing "exit" or "logout."

   For example:

   TYPE: exit
   PRESS: [Enter]

   Note: If a menu appears, choose the option to disconnect, or quit.

CHAPTER 2

Applications Software
Tools for Thinking & Working

Concepts You Should Know

After reading this chapter, you should be able to:

1. Distinguish between the two principal kinds of software: applications and systems.

2. Discuss the ethics of copying software.

3. Describe the four types of applications software: entertainment, education and reference, productivity, and business and specialized.

4. Explain key features shared by many types of applications software packages.

5. Identify the key functions of word processors, spreadsheets, database managers, graphics programs, communications programs, desktop accessories and personal information managers, integrated programs and suites, groupware, and Internet Web browsers.

6. Briefly describe the key functions of programs for desktop publishing, personal finance, project management, computer-aided design, drawing and painting, and of hypertext.

T hink of it as a map to the buried treasures of the Information Age."

That's how one writer described a particular kind of software named Mosaic when it first came out.[1] Mosaic is designed to help computer users find their way around the Internet. The global "network of networks," the Internet is rich in information but can be baffling to navigate. The developers of Mosaic had tried to remove that difficulty. Indeed, they had hoped their program might be the first "killer app"—killer application—of network computing. That is, it would be a breakthrough development that would help millions of people become comfortable using computer networks, a technology formerly used by only a relative few.

However, Mosaic was not to become the software that would make the Internet available to everyone, being overtaken in a matter of months by another program called Netscape Navigator. Indeed, as this is written, developers are engaged in a titanic struggle to come up with the defining tool that will simplify users' abilities to summon text, as well as sound and images, from among the Internet's many information sources.

Nevertheless, the search for highly useful applications shows how truly important software is. Without software, a computer on a desk is about as useful as half of a pair of bookends. Furthermore, the easier the software is to use, the greater the number of people who will use the hardware.

How to Think About Software

Preview & Review: Software consists of the step-by-step instructions that tell the computer how to perform a task.

Software is of two types, systems software and applications software.

Improvements to a software package may come out as a new version, representing a major upgrade, or as a new release, a minor upgrade.

Mosaic and Netscape Navigator illustrate a trend: software is continually getting easier to use. At one time—and even now, with some software—every user had to learn cryptic commands such as "format a: /n:9 /t:40." Now software is available that lets computer users perform operations by simply pointing to words and images on the display screen and clicking a mouse button. Needless to say, ease of use would probably greatly influence your choice of software.

The Most Popular Uses of Software

Let's get right to the point: What do most people use software for? The answer hasn't changed in a decade. If you don't count games, by far the most popular applications are (1) *word processing* and (2) *spreadsheets*, according to the Software Publishers Association. Moreover, studies show, most people use only a few basic features of these programs, and they use them for rather simple tasks. For example, 70% of all documents produced with word processing software are one-page letters, memos, or simple reports. And 70% of the time people use spreadsheets simply to add up numbers.[2]

This is important information. If you are this type of user, you may have no more need for fancy software and hardware than an ordinary commuter has for an expensive Italian race car. On the other hand, you may be in a profession in which you need to become a "power user," learning almost all the software features available in order to keep ahead in your career.

The Two Kinds of Software: Applications & Systems

As we've said, *software*, or *programs*, **consists of the step-by-step instructions that tell the computer how to perform a task.** Software is of two types. (■ *See Panel 2.1.*)

* **Applications software:** *Applications software* **is software that can perform useful work on general-purpose tasks,** such as word processing and creating spreadsheets.

* **Systems software:** As the user, you interact with the applications software. In turn, *systems software* **enables the applications software to interact with the computer and helps the computer manage its internal resources,** and you can instruct it in some of those tasks.

Software may be either *custom-written* or *packaged:*

* **Custom-written software:** *Custom-written software* **is software designed for a particular customer.** It is the kind of software written by a computer programmer to fulfill a highly specialized task. Unless you go on to become a programmer, you will probably not be required to know the intricacies of this kind of software.

* **Packaged software:** *Packaged software,* **or a** *software package,* **is an "off-the-shelf" program available on disk for use by the general public,** such as word processing or spreadsheets. This is the kind of software discussed in this chapter.

If you buy a new microcomputer in a store, you will find that some software has already been installed on it—that is, is "bundled" with it. This includes systems software and various types of applications software that are compatible with the systems software. We discuss systems software in Chapter 3. At this point, however, you are no doubt more concerned about what you can use a computer *for*. In this chapter, therefore, we describe applications software.

Versions & Releases

Every year or so, software developers find ways to enhance their products and put forth a new *version* or new *release:*

* **Version:** **A** *version* **is a major upgrade in a software product.** Versions are usually indicated by numbers such as 1.0, 2.0, 3.0, and so forth. The higher the number preceding the decimal point, the more recent the version.

* **Release:** **A** *release* **is a minor upgrade.** Releases are usually indicated by a change in number after the decimal point—3.0, then 3.1, then perhaps 3.11, then 3.2, and so on.

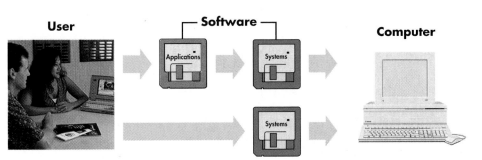

User **Software** **Computer**

Applications Systems Systems

PANEL 2.1

Applications software and systems software

You interact with both the applications and systems software. The systems software interacts with the computer.

Some software developers have departed from this system. Microsoft, for instance, decided to call its new operating system, launched in 1995, "Windows 95" instead of "Windows 4.0." (*Note:* When you buy a new version of a software product, make sure it's compatible with the existing programs and files on your computer.)

Before we continue with the details about various types of applications software, we need to raise the ethical issue of copying intellectual property, including software.

Ethics & Intellectual Property Rights: When Can You Copy?

Preview & Review: Intellectual property consists of the products of the human mind. Such property can be protected by copyright, the exclusive legal right that prohibits copying it without the permission of the copyright holder.

Software piracy, network piracy, and plagiarism violate copyright laws.

Public domain software, freeware, and shareware can be legally copied, which is not the case with proprietary software.

Information technology has presented legislators and lawyers—and you—with some new ethical questions regarding rights to intellectual property. *Intellectual property* consists of the products, tangible or intangible, of the human mind. There are three methods of protecting intellectual property. They are *patents* (as for an invention), *trade secrets* (as for a formula or method of doing business), and *copyrights* (as for a song or a book).

What Is a Copyright?

Of principal interest to us is copyright protection. **A *copyright* is the exclusive legal right that prohibits copying of intellectual property without the permission of the copyright holder.** Copyright law protects books, articles, pamphlets, music, art, drawings, movies—and, yes, computer software. Copyright protects the *expression* of an idea but not the idea itself. Thus, others may copy your idea for, say, a new shoot-'em-up videogame but not your particular variant of it. Copyright protection is automatic and lasts a minimum of 50 years; you do not have to register your idea with the government (as you do with a patent) in order to receive protection.

These matters are important because the Digital Age has made the act of copying far easier and more convenient than in the past. Copying a book on a photocopier might take hours, so people felt they might as well buy the book. Copying a software program onto another floppy disk, however, might take just seconds.

In addition, current copyright law doesn't specifically protect copyrighted material online. Says one article:

> Copyright experts say laws haven't kept pace with technology, especially digitization, the process of converting any data—sound, video, text—into a series of ones and zeros that are then transmitted over computer networks. Using this technology, it's possible to create an infinite number of copies of a book, a record, or a movie and distribute them to millions of people around the world at very little cost. Unlike photocopies of books or pirated audiotapes, the digital copies are virtually identical to the original.[3]

Piracy, Plagiarism, & Ownership of Images & Sounds

Three copyright-related matters deserve our attention: software and network piracy, plagiarism, and ownership of images and sounds.

- **Software and network piracy:** It may be hard to think of yourself as a pirate (no sword or eyepatch) when all you've done is make a copy of some commercial software for a friend. However, from an ethical standpoint, an act of piracy is like shoplifting the product off a store shelf—even if it's for a friend.

  Piracy is theft or unauthorized distribution or use. **Software piracy is the unauthorized copying of copyrighted software.** One way is to copy a program from one diskette to another. A second way is to download (transfer) a program from a network and make a copy of it. **Network piracy is using electronic networks for the unauthorized distribution of copyrighted materials in digitized form.** Record companies, for example, have protested the practice of computer users' sending unauthorized copies of digital recordings over the Internet.[4] Both types of piracy are illegal.

  The easy rationalization is to say that "I'm just a poor student, and making this one copy or downloading only one digital recording isn't going to cause any harm." But it is the single act of software piracy multiplied millions of times that is causing the software publishers a billion-dollar problem. They point out that the loss of revenue cuts into their budget for offering customer support, upgrading products, and compensating their creative people. Piracy also means that software prices are less likely to come down; if anything, they are more likely to go up.

- **Plagiarism:** *Plagiarism* is the expropriation of another writer's text, findings, or interpretations and presenting it as one's own. Information technology puts a new face on plagiarism in two ways. On the one hand, it offers plagiarists new opportunities to go far afield for unauthorized copying. On the other hand, the technology offers new ways to catch people who steal other people's material.

  Electronic online journals are not limited to a specific number of pages, and so they can publish papers that attract a small number of readers. In recent years, there has been an explosion in the number of such journals and of their academic and scientific papers. This proliferation may make it harder to detect when a work has been plagiarized, since few readers will know if a similar paper has been published elsewhere.[5]

  Yet information technology may also be used to identify plagiarism. Scientists have used computers to search different documents for identical passages of text. In 1990, two "fraud busters" at the National Institutes of Health alleged after a computer-based analysis that a prominent historian and biographer had committed plagiarism in his books. The historian, who said the technique turned up only the repetition of stock phrases, was later exonerated in a scholarly investigation.[6]

- **Ownership of images and sounds:** Computers, scanners, digital cameras, and the like make it possible to alter images and sounds to be almost anything you want. What does this mean for the original copyright holders? An unauthorized sound snippet of James Brown's famous howl can be electronically transformed by digital sampling into the background music for dozens of rap recordings.[7] Images can be appropriated by scanning them into a computer system, then altered or placed in a new context.

Software Piracy Estimates

More than half of all software in existence today is lost to piracy. Top countries reporting software losses in 1994.

Country	Dollars lost	% illegal
United States	$2,876,922,400	35%
Japan	2,075,809,729	67
Germany	1,874,741,352	50
France	771,460,734	57
Brazil	550,936,140	95
Korea	545,926,907	78
United Kingdom	543,516,297	43
Russian Federation	540,564,400	94
People's Republic of China	526,740,300	98
Italy	404,382,943	58

Because the line between artistic license and copyright infringement isn't always clear cut, a growing number of artists who use recycled material are taking steps to protect themselves from lawsuits.

These are the general issues you need to consider when you're thinking about how to use someone else's intellectual property in the Digital Age. Now let's see how software fits in.

Public Domain Software, Freeware, & Shareware

No doubt most of the applications programs you will study in conjunction with this book will be commercial software packages, with brand names such as Microsoft Word or Lotus 1-2-3. However, there are a number of software products—many available over communications lines from the Internet—that are available to you as *public domain software, freeware,* or *shareware.*

- **Public domain software:** ***Public domain software* is software that is not protected by copyright and thus may be duplicated by anyone at will.** Public domain programs—usually developed at taxpayer expense by government agencies—have been donated to the public by their creators. They are often available through sites on the Internet (or electronic bulletin boards) or through computer users groups. A users group is a club, or group, of computer users who share interests and trade information about computer systems.

  You can duplicate public domain software without fear of legal prosecution. (Beware: Downloading software through the Internet may introduce some problems—bad code called *viruses*—into your system. We discuss this problem, and how to prevent it, in Chapter 8.)

- **Freeware:** ***Freeware* is software that is available free of charge.** Freeware is distributed without charge, also usually through the Internet or computer users groups.

  Why would any software creator let the product go for free? Sometimes developers want to see how users respond, so they can make improvements in a later version. Sometimes it is to further some scholarly purpose, such as to create a standard for software on which people are apt to agree because there is no need to pay for it.

  Freeware developers often retain all rights to their programs, so that technically you are not supposed to duplicate and distribute it further. Still, there is no problem about your making several copies for your own use.

- **Shareware:** ***Shareware* is copyrighted software that is distributed free of charge but requires users to make a contribution in order to receive technical help, documentation, or upgrades.** Shareware, too, is distributed primarily through communications connections such as the Internet.

  Is there any problem about making copies of shareware for your friends? Actually, the developer is hoping you will do just that. That's the way the program gets distributed to a lot of people—some of whom, the software creator hopes, will make a "contribution" or pay a "registration fee" for advice or upgrades.

  Though copying shareware is permissible, because it is copyrighted you cannot use it as the basis for developing your own program in order to compete with the developer.

Proprietary Software & Software Licenses

Proprietary software **is software whose rights are owned by an individual or business,** usually a software developer. The ownership is protected by the copyright, and the owner expects you to buy a copy in order to use it. The software cannot legally be used or copied without permission.

Software manufacturers don't sell you the software so much as sell you a license to become an authorized user of it. What's the difference? In paying for a *software license,* **you sign a contract in which you agree not to make copies of the software to give away or for resale.** That is, you have bought only the company's permission to use the software and not the software itself. This legal nicety allows the company to retain its rights to the program and limits the way its customers can use it.[8] The small print in the licensing agreement allows you to make one copy (working copy or archival copy) for your own use.

The Software Police

Industry organizations such as the Software Publishers Association (hotline for reporting illegal copying: 800-388-7478) are going after software pirates large and small. Commercial software piracy is now a felony, punishable by up to five years in prison and fines of up to $250,000 for anyone convicted of stealing at least ten copies of a program or more than $2500 worth of software. Campus administrators are getting tougher with offenders and are turning them over to police.

The Four Types of Applications Software

Preview & Review: The four types of applications software may be considered to be (1) entertainment software, (2) education and reference software, (3) basic productivity software, and (4) business and specialized software.

Basic productivity tools include word processing, spreadsheet, database manager, presentation graphics, communications, desktop accessories and personal information managers, integrated programs and suites, groupware, and Internet Web browsers.

Software can change the way we act, even the way we think. Some readers may intuitively understand this because they grew up playing videogames. Indeed, some observers hold that videogames are not quite the time wasters we have been led to believe. However, these forms of entertainment are only a way station to something else. Videogames are training wheels for using more sophisticated software that can help us learn better and be more productive.

There are four types of applications software. (■ *See Panel 2.2.)* They are:

- Entertainment software
- Education and reference software
- Basic productivity software
- Business and specialized software

Entertainment Software: The Serious Matter of Videogames

Whatever else may come about during the convergence of computers and communications, you can bet one kind of software will be available: entertainment—videogames in particular. A $6.5-billion-a-year industry in the

PANEL 2.2

**Applications software:
The four types**

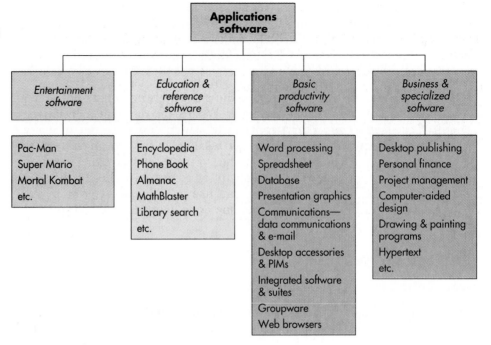

United States, *videogames* are interactive electronic games that may be played at home through a television set or personal computer or in entertainment arcades of the sort found in shopping malls.[9]

It was Pong—an electronic version of table tennis introduced by Atari in 1972—that popularized computers in the home. "Pong was the first time people saw computers as friendly and approachable," states one technology writer. "It launched a videogame boom that made thousands of kids want to become computer programmers, and prepared an entire generation for interaction with a blinking and buzzing computer screen."[10]

Pong was followed by Space Invaders and Pac-Man, and then by Super Mario, which led to Mortal Kombat I and II. In 1986 Nintendo began to reshape the market when it introduced 8-bit entertainment systems. *Bit numbers* measure how much data a computer chip can process at one time. Bit (✓ p. 6) numbers are important because the higher the bit number, the greater the screen resolution (clarity), the more varied the colors, and the more complex the games.[11] Since then, videogame hardware—which, after all, is just computer hardware—has increased in power just as microcomputers have. In the 1990s, videogame hardware manufacturers—Sega, 3DO, Atari—upped the ante to 16 bits, then 32 bits, until finally 64-bit machines were appearing on the market.

Educational & Reference Software

Because of the popularity of videogames, many educational software companies have been blending educational content with action and adventure—as in MathBlaster or the problem-solving game Commander Keen. They hope this marriage will help students be more receptive to learning. After all, as one writer points out, players of Nintendo's Super Mario Brothers must "become intimately acquainted with an alien landscape, with characters, artifacts, and rules completely foreign to ordinary existence. . . . Children assimilate this essentially useless information with astonishing speed."[12] Why not, then, design software that would educate as well as entertain?

Computers alone won't boost academic performance, but they can have a positive effect on student achievement in all major subject areas, preschool through college, according to an independent consulting firm, New York's Interactive Educational Systems Design. Skills improve when students use programs that are self-paced or contain interactive video. This is particularly true for low-achieving students. The reason, says a representative of the firm, which analyzed 176 studies done over the past 5 years, is that this kind of educational approach is "a different arena from the one in which they failed, and they have a sense of control."[13]

In addition to educational software, library search and reference software have become popular. For instance, there are CD-ROMs with encyclopedias, phone books, voter lists, mailing lists, maps, home-remodeling how-to information, and reproductions of famous art. With the CD-ROM encyclopedia Microsoft Encarta, for example, you can search for, say, music in 19th-century Russia, then listen to an orchestral fragment from Tchaikovsky's *1812 Overture.*

Basic Productivity Software

Basic productivity software consists of programs found in most offices and probably on all campuses, on personal computers and on larger computer systems. Their purpose is simply to make users more productive at performing general tasks.

The most popular kinds of productivity tools are:

- Word processing software
- Spreadsheet software
- Database software
- Presentation graphics software
- Communications software—both data communications and e-mail
- Desktop accessories and personal information managers
- Integrated software and suites
- Groupware
- Internet Web browsers

We will describe these common productivity software tools shortly.

Business & Specialized Software

Whatever your occupation, you will probably find it has specialized software available to it. This is so whether your career is as an architect, building contractor, chef, dairy farmer, dance choreographer, horse breeder, lawyer, nurse, physician, police officer, tax consultant, or teacher.

Some business software is of a general sort used in all kinds of enterprises, such as accounting software, which automates bookkeeping tasks, or payroll software, which keeps records of employee hours and produces reports for tax purposes. Other software is more specialized. Some programs help lawyers or advertising people, for instance, keep track of hours spent on particular projects for billing purposes. Other programs help construction estimators pull together the costs of materials and labor needed to estimate the costs of doing a job.

Later in this chapter we describe the following kinds of specialized software: *desktop publishing, personal finance, project management, computer-aided design, drawing and painting programs,* and *hypertext.*

The User Interface & Other Basic Features

Preview & Review: Applications software packages share some basic features and functions. They use special-purpose keys, function keys, and a mouse to issue commands and choose options. Their graphical user interfaces use menus, Help screens, windows, icons, and dialog boxes to make it easy for people to use the program.

Features of the Keyboard

We will describe the keyboard as an input device in Chapter 5. Here, however, we need to explain some aspects of the keyboard because it and the mouse are the means for manipulating software.

Besides a typewriter-like layout of letter, number, and punctuation keys and often a calculator-style numeric keypad, computer keyboards have special-purpose and function keys.

- **Special-purpose keys:** *Special-purpose keys* **are used to enter, delete, and edit data and to execute commands.** An example is the Esc (for "Escape") key. The most important is the Enter key, which you will use often to tell the computer to execute commands entered with other keys. *Commands* are instructions that cause the software to perform specific actions. For example, pressing the Esc key commands the computer, via software instructions, to cancel an operation or leave ("escape from") the current mode of operation.

  Special-purpose keys are generally used the same way regardless of the applications software package being used. Most IBM-style keyboards include the following special-purpose keys: Esc, Ctrl, Alt, Del, Ins, Home, End, PgUp, PgDn, Num Lock, and a few others. (*Ctrl* means Control, *Del* means Delete, *Ins* means Insert, for example.)

- **Function keys:** *Function keys,* **labeled F1, F2, and so on, are positioned along the top or left side of the keyboard. They are used for commands specific to the software being used.** For example, one applications software package may use F6 to exit a file, whereas another may use F6 to underline a word.

  Many software packages come with templates that you can attach to the keyboard. Like the explanation of symbols on a roadmap, the template explains the purpose of each function key and certain combinations of keys. For example, in one word processing program, pressing Alt and F6 at the same time means "position these lines flush right on the page."

- **Macros:** Sometimes you may wish to reduce the number of keystrokes required to execute a command. To do this, you use a macro. **A *macro* is a single keystroke or command—or a series of keystrokes or commands— used to automatically issue a longer, predetermined series of keystrokes or commands.** Thus, you can consolidate several keystrokes for a command into only one or two keystrokes. The user names the macro and stores the corresponding command sequence; once this is done, the macro can be used repeatedly.

  Although many people have no need for macros, others who find themselves continually repeating complicated patterns of keystrokes say they are quite useful.

The User Interface: GUIs, Menus, Help Screens, Windows, Icons, & Dialog Boxes

The first thing you look at when you call up any applications software on the screen is the user interface. **The *user interface* is the part of the software that displays information and presents on the screen the various commands by which you communicate with it.** The type of user interface is usually determined by the systems software (discussed in the next chapter). However, because this is what you see on the screen before you can begin using the applications software, we will briefly describe it here.

[handwritten margin note: this is the part of the operating system that allows you to communicate + interact with it.]

Some user interfaces require that you indicate your commands by typing in characters and text. However, the kind of interface now used by most people is the graphical user interface. **With a *graphical user interface*, or *GUI* (pronounced "gooey"), you may use graphics (images) and menus as well as keystrokes to choose commands, start programs, and see lists of files and other options.**

Common features of GUIs are *menus, Help screens, windows, icons, buttons,* and *dialog boxes.* (■ *See Panel 2.3, p. 52.*)

- **Menus:** **A *menu* is a list of available commands presented on the screen.** Menus may appear as menu bars or pull-down menus.

  A *menu bar* is a line or two of command options across the top or bottom of the screen. Examples of commands, which you activate with a mouse or with key combinations, are File, Edit, and Print.

  A *pull-down* menu is a list of command options that "drops down" from a selected menu bar item at the top of the screen. For example, you might use the mouse to "click on" (activate) a command (for example, File) on the menu bar, which in turn would yield a pull-down menu offering further options. Choosing one of these options may produce further menus called *pop-up menus,* which seem to appear out of nowhere on the screen.

  A particularly useful type of menu is the **Help menu, or *Help screen*, which offers assistance on how to perform various tasks,** such as printing out a document. Having a set of Help screens is like having a built-in electronic instruction manual.

- **Windows:** A particularly interesting feature of GUIs is the use of windows. **A *window* is a rectangle that appears on the screen and displays information from a particular part of a program.** A display screen may show more than one window—for instance, one showing information from a word processing program, another information from a spreadsheet.

  A window (small w) should not be confused with *Microsoft Windows* (capital W), which is the most popular form of systems software. However, as you might expect, Windows features extensive use of windows.

- **Icons:** **An *icon* is a picture used in a GUI to represent a command, a program, or a task.** For example, a picture of a floppy disk might represent the command "Save (store) this document." Icons are activated by a mouse or other pointing device.

- **Buttons:** **A *button* is a simulated on-screen button (kind of icon) that is activated ("pushed") by a mouse or other pointing device to issue a command,** such as "Print document."

- **Dialog box:** **A *dialog box* is a box that appears on the screen and displays a message requiring a response from you,** such as pressing Y for "Yes" or N for "No" or typing in the name of a file.

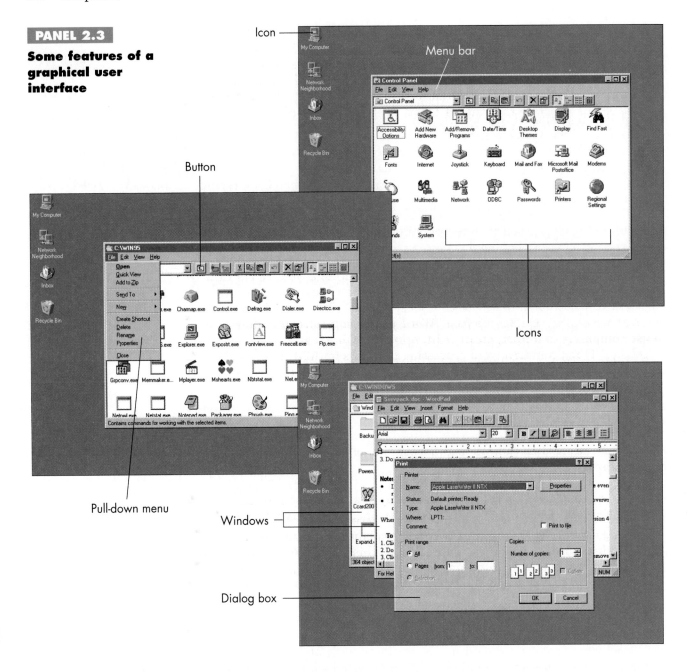

PANEL 2.3

Some features of a graphical user interface

Tutorials & Documentation

How are you going to learn a given software program? Most commercial packages come with tutorials and documentation.

- **Tutorials:** *A tutorial is an instruction book or program that takes you through a prescribed series of steps to help you learn how to use the product.* For instance, our publisher offers several how-to books, the Irwin Advantage Series, that enable you to learn different kinds of software. Tutorials can also be on-screen, provided as part of the software package.

- **Documentation:** *Documentation is a user manual (book) or reference manual that is a narrative and graphical description of a program.* Documentation may be instructional, but features and functions are usually grouped by category for reference purposes. For example, in word processing documentation, all cut-and-paste features are grouped together so you can easily look them up if you have forgotten how to perform them.

Often you can ask your software for directions on how to use the software. That is, some software makers equip their programs with features that allow you to ask questions in plain English. Thus, if you type "How do I add up the numbers in this column?" the software will respond by directing you to an interactive tutor or "coach" that can help you through the procedure.

Now let us consider the various forms of applications software used as productivity tools, plus a few specialized tools.

Word Processing

Preview & Review: Word processing software allows you to use computers to format, create, edit, print, and store text material.

One of the first typewriter users was Mark Twain. However, the typewriter, that long-lived machine, has gone to its reward. Indeed, if you have a manual typewriter, it is becoming as difficult to get it repaired as it is to find a blacksmith. What, then, has replaced the typewriter? The answer is the computer and word processing software. **Word processing software allows you to use computers to format, create, edit, print, and store text material.** (■ *See Panel 2.4.*) Three common word processing programs for IBM-style Windows computers are Microsoft Word, WordPerfect, and Ami Pro. For Macintoshes they are Word and MacWrite.

fceps

Formatting Documents

Formatting means determining the appearance of a document. There are many choices here.

5

- **Type:** You can decide what *typeface* and *type size* you wish to use. You can specify what parts of it should be <u>underlined</u>, *italic,* or **boldface.**
- **Spacing and columns:** You can choose whether you want the lines to be *single-spaced* or *double-spaced* (or something else). You can specify whether you want text to be *one column* (like this page), *two columns* (like many magazines and books), or *several columns* (like newspapers).

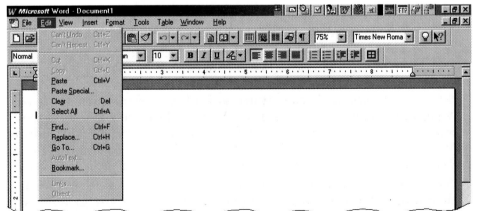

PANEL 2.4

Word processing

This Microsoft Word 7.0 for Windows program has a pull-down Edit menu that offers, among other things, several options for moving text around in a document and finding and replacing words.

- **Margins and justification:** You can indicate the dimensions of the *margins*—left, right, top, and bottom—around the text.

  You can specify whether the text should be *justified* or not. *Justify* means to align text evenly between left and right margins, as, for example, is done with most newspaper columns and this text. *Left-justify* means to not align the text evenly on the right side, as in many business letters ("ragged right").

- **Pages, headers, footers:** You can indicate *page numbers* and *headers* or *footers*. A *header* is common text (such as a date or document name) that is printed at the top of every page. A *footer* is the same thing printed at the bottom of every page.

- **Other formatting:** You can specify *borders* or other decorative lines, *shading*, *tables*, and *footnotes*. You can even pull in ("import") *graphics* or drawings from files in other software programs.

It's worth noting that word processing programs (and indeed most forms of applications software) come from the manufacturers with *default settings*. *Default settings* are the settings automatically used by a program unless the user specifies otherwise, thereby overriding them. Thus, for example, most word processing programs will automatically prepare a document single-spaced, left-justified, with 1-inch right and left margins unless you alter these default settings.

Creating Documents

Creating a document means entering text, using the keyboard. Word processing software has three features that you will not encounter with a typewriter—the *cursor, scrolling,* and *word wrap.* (■ See Panel 2.5.)

- **Cursor:** **The *cursor* is the movable symbol on the display screen that shows you where you may enter data or commands next.** The symbol is often a blinking rectangle or I-beam. You can move the cursor on the screen using the keyboard's directional arrow keys or an electronic mouse.

- **Scrolling:** ***Scrolling* is the activity of moving quickly upward or downward through the text or other screen display.** A standard computer screen displays only 20–22 lines of standard-size text. Of course, most documents are longer than that. Using the directional arrow keys or a mouse, you can move ("scroll") through the display screen and into the text above and below it.

- **Word wrap:** *Word wrap* automatically continues text on the next line when you reach the right margin. That is, the text "wraps around" to the next line. You do not need to press a carriage-return key when you reach the right margin, as you would on a typewriter.

Editing Documents

Editing is the act of making alterations in the content of your document. Some features of editing that can't be found on a typewriter are *insert and delete, undelete, search and replace, block and move, spelling checker, grammar checker,* and *thesaurus.*

- **Insert and delete:** *Inserting* is the act of adding to the document. You simply place the cursor wherever you want to add text and start typing; the existing characters will move aside.

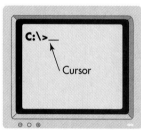

Cursor

Scrolling

PANEL 2.5

Cursor and scrolling

Deleting is the act of removing text, usually using the Delete or Back-space keys.

The *Undelete command* allows you to change your mind and restore text that you have deleted. Some word processing programs offer as much as 100 layers of "undo," allowing users who delete several blocks of text, but then change their minds, to reinstate one or more of the blocks.

- **Search and replace:** **The *Search command* allows you to find any word, phrase, or number that exists in your document. The *Replace command* allows you to automatically replace it with something else.**

- **Block and move:** Typewriter users were accustomed to using scissors and glue to "cut and paste" to move a paragraph or block of text from one place to another in a manuscript. With word processing, you can exercise the ***Block command* to indicate the beginning and end of the portion of text you want to move.** Then you can use the ***Move command* to move it to another location in the document.**

  You can also use the *Copy command* to copy the block of text to a new location while also leaving the original block where it is.

- **Spelling checker, grammar checker, thesaurus:** Many writers auto-matically run their completed documents through a ***spelling checker, which tests for incorrectly spelled words.*** (Some programs, such as Microsoft Word 6.0, have an "Auto Correct" function that automatically fixes such common mistakes as transposed letters—"teh" instead of "the.") Another feature is a ***grammar checker, which flags poor grammar, wordiness, incomplete sentences, and awkward phrases.***

  If you find yourself stuck for the right word while you're writing, you can call up an on-screen ***thesaurus, which will present you with the appro-priate word or alternative words.***

Printing Documents

Most word processing software gives you several options for printing. For example, you can print *several copies* of a document. You can print *indi-vidual pages* or *a range of pages.* You can even preview a document before printing it out. *Previewing (print previewing)* means viewing a document on screen to see what it will look like in printed form before it's printed. Whole pages are displayed in reduced size.

Saving Documents

Saving means to store, or preserve, the electronic files of a document per-manently on diskette, hard disk, or magnetic tape. Saving is a feature of nearly all applications software, but anyone accustomed to writing with a typewriter will find this activity especially valuable. Whether you want to make small changes or drastically revise your word processing document, having it stored in electronic form spares you the arduous chore of having to retype it from scratch. You need only call it up from disk or tape and make just those changes you want, then print it out again.

Spreadsheets

Preview & Review: Spreadsheet software allows users to create tables and financial schedules by entering data into rows and columns arranged as a grid on a display screen. If one (or more) numerical value or formula is changed, the software automatically calculates the effect of the change on the rest of the spreadsheet.

Spreadsheet software also allows users to create analytical graphics charts to present data.

What is a spreadsheet? Traditionally, it was simply a grid of rows and columns, printed on special green paper, that was used by accountants and others to produce financial projections and reports. A person making up a spreadsheet often spent long days and weekends at the office penciling tiny numbers into countless tiny rectangles. When one figure changed, all the rest of the numbers on the spreadsheet had to be recomputed—and ultimately there might be wastebaskets full of jettisoned worksheets.

In the late 1970s, Daniel Bricklin was a student at the Harvard Business School. One day he was staring at columns of numbers on a blackboard when he got the idea for computerizing the spreadsheet. The result, VisiCalc, was the first of the electronic spreadsheets. **An *electronic spreadsheet*, also called simply a *spreadsheet*, allows users to create tables and financial schedules by entering data into rows and columns arranged as a grid on a display screen.**

The electronic spreadsheet quickly became the most popular small-business program. Unfortunately for Bricklin, VisiCalc was shortly surpassed by Lotus 1-2-3, a sophisticated program that combines the spreadsheet with database and graphics programs. Today the principal spreadsheets are Microsoft Excel, Lotus 1-2-3, and Quattro Pro.

Principal Features

The arrangement of a spreadsheet is as follows. (■ *See Panel 2.6.*)

- **Columns, rows, and labels:** *Column headings* appear across the top ("A" is the name of the first column, "B" the second, and so on). *Row headings* appear down the left side ("1" is the name of the first row, "2" the second, and so forth). Labels are any descriptive text, such as APRIL, PHONE, or GROSS SALES.

- **Cells, cell addresses, values, and spreadsheet cursor:** **The place where a row and a column intersect is called a *cell*, and its position is called a *cell address*.** For example, "A1" is the cell address for the top left cell, where column A and row 1 intersect. A number entered in a cell is called a *value*. The values are the actual numbers used in the spreadsheet—dollars, percentages, grade points, temperatures, or whatever. A *cell pointer*, or *spreadsheet cursor*, indicates where data is to be entered. The cell pointer can be moved around like a cursor in a word processing program.

- **Formulas, functions, and recalculation:** Now we come to the reason the electronic spreadsheet has taken offices by storm. **Formulas are instructions for calculations.** For example, a formula might be @SUM(A5..A15), meaning "Sum (add) all the numbers in the cells with cell addresses A5 through A15."

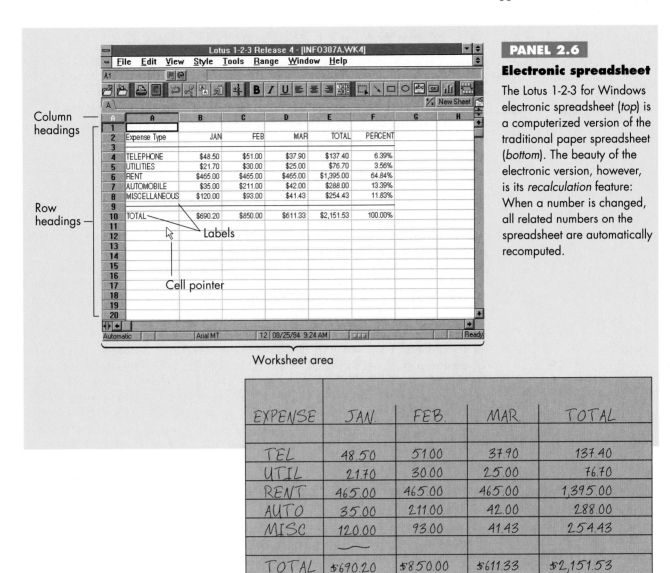

PANEL 2.6

Electronic spreadsheet

The Lotus 1-2-3 for Windows electronic spreadsheet (*top*) is a computerized version of the traditional paper spreadsheet (*bottom*). The beauty of the electronic version, however, is its *recalculation* feature: When a number is changed, all related numbers on the spreadsheet are automatically recomputed.

Functions are stored formulas that perform common calculations. For instance, a function might average a range of numbers or round off a number to two decimal places.

After the values have been plugged into the spreadsheet, the formulas and functions can be used to calculate outcomes. What is revolutionary, however, is the way the spreadsheet can easily do recalculation. **Recalculation is the process of recomputing values *automatically*, either as an ongoing process as data is being entered or afterward, with the press of a key.** With this simple feature, the hours of mind-numbing work required in manually reworking paper spreadsheets became a thing of the past.

• **The "what if?" world:** The recalculation feature has opened up whole new possibilities for decision making. As a user, you can create a plan, put in formulas and numbers, and then ask yourself, "What would happen if we change that detail?"—and immediately see the effect on the bottom line. You could use this if you're considering buying a new car. Any number of things can be varied: total price ($10,000? $15,000?), down payment ($2,000? $3,000?), interest rate on the car loan (7%? 8%?), or number of months to pay (36? 48?). You can keep changing the "what if" possibilities until you arrive at a monthly payment figure that you're comfortable with.

Analytical Graphics: Creating Charts

A nice feature of spreadsheet packages is the ability to create analytical graphics. **Analytical graphics, or business graphics, are graphical forms that make numeric data easier to analyze** than when it is in the form of rows and columns of numbers, as in electronic spreadsheets. Whether viewed on a monitor or printed out, analytical graphics help make sales figures, economic trends, and the like easier to comprehend and analyze.

The principal examples of analytical graphics are *bar charts*, *line charts*, and *pie charts*. (■ *See Panel 2.7.*) Quite often these charts can be displayed or printed out so that they look three-dimensional. Spreadsheets can even be linked to more exciting graphics, such as digitized maps.

Database Software

Preview & Review: A database is a computer-based collection of interrelated files. Database software is a program that controls the structure of a database and access to the data.

In its most general sense, a database is any electronically stored collection of data in a computer system. In its more specific sense, **a *database* is a collection of interrelated files** in a computer system. These computer-based files are organized according to their common elements, so that they can be retrieved easily. (Databases are covered in detail in Chapter 7.) Sometimes called a *database manager* or *database management system (DBMS)*, **database software is a program that controls the structure of a database and access to the data.**

The Benefits of Database Software

Because it can access several files at one time, database software is much better than the old file managers (also known as flat-file management systems) that used to dominate computing. A *file manager* is a software package that can access only one file at a time. With a file manager, you could call up a list of, say, all students at your college majoring in English. You

---

PANEL 2.7

Analytical graphics

Bar charts, line graphs, and pie charts are used to display numerical data in graphical form.

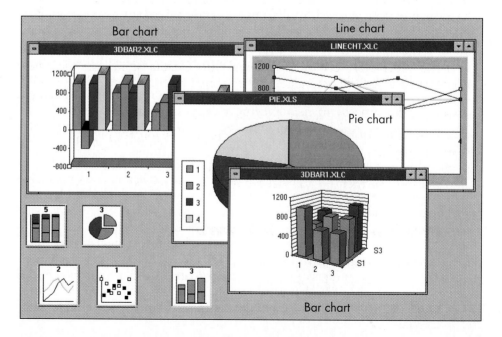

could also call up a separate list of all students from Wisconsin. But you could not call up a list of English majors from Wisconsin, because the relevant data is kept in separate files. Database software allows you to do that.

Databases are a lot more interesting than they used to be. Once they included only text. The Digital Age has added new kinds of information—not only documents but also pictures, sound, and animation. It's likely, for instance, that your personnel record in a future company database will include a picture of you and perhaps even a clip of your voice. If you go looking for a house to buy, you will be able to view a real estate agent's database of video clips of homes and properties without leaving the realtor's office.[14] Today the principal database software packages are dBASE, Access, Paradox, Filemaker Pro for Windows, FoxPro for Windows, Q&A for Windows, and Approach for Windows.

Databases have gotten easier to use, but they still can be difficult to set up. Even so, the trend is toward making such programs easier for both database creators and database users.

Principal Features of Database Software

Some features of databases are as follows: *OSSC*

- **Organization of a database:** A database is organized—from smallest to largest items—into *fields*, *records*, and *files*. *frf*

   A *field* is a unit of data consisting of one or more characters. An example of a field is your name, your address, or your driver's license number.

   A *record* is a collection of related fields. An example of a record would be your name *and* address *and* driver's license number.

   A *file* is a collection of related records. An example of a file could be one in your state's Department of Motor Vehicles. The file would include everyone who received a driver's license on the same day, including their names, addresses, and driver's license numbers.

- **Select and display:** The beauty of database software is that you can locate records in the file quickly. For example, your college may maintain several records about you—one at the registrar's, one in financial aid, one in the housing department, and so on. Any of these records can be called up on a computer display screen for viewing and updating. Thus, if you move, your address field will need to be changed in all records. The database is quickly corrected by finding your name field. Once the record is displayed, the address field can be changed.

- **Sort:** With database software you can easily change the order of records in a file. Normally, records are entered into a database in the order they occur, such as by the date a person registered to attend college. However, all these records can be sorted in different ways. For example, they can be rearranged by state, by age, or by Social Security number.

- **Calculate and format:** Many database programs contain built-in mathematical formulas. This feature can be used, for example, to find the grade-point averages for students in different majors or in different classes. Such information can then be organized into different formats and printed out.

Presentation Graphics Software

Preview & Review: Presentation graphics software allows people to create graphical representations of data to present to other people. This type of graphics is more sophisticated than the analytical graphics produced by spreadsheet packages.

Computer graphics can be highly complicated, such as those used in special effects for movies (such as *Toy Story* or *Jurassic Park)*. Here we are concerned with just one kind of graphics called presentation graphics.

***Presentation graphics* are graphics used to communicate or make a presentation of data to others,** such as clients or supervisors. Presentations may make use of bar, line, and pie charts, but they usually look much more sophisticated, using, for instance, different texturing patterns (speckled, solid, cross-hatched), color, and three-dimensionality. (■ *See Panel 2.8.)* Examples of well-known presentation graphics packages are Microsoft PowerPoint, Aldus Persuasion, Lotus Freelance Graphics, and SPC Harvard Graphics. In general, these graphics are presented as *slides*, which can be projected on a screen or displayed on a large monitor.

Communications Software

Preview & Review: Communications software manages the transmission of data between computers. It also enables users to send and receive electronic mail.

In the past, many microcomputer users felt they had all the productivity they needed without ever having to hook up their machines to a telephone. One of the major themes of this book, however, is that having communications capabilities vastly extends your range. This great leap forward is made possible with communications software. Two types of communications software are *data communications software* and *electronic mail software.*

Data Communications Software

***Data communications software* manages the transmission of data between computers.** For most microcomputer users this sending and receiving of data is by way of a modem and a telephone line. As described in Chapter 1 (✓ p. 6), a *modem* is an electronic device that allows computers to communicate with each other over telephone lines. The modem translates the digital signals of the computer into analog signals that can travel over telephone lines to another modem, which translates the analog signals back to digital. When you buy a modem, you often get communications software with it. Popular microcomputer communications programs are Crosstalk and Procomm Plus.

PANEL 2.8

Presentation graphics

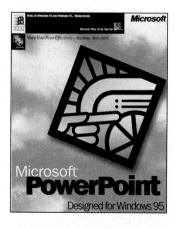

Electronic Mail Software

Electronic mail (e-mail) software **enables users to send letters and documents from one computer to another.** Many organizations have "electronic mailboxes." If you were a sales representative, for example, such a mailbox would allow you to transmit a report you created on your word processor to a sales manager in another area. Or you could route the same message to a number of users on a distribution list.

Desktop Accessories & Personal Information Managers

Preview & Review: Desktop accessory software provides an electronic version of tools or objects commonly found on a desktop: calendar, clock, card file, calculator, and notepad.

Personal information manager (PIM) software combines some features of word processing, database manager, and desktop accessory programs to organize specific types of information, such as address books.

Pretend you are sitting at a desk in an old-fashioned office. You have a calendar, clock, calculator, Rolodex-type address file, and notepad. Most of these items could also be found on a student's desk. How would a computer and software improve on this arrangement?

Many people find ready uses for types of software known as *desktop accessories* and *personal information managers (PIMs)*.

Desktop Accessories

A *desktop accessory,* or *desktop organizer,* **is a software package that provides an electronic version of tools or objects commonly found on a desktop: calendar, clock, card file, calculator, and notepad.**

Some desktop-accessory programs come as standard equipment with some systems software (such as with Microsoft Windows). Others, such as Borland's SideKick or Lotus Agenda, are available as separate programs to run in your computer's main memory at the same time you are running other software. Some are principally *scheduling and calendaring programs;* their main purpose is to enable you to do time and event scheduling.

Personal Information Managers

A more sophisticated program is the **personal information manager (PIM), a combination word processor, database, and desktop accessory program that organizes a variety of information.** Examples of PIMs are Commence, Dynodex, Ecco, Lotus Organizer, and Franklin Planner.

Lotus Organizer, for example, looks much like a paper datebook on the screen—down to simulated metal rings holding simulated paper pages. The program has screen images of section tabs labeled Calendar, To Do, Address, Notepad, Planner, and Anniversary. The Notepad section lets users enter long documents, including text and graphics, that can be called up at any time. Whereas Lotus Organizer resembles a datebook, the PIM called Dynodex resembles an address book, with spaces for names, addresses, phone numbers, and notes.

Integrated Software & Suites

Preview & Review: Integrated software packages combine the features of several applications programs—for example, word processing, spreadsheet, database manager, graphics, and communications—into one software package.

What if you want to take data from one program and use it in another—say, call up data from a database and use it in a spreadsheet? You can try using separate software packages, but one may not be designed to accept data from the other. Two alternatives are the collections of software known as *integrated software* and *software suites.*

Integrated Software: "Works" Programs

Integrated software packages **combine the features of several applications programs—such as word processing, spreadsheet, database, graphics, and communications—into one software package.** These so-called "works" collections—the principal representatives are AppleWorks, ClarisWorks, Lotus Works, Microsoft Works, and PerfectWorks—give good value because the entire bundle often sells for $100 or less.

Integrated software packages are less powerful than separate programs used alone, such as a word processing or spreadsheet program used by itself. But that may be fine, because single-purpose programs may be more complicated and demand more computer resources than necessary.

Software Suites: "Office" Programs

Software suites, **or simply** *suites,* **are applications—like spreadsheets, word processing, graphics, communications, and groupware—that are bundled together and sold for a fraction of what the programs would cost if bought individually.** The principal suites, sometimes called "office" programs, are Microsoft Office from Microsoft, SmartSuite from Lotus, and PerfectOffice from Corel.

"Bundled" and "unbundled" are jargon words frequently encountered in software and hardware merchandising. *Bundled* means that components of a system are sold together for a single price. *Unbundled* means that a system has separate prices for each component.

Although cost is what makes suites attractive to many corporate customers, they have other benefits as well. Software makers have tried to integrate the "look and feel" of the separate programs within the suites to make them easier to use. "The applications mesh more smoothly in the package form," says one writer, "and the level of integration is increasing. More and more, they use the same commands and similar icons in the spreadsheet, word processor, graphics, and other applications, making them easier to use and reducing the training time."[15]

A tradeoff, however, is that such packages require a lot of hard-disk storage capacity. Microsoft Office 95, for instance, comes on 24 or more floppy disks (there is also a CD-ROM version) and occupies *at least 89 megabytes* of hard disk space—quite a lot if your hard disk holds only 200 megabytes.

Groupware

Preview & Review: Groupware is software used on a network that serves a group of users working together on the same project.

Most microcomputer software is written for people working alone. **Groupware is software that is used on a network and serves a group of users working together on the same project.** Groupware improves productivity by keeping you continually notified about what your colleagues are thinking and doing, and they about you. "Like e-mail," one writer points out, "groupware became possible when companies started linking PCs into networks. But while e-mail works fine for sending a message to a specific person or group—communicating one-to-one or one-to-many—groupware allows a new kind of communication: many-to-many."[16]

Groupware is essentially of four types:[17]   *BWMS*

- **Basic groupware:** Exemplified by Lotus Notes and Microsoft Exchange, this kind of groupware uses an enormous database containing work records, memos, and notations and combines it with a messaging system. Thus, a company like accounting giant Coopers & Lybrand uses Lotus Notes to let co-workers organize and share financial and tax information. It can also be used to relay advice from outside specialists, speeding up audits and answers to complex questions from clients.[18]
- **Workflow software:** Workflow software, exemplified by ActionWorkflow System and ProcessIt, helps workers understand and redesign the steps that make up a particular process. It also routes work automatically among employees and helps organizations reduce paper-jammed bureaucracies.
- **Meeting software:** An example of meeting software is Ventana's GroupSystems V, which allows people to have computer-linked meetings. With this software, people "talk," or communicate, with one another at the same time by typing on microcomputer keyboards. As one writer describes it, "Because people read faster than they speak, and don't have to wait for others to finish talking, the software can dramatically speed progress toward consensus."[19]
- **Scheduling software:** Scheduling software uses a microcomputer network to coordinate co-workers' electronic datebooks or appointment calendars so they can figure out a time when they can all get together. An example is Network Scheduler 3 from Powercore.

Internet Web Browsers

Preview & Review: Web browsers are software programs that allow people to view information at Web sites in the form of colorful, on-screen magazine-style "pages" with text, graphics, and sound.

The Internet, that network of thousands of interconnected networks, "is just a morass of data, dribbling out of [computers] around the world," says one writer. "It is unfathomably chaotic, mixing items of great value with cybertrash." This is why so-called *browsers* have caught people's imaginations, he states. "A browser cuts a path through the tangled growth and even creates a form of memory, so each path can be retraced."[20]

We cover the Internet in detail elsewhere (especially Chapter 8). Here let us consider just a part of a part of it, one that you may find particularly useful.

The World Wide Web

The most exciting part of the Internet is probably that fast-growing region or subset of it known as the World Wide Web. The *World Wide Web*, or simply *"the Web,"* consists of hundreds of thousands of intricately interlinked sites called "home pages" set up for on-screen viewing in the form of colorful magazine-style "pages" with text, images, and sound.

To be connected to the World Wide Web, you need an automatic setup with an online service or Internet access provider (described in the Chapter 4 Experience Box and also in Chapter 8), who will then give you a "browser" for actually exploring the Web. (The reverse is also true: If you buy some Web browsers, they will help you find an access provider.) **A *Web browser*, or simply *browser*, is software that enables you to "browse through" and view Web sites.** You can move from page to page by "clicking on" or selecting an icon or by typing in the address of the page. The accompanying drawing explains what the parts of a Web electronic address mean. (■ *See Panel 2.9.*)

There are a great many browsers, including some unsophisticated ones offered by Internet access providers and some by the large commercial online services such as America Online, CompuServe, and Prodigy. However, the recent battle royal to find the "killer app" browser has been between Netscape, which produces Navigator, and Microsoft, which developed a browser called Internet Explorer. The major online services now offer Netscape Navigator in addition to other browsers.

Search Tools on the Web: Directories & Indexes

Once you're on your browser, you need to know how to find what you're looking for. Search tools are of two basic types—*directories* and *indexes*.

PANEL 2.9

What's a Web browser?

This "tool bar" from Netscape's Navigator shows what you see at the top of each home page (Web site). From here you can move from one page to the next.

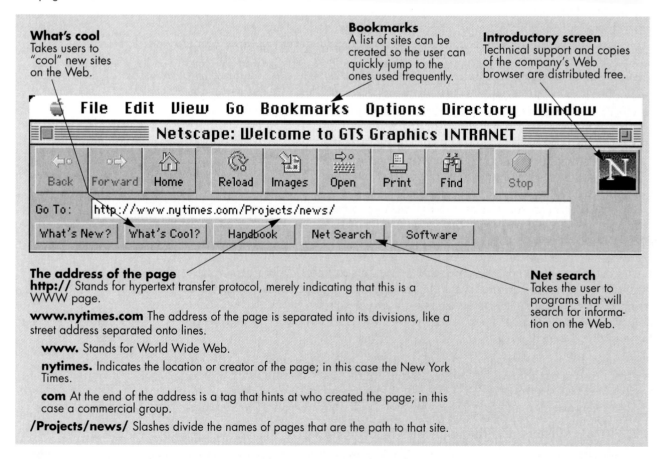

What's cool
Takes users to "cool" new sites on the Web.

Bookmarks
A list of sites can be created so the user can quickly jump to the ones used frequently.

Introductory screen
Technical support and copies of the company's Web browser are distributed free.

File   Edit   View   Go   Bookmarks   Options   Directory   Window

Netscape: Welcome to GTS Graphics INTRANET

Back   Forward   Home   Reload   Images   Open   Print   Find   Stop

Go To: http://www.nytimes.com/Projects/news/

What's New?   What's Cool?   Handbook   Net Search   Software

Net search
Takes the user to programs that will search for information on the Web.

The address of the page
http:// Stands for hypertext transfer protocol, merely indicating that this is a WWW page.

www.nytimes.com The address of the page is separated into its divisions, like a street address separated onto lines.

www. Stands for World Wide Web.

nytimes. Indicates the location or creator of the page; in this case the New York Times.

com At the end of the address is a tag that hints at who created the page; in this case a commercial group.

/Projects/news/ Slashes divide the names of pages that are the path to that site.

- **Directories:** *Web directories* are search tools classified by topic. One of the foremost examples is Yahoo *(http://www.yahoo.com)*, which provides you with an opening screen offering 14 general categories.
- **Indexes:** *Web indexes* allow you to find specific documents through keyword searches. An example of one useful index tool is Lycos *(http://www.lycos.com)*.

Specialized Software

Preview & Review: Specialized software tools include programs for desktop publishing, personal finance, project management, computer-aided design, drawing and painting, and hypertext.

After learning to use some of the productivity software just described, you may wish to extend your range by becoming familiar with more specialized programs. For example, you might first learn word processing and then move on to desktop publishing, the technology used to prepare much of today's printed information. Or you might learn spreadsheet programs and then go on to master personal-finance, tax, and investment software. Let us consider some of these specialized tools. We describe the following, although these are but a handful of the thousands of programs available:

- Desktop-publishing programs
- Personal finance programs
- Project management programs
- Computer-aided design
- Painting and drawing programs
- Hypertext

Desktop Publishing

Once you've become comfortable with a word processor, could you then go on and learn to do what Margaret Trejo did? When Trejo, then 36, was laid off from her job in 1987 because her boss couldn't meet the payroll, she was stunned. "Nothing like that had ever happened to me before," she said later. "But I knew it wasn't a reflection on my work. And I saw it as an opportunity."[21]

Today Trejo Production is a successful desktop-publishing company in Princeton, New Jersey, using Macintosh equipment to produce scores of books, brochures, and newsletters. "I'm making twice what I ever made in management positions," says Trejo, "and my business has increased by 25% every year."

Not everyone can set up a successful desktop-publishing business, because many complex layouts require experience, skill, and a knowledge of graphic design. Indeed, use of these programs by nonprofessional users can lead to rather unprofessional-looking results. Nevertheless, the availability of microcomputers and reasonably inexpensive software has opened up a career area formerly reserved for professional typographers and printers. **Desktop publishing, abbreviated *DTP*, involves using a microcomputer and mouse, scanner, laser printer, and DTP software for mixing text and graphics to produce high-quality printed output.** Often the laser printer is used primarily to get an advance look before the completed job is sent to a typesetter for even higher-quality output. Principal desktop-publishing programs are Aldus Page-Maker, Ventura Publisher, Quark XPress, and First Publisher. Microsoft Publisher is a low-end DTP package. Some word processing programs, such as Word and WordPerfect, also have many DTP features.

Desktop publishing has the following characteristics:

- **Mix of text with graphics:** Unlike traditional word processing programs, desktop-publishing software allows you to manage and merge text with graphics. Indeed, while laying out a page on screen, you can make the text "flow," liquid-like, around graphics such as photographs.

  Software used by many professional typesetters shows display screens full of formatting codes rather than what you will see when the job is printed out. By contrast, DTP programs can display your work in WYSI-WYG form. *WYSIWYG* (pronounced "wizzy-wig") stands for "What You See Is What You Get." It means that the text and graphics appear on the display screen exactly as they will print out.

- **Varied type and layout styles:** DTP programs provide a variety of *fonts, or typestyles,* from readable Times Roman to staid Tribune to wild Jester and Scribble. You can also create all kinds of rules, borders, columns, and page numbering styles.

- **Use of files from other programs:** Most DTP programs don't have all the features of full-fledged word processing or computerized drawing and painting programs. Thus, text is usually composed on a word processor, artwork is created with drawing and painting software, and photographs are scanned in using a scanner. Prefabricated art may also be obtained from disks containing *clip art,* or "canned" images that can be used to illustrate DTP documents. The DTP program is used to integrate all these files. (■ *See Panel 2.10.*) You can look at your work on the display screen as one page or as two facing pages in reduced size. Then you can see it again after it is printed out on a printer.

PANEL 2.10

How DTP uses other files

Text is composed on a word processor, graphics are drawn with drawing and painting programs, and photographs and other artwork are scanned in with a scanner. Data from these files is integrated using desktop-publishing software, then printed out on a laser printer.

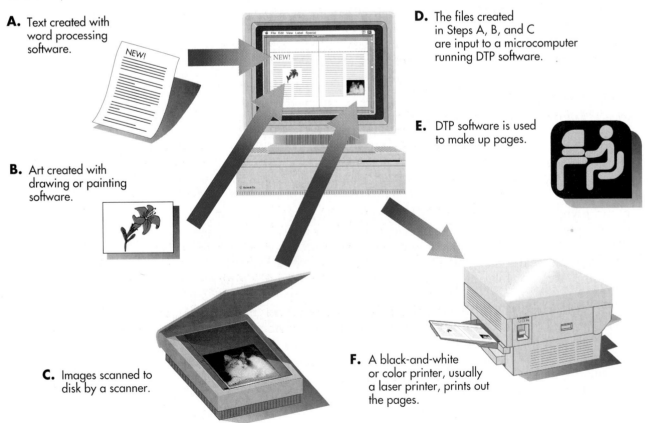

A. Text created with word processing software.

B. Art created with drawing or painting software.

C. Images scanned to disk by a scanner.

D. The files created in Steps A, B, and C are input to a microcomputer running DTP software.

E. DTP software is used to make up pages.

F. A black-and-white or color printer, usually a laser printer, prints out the pages.

- **Page description language:** Once you have finished your composition and layout, you can send the document to the printer. Much of the shaping of text characters and graphics is done within the printer rather than in the computer. For instance, instead of sending the complete image of a circle from the computer to the printer, you send a command to the printer to draw a circle. **A *page description language* is software used to describe to the printer the shape and position of letters and graphics.** An example of a page description language is Adobe's PostScript, which is used with Aldus PageMaker.

Personal Finance Programs

Personal finance software **lets you keep track of income and expenses, write checks, and plan financial goals.** Whether or not you learn how to use electronic spreadsheet programs, you'll probably find it useful to use personal finance software. Such programs don't promise to make you rich, but they can help you manage your money, maybe even get you out of trouble.

Many personal finance programs include a calendar and a calculator, but the principal features are the following:

- **Tracking of income and expenses:** The programs allow you to set up various account categories for recording income and expenses, including credit card expenses.
- **Checkbook management:** All programs feature checkbook management, with an on-screen check writing form and check register that look like the ones in your checkbook. Checks can be purchased to use with your computer printer.
- **Reporting:** All programs compare your actual expenses with your budgeted expenses. Some will compare this year's expenses to last year's.
- **Income tax:** All programs offer tax categories, for indicating types of income and expenses that are important when you're filing your tax return.
- **Other:** Some of the more versatile personal finance programs also offer financial-planning and portfolio-management features.

Quicken (there are versions for DOS, Windows, and Macintosh) seems to have generated a large following, but other personal finance programs exist as well. They include Kiplinger's CA-Simply Money, Managing Your Money, Microsoft Money, and WinCheck. Some offer enough features that you could use them to manage a small business.

In addition, there are tax software programs, which provide virtually all the forms you need for filing income taxes. Tax programs make complex calculations, check for mistakes, and even unearth deductions you didn't know existed. (Principal tax programs are Andrew Tobias' TaxCut, Kiplinger TaxCut, TurboTax/MacInTax, Personal Tax Edge, and CA-Simply Tax.) Finally, there are investment software packages, such as StreetSmart from Charles Schwab and Online Xpress from Fidelity, as well as various retirement planning programs.

Project Management Software

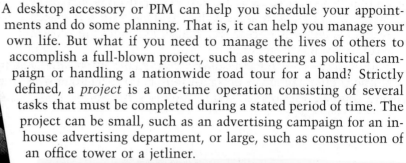

A desktop accessory or PIM can help you schedule your appointments and do some planning. That is, it can help you manage your own life. But what if you need to manage the lives of others to accomplish a full-blown project, such as steering a political campaign or handling a nationwide road tour for a band? Strictly defined, a *project* is a one-time operation consisting of several tasks that must be completed during a stated period of time. The project can be small, such as an advertising campaign for an in-house advertising department, or large, such as construction of an office tower or a jetliner.

Project management software is a program used to plan, schedule, and control the people, costs, and resources required to complete a project on time. For instance, the associate producer on a feature film might use such software to keep track of the locations, cast and crew, materials, dollars, and schedules needed to complete the picture on time and within budget. The software would show the scheduled beginning and ending dates for a particular task—such as shooting all scenes on a certain set—and then the date that task was actually completed. Project management software is also used to manage the development of the many components of multimedia projects. Examples of project management software are Harvard Project Manager, Microsoft Project for Windows, Project Scheduler 4, SuperProject, and Time Line.

Computer-Aided Design

Computers have long been used in engineering design. **Computer-aided design (CAD) programs are software programs for the design of products and structures.** CAD programs, which are now available for microcomputers, help architects design buildings and work spaces and engineers design cars, planes, and electronic devices. One advantage of CAD software is that the product can be drawn in three dimensions and then rotated on the screen so the designer can see all sides. (■ *See Panel 2.11.*) Examples of CAD programs for beginners are Autosketch, EasyCAD2 (Learn CAD Now), and TurboCAD.

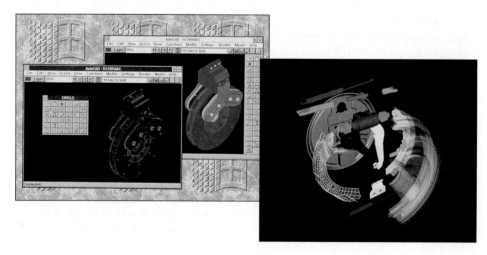

A variant on CAD is ***CADD, for computer-aided design and drafting, software that helps people do drafting.*** CADD programs include symbols (points, circles, straight lines, and arcs) that help the user put together graphic elements, such as the floor plan of a house. Examples are Autodesk's AutoCAD and Intergraph's Microstation.

CAD/CAM—for computer-aided design/computer-aided manufacturing—software allows products designed with CAD to be input into an automated manufacturing system that makes the products. For example, CAD, and its companion, CAM, brought a whirlwind of enhanced creativity and efficiency to the fashion industry. Some CAD systems, says one writer, "allow designers to electronically drape digital-generated mannequins in flowing gowns or tailored suits that don't exist, or twist imaginary threads into yarns, yarns into weaves, weaves into sweaters without once touching needle to garment."[22] The designs and specifications are then input into CAM systems that enable robot pattern-cutters to automatically cut thousands of patterns from fabric, with only minimal waste. Whereas previously the fashion industry worked about a year in advance of delivery, CAD/CAM has cut that time to 8 months—a competitive edge for a field that feeds on fads.

Drawing & Painting Programs

It may be no surprise to learn that commercial artists and fine artists have begun to abandon the paintbox and pen-and-ink for software versions of palettes, brushes, and pens. The surprise, however, is that an artist can use mouse and pen-like stylus to create computer-generated art as good as that achievable with conventional artist's tools. More surprising, even *nonartists* can be made to look good with these programs.

There are two types of computer art programs: drawing and painting.

- **Drawing programs:** A ***drawing program* is graphics software that allows users to design and illustrate objects and products.** CAD and drawing programs are similar. However, CAD programs provide precise dimensioning and positioning of the elements being drawn, so that they can be transferred later to CAM programs. Also, CAD programs lack the special effects for illustrations that come with drawing programs.[23] Some drawing programs are CorelDraw, Illustrator, Freehand, and Sketcher.

- **Painting programs:** Whereas drawing programs are generally gray-scale programs, painting programs add color. ***Painting programs* are graphics programs that allow users to simulate painting on screen.** A mouse or a tablet stylus is used to simulate a paintbrush. The program allows you to select "brush" sizes, as well as colors from a color palette.

  The difficulty with using painting programs is that a powerful computer system is needed because color images take up so much main memory and disk storage space. In addition, these programs require sophisticated color printers.

Hypertext

***Hypertext* is software that allows users to have fast and flexible access to information in large documents, constructing associations among data items as needed.** Hypertext works the way people think, allowing them to link facts into sequences of information in ways that resemble those that people use to obtain new knowledge.

Hypertext is often used in Help systems (✓ p. 51). Another well-known example of the use of hypertext is that found in HyperCard, introduced for the Apple Macintosh in 1987. HyperCard is based on the concept of *cards*

`http://www.fieryfoods.com`
(Click the "Little-Guy-Icons" for your favorite regional foods. Click the "Compass" for your preferred heat level.)

Hot sauces, championship chili mixes, Habañero salsas, jerked meat marinades, scorching snacks, Jalapeño concoctions and—ahhh, yet more hot sauces. Yep, if you're seeking a chile-head's paradise, a pepper phreak's dream, drop in and partake of Bob's Fiery Foods.

What do we bring to the table?

You'll find the hottest foods from <u>Jamaica</u>, <u>Trinidad</u>, <u>Tobago</u>, <u>Dominica</u>, <u>Costa Rica</u>, <u>Belize</u>, <u>Mexico</u>, <u>Cajun Country</u>, <u>Florida</u>, the <u>Eastern U.S.</u>, the <u>Great American West</u>, and even special goodies stored away in <u>Bob's Pantry</u>.

A hot sauce is a hot sauce. Right? Wrong.

Because we're not talking that ubiquitous skinny bottle of red hot you see among your grocer's generic offerings. We're into a different crowd: quirky micro-brands emblazoned with warnings such as *after death, insanity sauce* and *last rites*. Replete with the world's hottest peppers and spices. Sauces imparting bizarre, memorable taste and aroma. Woo, boy.

Sadomasochists welcome at Bob's.

You'll find that Bob's Fiery Foods offers a whole new set of possibilities. Touch your taste buds to a single drop of *Dave's Gourmet Insanity Sauce*, *Endorphin Rush* or *Mad Dog Inferno* and you'll likely exceed your threshold of pain in seconds. You simply have no way of knowing how hot hot is, until you sample our most forbidden pleasures, sauces blended with pure pepper extract.

PANEL 2.12

Hypertext

Hidden hypertext codes provide links among related web sites. Terms linked to other sites are underlined.

and *stacks* of cards—just like notecards, only they are electronic. A card is a screenful of data that makes up a single record; cards are organized into related files called stacks. On each card there may be one or more *buttons*, which, when clicked on by a mouse, can pull up another card or stack. By clicking buttons, you can make your way through the cards and stacks to find information or discover connections between ideas.

Recently, hypertext has come into its own as a means of accessing Web sites (discussed in Chapter 8). The hidden codes of hypertext allow users to use a mouse to click on a highlighted or underlined word or phrase to automatically access a new but related site. (■ *See Panel 2.12.*)

Onward

In this chapter we described the more elementary forms of software. Still to be discussed, however, are some of the truly exciting software developments of the Digital Age. Later in the book, we describe software technology such as virtual reality, multimedia, simulation, expert systems, and information-seeking "agents," as well as software affecting the Internet.

Getting Started With Computers in College & Going Online

You may still hear the sounds of late-night typing in a college residence hall. However, it's certainly not the smart way to work anymore. Indeed, coping with a typewriter actually detracts from learning. You're worrying about making mistakes and avoiding retyping a whole paper rather than concentrating on educational principles. Using a computer to write your papers not only makes life easier, it also opens up new areas of freedom and knowledge and helps prepare you for the future.

Thus, students who come to college with a personal computer as part of their luggage are certainly ahead of the game. If you don't have one, however, there are other options.

If You Don't Own a Personal Computer

If you don't have a PC, you can probably borrow someone else's sometimes. However, if you have a paper due the next day, you may have to defer to the owner, who may also have a deadline. When borrowing, then, you need to plan ahead and allow yourself plenty of time.

Virtually every campus now makes computers available to students, either at minimal cost or essentially for free as part of the regular student fees. This availability may take two forms:

- *Library or computer labs:* Even students who have their own PCs may sometimes want to use the computers available at the library or campus computer lab. These may have special software or better printers that they don't have themselves.

- *Dormitory computer centers or dorm-room terminals:* Some campuses provide dormitory-based computer centers (for example, in the basement). Even if you have your own PC, it's nice to know about these for backup purposes.

  More and more campuses are also providing computers or terminals within students' dormitory rooms. These are usually connected by a campuswide local area network (LAN) to lab computers and administrative systems. Often, however, they also allow students to communicate over phone lines to people in other states.

If You Do Own a Personal Computer

Perhaps someone gave you a personal computer, or you acquired one, before you came to college. It will probably be one of three types: (1) an IBM or IBM-compatible, such as a Compaq, Packard Bell, AST, Radio Shack, Zenith, or Dell; (2) an Apple Macintosh or compatible; (3) other, such as an Apple II or Commodore.

If all you need to do is write term papers, nearly any microcomputer, new or used, will do. Indeed, you may not even need to have a printer, if you can find other ways to print out things. The University of Michigan, for instance, offers "express stations" or "drive-up windows." These allow students to use a diskette (floppy disk) or connect a computer to a student-use printer to print out their papers. Or, if a friend has a compatible computer, you can ask to borrow it and the printer for a short time to print your work.

You should, however, take a look around you to see if your present system is appropriate for your campus and your major.

- *The fit with your campus:* Some campuses are known as "IBM" (or IBM-compatible) schools, others as "Mac" (Macintosh) schools. Apple IIs and Commodores, still found in elementary and high schools, are not used much at the college level.

  Why should choice of machine matter? The answer is that diskettes generally can't be read interchangeably among the two main types of microcomputers. Thus, if you own the system that is out of step for your campus, you may find it difficult to swap files or programs with others. Nor will you be able to borrow their equipment to finish a paper if yours breaks down. (There are some conversion programs, but these take time and may not be readily available.)

  Most campuses favor either Macintoshes or IBMs (and IBM-compatibles). You should call the dean of students or otherwise ask around to find which system is most popular.

- *The fit with your major:* Speech communications, foreign language, physical education, political science, biology, and English majors probably don't need a fancy computer system (or even any system at all). Business, engineering, architecture, and journalism majors may have special requirements. For instance, an architecture major doing computer-aided design (CAD) projects or a journalism major doing desktop publishing will need reasonably powerful systems. A history or nursing major, who will mainly be writing papers, will not.

  Of course you may be presently undeclared or undecided about your major. Even so, it's a good idea to find out what kinds of equipment and programs are being used in the majors you are contemplating.

How to Buy a Personal Computer

Buying a personal computer, like buying a car, often requires making a tradeoff between power and expense.

Power Many computer experts try to look for a personal computer system with as much power as possible. The word *power* has different meanings when describing software and hardware:

- *Powerful software:* Applied to software, "powerful" means that the program is *flexible*. That is, it can do many different things. For example, a word processing program that can print in different typestyles (fonts) is more powerful than one that prints in only one style.

- *Powerful hardware:* Applied to hardware, "powerful" means that the equipment (1) is *fast* and (2) has *great capacity*.

     A fast computer will process data more quickly than a slow one. With an older computer, for example, it may take several seconds to save, or store on a disk, a 50-page term paper. On a newer machine, it might take less than a second.

     A computer with great capacity can run complex software and process voluminous files. *This is an especially important matter if you want to be able to run the latest releases of software.*

Will computer use make up an essential part of your major, as it might if you are going into engineering, business, or graphic arts? If so, you may want to try to acquire powerful hardware and software. People who really want (and can afford) their own desktop publishing system might buy a new Macintosh Power PC with laser printer, scanner, and PageMaker or Quark software. This might well cost $8000. Most students, of course, cannot afford anything close to this.

Expense If your major does not require a special computer system, a microcomputer can be acquired for relatively little. You can probably buy a used computer, with software thrown in, for under $500 and a printer for under $200.

What's the *minimum* you should get? Probably an IBM-compatible or Macintosh system, with 4 megabytes of memory (✓ p. 16) and two diskette drives or one diskette and one hard-disk drive. However, up to 16 megabytes of memory is preferable if you're going to run many of today's programs. Dot-matrix printers are still in use on many campuses (24-pin printers are preferable to 9-pin). To be sure, the more expensive laser and inkjet printers produce a better image. However, you can always use the dot-matrix for drafts and print out the final version on a campus student-use printer.

Where to Buy New Fierce price wars among microcomputer manufacturers and retailers have made hardware more affordable. One reason IBM-compatibles have become so widespread is that non-IBM manufacturers early on were able to copy, or "clone," IBM machines and offer them at cut-rate prices. For a long time, Apple Macintoshes were considerably more expensive. In part this was because other manufacturers were unable to offer inexpensive clones. In recent times, however, Apple has felt the pinch of competition and has dropped its prices. It also has licensed parts of its technology to others so that we are now seeing Macintosh "clones."

There are several sources for inexpensive new computers, such as student-discount sources, computer superstores, and mail-order houses.

When buying hardware, look to see if software, such as word processing or spreadsheet programs, comes bundled with it.

Because computers are somewhat fragile, it's not unusual for them to break down, some even when newly purchased. Indeed, nearly 25% of 45,000 PC users surveyed by one computer magazine reported some kind of problem with new computers.[24] (The failure rates were: hard drive—21%, motherboard—20%, monitor—12%, diskette drive—11%, and power supply—10%.) The PCs (Apple was not included) that had the fewest problems included those from AST Research, Compaq, Epson, Hewlett-Packard, IBM, and NCR (AT&T). Most troublesome were low-cost PCs.

Where to Buy Used Buying a used computer can save you a minimum of 50%, depending on its age. If you don't need the latest software, this can often be the way to go. The most important thing is to buy *recognizable* brand names, examples being Apple and IBM or well-known IBM-compatibles: Compaq, Hewlett-Packard, NCR, Packard Bell, Tandy, Toshiba, Zenith. Obscure or discontinued brands may not be repairable.

Among the sources for used computers are the following:

- *Retail sources:* A look in the telephone-book Yellow Pages under "Computers, Used" will produce several leads. Authorized dealers (of IBM, Apple, Compaq, and so on) may shave prices on demonstration (demo) or training equipment.

- *Used-computer brokers:* There are a number of used-computer brokers, such as American Computer Exchange, Boston Computer Exchange, Damark, and National Computer Exchange.

- *Individuals:* Classified ads in local newspapers, shopper throwaways, and (in some localities) free computer newspapers/magazines provide listings of used computer equipment. Similar listings may also appear on electronic bulletin board systems (BBSs).

  One problem with buying from individuals is that they may not feel obligated to take the equipment back if something goes wrong. Thus, you should inspect the equipment carefully. For a small fee, a computer-repair shop can check out the hardware for damage before you buy it.

How much should you pay for a used computer? This can be tricky. Some sellers may not be aware of the rapid depreciation of their equipment and price it too high. The best bet is to look through back issues of the classified ads for a couple of newspapers in your area until you have a sense of what equipment may be worth.

Checklist Here are some decisions you should make before buying a computer:

- *What software will I need?* Although it may sound backward, you should select the software before the hardware. This is because you want to choose software that will perform the kind of work you want to do. First find the kind of programs you want—word processing, spreadsheets, communications, graphics, or whatever. Check out the memory and other hardware requirements for those programs. Then make sure you get a system to fit them.

  The advice to start with software before hardware has always been standard for computer buyers. However, it is becoming increasingly important as programs with extensive graphics come on the market. Graphics tend to require a lot of memory, hard-disk storage, and screen display area.

- *Do I want a desktop or a portable?* Look for a computer that fits your work style. For instance, you may want a portable if you spend a lot of time at the library. Some students even use portables to take notes in class. If you do most of your work in your room, you may find it more comfortable to have a desktop PC. Though not portable, the monitors of desktop computers are usually easier to read.

  It's possible to have both portability and a readable display screen. Buy a laptop, but also buy a monitor that you can plug the portable into. Computers are also available with "docking" systems that permit a portable to fit inside a desktop computer or monitor.

Also keep in mind that portable computers are more expensive to maintain than desktop computers, and portable keyboards are smaller.

- *Is upgradability important?* The newest software being released is so powerful (meaning flexible) that it requires increasingly more powerful hardware. That is, the software requires hardware that is faster and has greater main memory and storage capacity. If you buy an outdated used computer, you probably will not be able to *upgrade* it. That is, you will be unable to buy internal parts, such as additional memory, that can run newer software. This limitation may be fine if you expect to be able to afford an all-new system in a couple of years. If, however, you are buying new equipment right now, be sure to ask the salesperson how the hardware can be upgraded.

- *Do I want an IBM-style or a Macintosh?* Although the situation is changing, until recently the division between IBM and IBM-compatibles on the one hand and Apple Macintoshes on the other was fundamental. Neither could run the other's software or exchange files of data without special equipment and software. We mentioned that some campuses and some academic majors tend to favor one type of microcomputer over the other. Outside of college, however, the business world tends to be dominated by IBM and IBM-compatible machines. In a handful of areas—graphic arts and desktop publishing, for example—Macintoshes are preferred.

Getting Started Online

Computer networks have transformed life on campuses around the country, becoming a cultural and social force affecting everybody, not just nerds and wonks. How do you join this vast world of online information and interaction?

Hardware & Software Needed Besides a microcomputer with a hard disk (any made in the last 5 years will do), you need a modem, to send messages from one computer to another via a phone line. The speed at which a modem can transmit data is generally measured in bps (bits per second). A slow modem, 2400 bps, can be bought for less than $50. For less than $100 you can get a faster modem—9600, 14,400, or 28,800 bps—which allows data to flow faster and reduces your phone charges. About 40% of personal computers these days have a modem. If yours doesn't, you can have a store install an internal modem as an electronic circuit board on the inside of the computer, or buy an external modem, which appears as a box outside the computer.

To go online, you'll need communications software, which may come bundled with any computer you buy or is sold on floppy disks in computer stores. (Popular brands are Microphone and White Knight for Macintosh computers. IBM-style computers running Windows and DOS may use ProComm Plus, HyperAccess, QModem Pro, or Crosstalk.) However, many modems come with communications software when you buy them. Or, if you sign up for an online service, it will supply the communications program you need to use its network.

Your modem will connect to a standard telephone wall jack. (When your computer is in use, it prevents you from using the phone line, and callers trying to reach you will hear a busy signal.)

Getting Connected: Starting with an Online Service Unless you already have access to a campus network, probably the easiest first step for using this equipment is to sign up with a commercial online service, such as one of the Big Three: America Online, CompuServe, or Prodigy. (Another is the Microsoft Network.)

You'll need a credit card in order to join, since online services charge a monthly fee, typically $10 to $20. Most also charge for the time spent online—from the time you dial and connect to the time you disconnect. Charges are billed to your credit card. Note: It's easy to get carried away and run up charges by staying online too long. However, you can keep your costs down by going online only during off-hours (evenings and weekends). You can also not do your reading (of articles, say) online but download (save to your hard disk) material unread, then read it later.

All online services have introductory offers that allow you a free trial period. You can get instructions and free start-up communications disk by phoning their toll-free 800 numbers. (■ *See Panel 2.13.*) You'll also find promotional offers at computer stores or promotional diskettes shrink-wrapped inside computer magazines on newsstands.

PANEL 2.13

The Big Three online services

Listed here are the leading online services, number of users in 1996, costs, phone numbers, and other details. Rates are subject to change. All have special introductory offers.

AMERICA Online	Number of users:	5 million
	Cost:	$9.95/month for 5 hours for all services, then $2.95/hour.
	How to connect:	800-827-6364; startup software, with 10 free hours of online time, available in computer stores or from America Online.
CompuServe	Number of users:	4.2 million
	Cost:	$9.95 for 5 hours, then $2.95/hour. 20–hour plan: $19.95, then $2.00/hour. Premium services cost extra.
	How to connect:	800-848-8199; membership kits with user's guide and software available through computer stores, mail-order outlets, or CompuServe. 28,800 bps modem.
PRODIGY Service	Number of users:	1.4 million
	Cost:	$9.95 for 5 hours, then $2.95/hour. 30–hour plan: $29.95, then $2.95/hour. Plus Service costs extra.
	How to connect:	Phone: 800-776-3449. Startup software is also available in computer stores.

SUMMARY

What It Is / What It Does	Why It's Important

analytical graphics *(p. 58, LO 5)* Also called *business graphics;* graphical forms representing numeric data. The principal examples are bar charts, line graphs, and pie charts. Analytical graphics programs are a type of applications software.

Numeric data is easier to analyze in graphical form than in the form of rows and columns of numbers, as in electronic spreadsheets.

applications software *(p. 43, LO 1)* Software that can perform useful work on general-purpose tasks—for example, word processing or spreadsheet software.

Applications software such as word processing, spreadsheet, database manager, graphics, and communications packages are used to increase people's productivity.

block command *(p. 55, LO 5)* Software command that allows the user to indicate beginning and end points of text to be moved or otherwise manipulated.

The block command makes it easy to rearrange and reformat documents.

button *(p. 51, LO 4)* Simulated on-screen button (kind of icon) that is activated ("pushed") by a mouse or other pointing device to issue a command.

Buttons make it easier for users to enter commands.

cell *(p. 56, LO 5)* In an electronic spreadsheet, the rectangle where rows and columns intersect.

The cell is the smallest working unit in a spreadsheet. Data and formulas are entered into the cells.

cell address *(p. 56, LO 5)* In an electronic spreadsheet, the position of a cell—for example, "A1," where column A and row 1 intersect.

Cell addresses provide location references for spreadsheet users.

clip art *(p. 66, LO 6)* "Canned" images that come on disks, which users can copy and place as desired to illustrate documents.

Clip art enables users to illustrate their word-processed or desktop-published documents without having to hire an artist.

computer-aided design (CAD) *(p. 68, LO 6)* Applications software programs for designing products and structures.

CAD programs help architects design buildings and work spaces and engineers design cars, planes, and electronic devices. With CAD software, a product can be drawn in three dimensions and then rotated on the screen so the designer can see all sides.

computer-aided design and drafting (CADD) *(p. 69, LO 6)* Applications software that helps people do drafting.

CADD programs include symbols (points, circles, straight lines, and arcs) that help the user put together graphic elements, such as the floor plan of a house.

computer-aided design/computer-aided manufacturing (CAD/CAM) *(p. 69, LO 6)* Applications software that allows products designed with CAD to be input into a computer-based manufacturing system (CAM) that makes the products.

CAD/CAM systems have greatly enhanced creativity and efficiency in many industries.

copyright *(p. 44, LO 2)* Body of law that prohibits copying of intellectual property without the permission of the copyright holder.

Copyright law aims to prevent people from taking credit for and profiting unfairly from other people's work.

cursor *(p. 54, LO 4)* Also called a *pointer;* the movable symbol on the display screen that shows the user where data may be entered next. The cursor is moved around with the keyboard's directional arrow keys or an electronic mouse.

All applications software packages use cursors to show users where their current work location is on the screen.

custom-written software *(p. 43, LO 1)* Applications software designed for a particular customer and written for a highly specialized task by a computer programmer.

When packaged applications software can't do the tasks a user requires, then software must be custom-written by a professional computer programmer.

database *(p. 58, LO 5)* Collection of interrelated files in a computer system that is created and managed by database manager software. These files are organized so that those parts with a common element can be retrieved easily.

Online database services provide users with enormous research resources. Businesses and organizations use databases to keep track of transactions and increase people's efficiency.

database software *(p. 58, LO 5)* Applications software for maintaining a database. It controls the structure of a database and access to the data.

Database manager software allows users to organize and manage huge amounts of data.

data communications software *(p. 60, LO 5)* Applications software that manages the transmission of data between computers.

Communications software is required to transmit data via modems in a communications system.

desktop accessory *(p. 61, LO 5)* Also called *desktop organizer;* software package that provides electronic counterparts of tools or objects commonly found on a desktop: calendar, clock, card file, calculator, and notepad.

Desktop accessories help users to streamline their daily activities.

desktop publishing (DTP) *(p. 65, LO 6)* Applications software that, along with a microcomputer, mouse, scanner, and laser printer (usually), is used to mix text and graphics, including photos, to produce high-quality printed output. Some word processing programs also have many DTP features. Text is usually composed first on a word processor, artwork is created with drawing and painting software, and photographs are scanned in using a scanner. Prefabricated art and photos may also be obtained from disks (CD-ROM and/or floppy) containing clip art.

Desktop publishing has reduced the number of steps, the time, and the money required to produce professional-looking printed projects.

dialog box *(p. 51, LO 4)* With graphical user interface (GUI) software, a box that appears on the screen and displays a message requiring a response from you—for example, Y for "Yes" or N for "No."

Dialog boxes are only one aspect of GUIs that make software easier for people to use.

documentation *(p. 52, LO 4)* User's manual or reference manual that is a narrative and graphical description of a program. Documentation may be instructional, but usually features and functions are grouped by category.

Documentation helps users learn software commands and use of function keys, solve problems, and find information about system specifications.

drawing programs *(p. 69, LO 6)* Applications software that allows users to design and illustrate objects and products.

Drawing programs and CAD are similar. However, drawing programs provide special effects that CAD programs do not.

electronic mail (e-mail) software *(p. 61, LO 5)* Software that enables computer users to send letters and documents from one computer to another.

E-mail allows businesses and organizations to quickly and easily send messages to employees and outside people without resorting to paper messages.

electronic spreadsheet *(p. 56, LO 5)* Also called *spreadsheet;* applications software that simulates a paper worksheet and allows users to create tables and financial schedules by entering data and/or formulas into rows and columns displayed as a grid on a screen. If data is changed in one cell, values in other cells specified in the spreadsheet will automatically recalculate.

The electronic spreadsheet became such a popular small-business applications program that it has been held directly responsible for making the microcomputer a widely used business tool.

font *(p. 66, LO 6)* Set of type characters in a particular typestyle and size.

Desktop publishing programs, along with laser printers, have enabled users to dress up their printed projects with many different fonts.

formula *(p. 56, LO 5)* In an electronic spreadsheet, instructions for calculations that are entered into designated cells. For example, a formula might be "Sum (add) all the numbers in the cells with cell addresses A5 through A15."

The use of formulas enables spreadsheet users to change data in one cell and have all the cells linked to it by formulas automatically recalculate their values.

freeware *(p. 46, LO 2)* Software that is available free of charge.

Freeware is usually distributed through the Internet. Users can make copies for their own use but are not free to make unlimited copies.

function keys *(p. 50, LO 4)* Computer keyboard keys that are labeled F1, F2, and so on; usually positioned along the top or left side of the keyboard.

Function keys are used to issue commands. These keys are used differently, depending on the software.

grammar checker *(p. 55, LO 5)* Software feature that flags poor grammar, wordiness, incomplete sentences, and awkward phrases.

A grammar checker allows users to improve their prose for both style and accuracy.

graphical user interface (GUI) *(p. 51, LO 4)* User interface that uses images to represent options. Some of these images take the form of icons, small pictorial figures that represent tasks, functions, or programs.

GUIs are easier to use than command-driven interfaces and menu-driven interfaces; they permit liberal use of the electronic mouse as a pointing device to move the cursor to a particular icon or place on the display screen. The function represented by the icon can be activated by pressing ("clicking") buttons on the mouse.

groupware *(p. 63, LO 5)* Applications software that is used on a network and serves a group of users working together on the same project.

Groupware improves productivity by keeping users continually notified about what colleagues are thinking and doing, and vice versa.

Help menu *(p. 51, LO 4)* Also called *Help screen;* offers on-screen instructions for using software. Help screens are accessed via a function key or by using the mouse to select Help from a menu.

Help screens provide a built-in electronic instruction manual.

hypertext *(p. 69, LO 6)* Applications software that allows users to link information in large documents, constructing associations among data items as needed.

Hypertext goes beyond the restrictive search-and-retrieval methods of traditional database systems and encourages people to follow their natural train of thought as they discover information. Hypertext is used to link Web pages.

icon *(p. 51, LO 4)* In a GUI, small pictorial figure that represents a task, function, or program.

The function represented by the icon can be activated by pointing at it with the mouse pointer and pressing ("clicking") on the mouse. The use of icons has simplified the use of computers.

integrated software *(p. 62, LO 5)* Applications software that combines several applications programs into one package—usually electronic spreadsheets, word processing, database management, graphics, and communications.

Integrated software packages offer greater flexibility than separate single-purpose programs.

macro *(p. 50, LO 4)* Software feature that allows a single keystroke or command to be used to automatically issue a predetermined series of keystrokes or commands.

Macros increase productivity by consolidating several command keystrokes into one or two.

menu *(p. 51, LO 4)* List of available commands displayed on the screen.

Menus are used in graphical-user interface programs to make software easier for people to use.

move command *(p. 55, LO 5)* Software command that allows users to move any highlighted, or blocked, items in a document to any other designated location.

See block command.

network piracy *(p. 45, LO 2)* The use of electronic networks for unauthorized distribution of copyrighted materials in digitized form.

If piracy is not controlled, people may not want to let their intellectual property and copyrighted material be dealt with in digital form.

packaged software *(p. 43, LO 1)* Also called *software package;* an "off-the-shelf" program available on disk for sale to the general public.

Most software used by general microcomputer users comes in software packages available at local stores.

What It Is / What It Does	Why It's Important

page description language *(p. 67, LO 6)* Software used in desktop publishing that describes the shape and position of characters and graphics to the printer.

Page description languages, used along with laser printers, gave birth to desktop publishing. They allow users to combine different types of graphics with text in different fonts, all on the same page.

painting programs *(p. 69, LO 6)* Applications programs that simulate painting using a mouse or tablet stylus like a paintbrush and that use colors. A powerful computer system is required to use these programs.

Painting programs can render sophisticated illustrations.

personal finance software *(p. 67, LO 6)* Applications software that helps users track income and expenses, write checks, and plan financial goals.

Personal finance software can help people manage their money more effectively.

personal information manager (PIM) *(p. 61, LO 5)* Applications software that combines a word processor, database, and desktop accessory program to organize a variety of information.

PIMs offer an electronic version of an appointment calendar, to-do list, address book, notepad, and similar daily office tools, all in one place.

plagiarism *(p. 45, LO 2)* Expropriation of another writer's text, findings, or interpretations and presenting them as one's own.

Information technology offers plagiarists new opportunities to go far afield for unauthorized copying, yet it also offers new ways to catch these people.

presentation graphics *(p. 60, LO 5)* Graphical forms used to communicate or make a presentation of data to others, such as clients or supervisors. Presentation graphics programs are a type of applications software.

Presentation graphics programs may make use of analytical graphics but look much more sophisticated, using texturing patterns, complex color, and dimensionality.

project management software *(p. 68, LO 6)* Applications software used to plan, schedule, and control the people, costs, and resources required to complete a project on time.

Project management software increases the ease and speed of planning and managing complex projects.

proprietary software *(p. 47, LO 2)* Software whose rights are owned by an individual or business.

Ownership of proprietary software is protected by copyright. This type of software must be purchased to be used. Copying is restricted.

public domain software *(p. 46, LO 2)* Software that is not protected by copyright and thus may be duplicated by anyone at will.

Public domain software offers lots of software options to users who may not be able to afford a lot of commercial software. Users may make as many copies as they wish.

recalculation *(p. 57, LO 5)* Feature of electronic spreadsheet software whereby values are re-computed automatically.

This feature allows changes to be made to spreadsheets while having all values dependent on the changed numbers automatically updated.

release *(p. 43, LO 1)* Refers to a minor upgrade in a software product.

Releases are usually indicated by a change in the number after a version's decimal point—such as 3.1, 3.2. The higher the number, the more recent the release.

replace command *(p. 55, LO 5)* Software command that allows users to automatically replace with a new item any existing item identified using the search command.

All occurrences of an item in a document can be replaced automatically, using just a single command.

scrolling *(p. 54, LO 4)* The activity of moving quickly upward or downward through text or other screen display, using directional arrow keys or mouse.

Normally a computer screen displays only 20–22 lines of text. Scrolling enables users to view an entire document, no matter how long.

search command *(p. 55, LO 5)* Software command that allows users to find any item known to exist in the document.

The search command saves users from having to read an entire document to find a particular item they want to check or change.

shareware *(p. 46, LO 2)* Copyrighted software that is distributed free of charge, usually over the Internet, but that requires users to make a contribution in order to receive technical help, documentation, or upgrades.

Along with public domain software and freeware, shareware offers yet another inexpensive way to obtain new software.

What It Is / What It Does	Why It's Important

software *(p. 43, LO 1)* Also called *programs;* step-by-step instructions that tell the computer how to perform a task. Software instructions are written by programmers. In most instances, the words *software* and *program* are interchangeable. Software is of two types: applications software and systems software.

Without software, hardware would be useless.

software license *(p. 47, LO 2)* Contract by which users agree not to make copies of proprietary software to give away or to sell.

Software manufacturers don't sell people software so much as sell them licenses to become authorized users of the software.

software piracy *(p. 45, LO 2)* Unauthorized copying of copyrighted software—for example, copying a program from one floppy disk to another or downloading a program from a network and making a copy of it.

Software piracy represents a serious loss of income to software manufacturers and is a contributor to high prices in new programs.

software suite *(p. 62, LO 5)* Several applications software packages—like spreadsheets, word processing, graphics, communications, and groupware— bundled together and sold for a fraction of what the programs would cost if bought individually.

Software suites can save users a lot of money.

special-purpose keys *(p. 50, LO 4)* Computer keyboard keys used to enter and edit data and execute commands— for example, Esc, Alt, and Ctrl.

All computer keyboards have special-purpose keys. The user's software program determines how these keys are used.

spelling checker *(p. 55, LO 5)* Word processing software feature that tests for incorrectly spelled words.

Although spelling checkers cannot flag words that are correctly spelled but incorrectly used (like "for" instead of "four"), they are helpful in assisting users to proofread documents before printing them out.

systems software *(p. 43, LO 1)* Software that controls the computer and enables it to run applications software. Systems software, which includes the operating system, allows the computer to manage its internal resources.

Applications software cannot run without systems software.

thesaurus *(p. 55, LO 5)* Word processing software feature that provides a list of similar words and alternative words for any word specified in a document.

The thesaurus feature helps users find the right word and avoid repetitiveness (using the same word again and again).

tutorial *(p. 52, LO 4)* Instruction book or program that takes users through a prescribed series of steps to help them learn the product.

Tutorials, which accompany applications software packages, enable users to practice new software in a graduated fashion, thereby saving them the time they would have used trying to teach themselves.

user interface *(p. 51, LO 4)* Part of a software program that presents on the screen the alternative commands by which you communicate with the system and that displays information.

Some user interfaces are easier to use than others. Most users prefer a graphical user interface.

version *(p. 43, LO 1)* Refers to a major upgrade in a software product.

Versions are usually indicated by numbers, such as 1.0, 2.0, 3.0, and so on. The higher the number before the decimal point, the more recent the version.

Web browser *(p. 64, LO 5)* Software that enables people to view Web sites on their computers.

Without browser software, users cannot use the part of the Internet called the World Wide Web.

window *(p. 51, LO 4)* Feature of graphical user interfaces; rectangle that appears on the screen and displays information from a particular part of a program.

Using the windows feature, an operating system (or operating environment) can display several windows on a computer screen, each showing a different application program such as word processing, spreadsheets, and graphics.

word processing software *(p. 53, LO 5)* Applications software that enables users to create, edit, revise, store, and print text material.

Word processing software allows a person to use a computer to easily create, edit, copy, save, and print documents such as letters, memos, reports, and manuscripts.

(Selected answers appear at the end of the book.)

Short-Answer Questions

1. What is the difference between freeware and shareware?

2. What is software piracy? Network piracy?

3. What are the four categories of applications software?

4. What would be a good use for database software?

5. Why is the World Wide Web one of the fastest-growing subsets of the Internet?

6. What do the abbreviations CAD, CADD, and CAM mean? What do these programs do?

7. What is the purpose of data communications software?

8. What is the difference between analytical graphics and presentation graphics?

Fill-in-the-Blank Questions

1. _database_ is a collection of related programs designed to enable users to perform work on general-purpose tasks.

2. _formatting_ offers capabilities that enable the user to create and edit documents easily.

3. If you need to develop a report that involves the use of extensive mathematical, financial, or statistical analysis, you would use _spreadsheet_ software.

4. _____ software enables you to combine near-typeset-quality text and graphics on the same page in a professional-looking document.

5. A _dialog_ is a box that appears on a GUI screen that requires a response from you.

6. In a computer system, a _GUI_ is organized into fields, records, and files.

7. _____ software packages combine the features of several applications into one software package.

8. _systems software_ enables applications software to interact with the computer.

9. A _Web browser_ is software that lets you move through sites on the World Wide Web.

Multiple-Choice Questions

1. Which of the following software types may be copied by anyone for any purpose without penalty?
   a. applications software
   b. systems software
   c. shareware
   d. public domain software
   e. groupware

2. The most popular uses of software are:
   a. word processing and personal information managers
   b. word processing and spreadsheet
   c. spreadsheet and database
   d. word processing and graphics
   e. word processing and desktop publishing

3. Which of the following types of software is free to use but requires that you pay a fee before you can receive technical help or documentation?
   a. freeware
   b. shareware
   c. proprietary software
   d. entertainment software
   e. none of the above

4. Which of the following is a pictorial representation of a command or task that is used in graphical user interfaces?
   a. window
   b. icon
   c. button
   d. dialog box
   e. none of the above

5. Which of the following isn't a feature of word processing software?
   a. what-if analysis
   b. thesaurus
   c. spell-checker
   d. justification
   e. headers and footers

6. Which of the following is a capability of data communications software?
   a. online connections
   b. use of financial services
   c. automatic dialing services
   d. remote access communications
   e. all of the above

7. Which of the following combines the capabilities of several applications into a single application?
   a. software suite
   b. shareware
   c. integrated software
   d. groupware
   e. none of the above

8. Which of the following is used on a network and serves multiple users working on the same project?
   a. software suite
   b. shareware
   c. integrated software
   d. groupware
   e. none of the above

True/False Questions

T (F) 1. Function keys are used the same way with all applications software packages.

(T) (F) 2. Tutorials are the same as documentation.

T F 3. A personal information manager combines some of the capabilities of a word processor, database, and spreadsheet.

T F 4. To price an unbundled computer system, you must price each system component individually.

T (F) 5. Hyperlinks of the World Wide Web allow users to browse through information by jumping from topic to topic.

T (F) 6. Spreadsheet software offers presentation graphics capabilities.

T F 7. Hypertext is commonly used in Help systems to allow users to go to related helpful information.

T F 8. A document appearing on the screen in WYSIWYG form must be printed before you see how it will look in final form.

Projects/Critical-Thinking Questions

1. Locate a department at your school or place of work that is using some custom-written software. What does this software do? Who uses it? Why couldn't it have been purchased off the shelf? How much did it cost? Do you think there is an off-the-shelf program that can be used instead? Why/why not?

2. Prepare a short report about how you would use an electronic spreadsheet to organize and manage your personal finances and to project outcomes of changes. What column headings (labels) would you use? Row headings? What formula relationships would you want to establish among the cells? (For example, if your tuition increased by $2000, how would that affect the monthly amount you set aside to buy a car or take a trip?)

net **Using the Internet**

Objective: *In this exercise we describe how to send and receive mail using the Pine mail program.*

Before you continue: *We assume that your Internet connection provides access to the Pine mail program. We also assume your Internet connection uses: (1) a command-line interface, (2) VT100 terminal emulation, and (3) the Unix operating system. If any of these assumptions are incorrect, you may still be able to perform this Internet exercise; however, some of the commands and screen output may be different. If necessary, ask your instructor or system administrator for assistance.*

1. Access your Internet account using your user name and password. The procedure you use depends on your particular system. Next, depending on your system, you may be prompted to choose a terminal emulation mode. If possible, choose VT100 terminal emulation. Otherwise, simply press **Enter** to bypass this option.

2. At this point you are either presented with a command prompt such as $, %, or > and a blinking cursor, or a menu. If a menu appears, choose the option that exits you to command-line, or shell, mode.

3. To start the Pine mail program:
   TYPE: pine
   PRESS: **Enter**
   The following menu should appear:

4. In this step you will use the Compose menu option to send a message to yourself. To choose the Compose option:
   TYPE: C
   The Compose Message screen should appear:

5. The cursor is currently positioned in the message header. Although you must type an e-mail address, or user name, into the "To:" information, the rest of the information (Cc, Attchmnt, and Subject) is optional. However it is considered good "netiquette" to include an appropriate subject line.

   The following is a sample e-mail address: kjones@is.college.edu (*Note:* If the person you're sending a message to uses the same computer system that you're logged into, you don't have to type "@" or any of the information to the right in the address.)

   To send a message to yourself:
   TYPE: *your e-mail address (user name)*

6. To move the cursor to the Subject area:
   PRESS: [Enter] *three times*

7. In the Subject area:
   TYPE: About Shareware
   PRESS: [Enter]

8. In the Message Text area, type the following message to yourself:
   TYPE: Shareware is copyrighted software that is widely available on the Internet. You can download the software for free and distribute it legally to other users. However, you must pay a fee to receive technical help, documentation, or upgrades.

   Your Compose Message screen may appear similar to the following:

9. As indicated on the bottom of the screen, to send the message you hold down [Ctrl] and type X.
   PRESS: [Ctrl] +X

10. At the "Send message?" prompt, you can simply press [Enter] to send the message.
    PRESS: [Enter]
    The Pine main menu should appear and the words "Message Sent" should appear above the commands at the bottom of the screen. At this point, wait about one minute before proceeding with the next step.

11. You should be viewing the main menu and the Folder List option should be highlighted in bold letters. To locate the message you sent to yourself:
    PRESS: [Enter] to select the Folder List option
    PRESS: [Enter] to display the contents of your Inbox
    The following screen shows the Inbox with three messages:

12. The message you sent to yourself should be highlighted. If it isn't, press [↓] until the message is highlighted. To view the message you sent to yourself, you can use the ViewMsg option that is listed on the bottom of the screen, or simply press [Enter].
    TYPE: V
    The screen should appear similar to the following:

13. To redisplay the main menu, choose the Main Menu option:
    TYPE: M

14. To quit the Pine mail program, you must use the Quit option:
    TYPE: Q
    PRESS: [Enter]

15. Now that you know how to send and receive e-mail, let's disconnect from the Internet. You can exit by typing "exit" or "logout."
    For example:
    TYPE: exit
    PRESS: [Enter]

    Note: If a menu appears, choose the option to disconnect, or quit.

CHAPTER 3

Systems Software
The Power Behind the Power

Concepts You Should Know

After reading this chapter, you should be able to:

1. Explain the difference between applications software and systems software.
2. Explain the three types of systems software.
3. Describe six functions of the operating system.
4. Define and describe the three types of user interfaces, including the GUI.
5. Explain the key features of the principal microcomputer operating systems.
6. List and describe the principal external utility programs.
7. Explain how new developments in communications could eliminate the need for users to concern themselves with systems software.

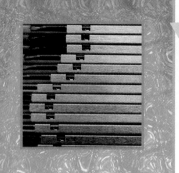

ou're gonna know in the first 10 minutes whether you are going to like that person."

That's how we react to meeting someone for the first time, says Tim Eckles, a consumer-product manager for a microcomputer maker. And, he says, many novices react to their first encounter with a computer in the same way.[1]

Not too many years ago, manufacturers could get away with sending out personal computers that required arcane commands to operate. That's because most customers were businesses, usually with some technical expertise on staff. Today, however, sales of microcomputers are exploding among non-business users, as people become attracted to multimedia capabilities (✓ p. 27) that add video and sound. These consumers may not have the motivation to figure out typical computer commands. "PC makers are afraid home users will pack up their new computers and send them back if they get too frustrated," says one journalist.[2]

Accordingly, computer makers are working hard to make sure that the first meeting consumers have with a computer is a friendly one. Indeed, a great deal of effort has gone into trying to make interfaces (✓ p. 50) as intuitive as possible for newcomers. In time, as interfaces are refined, computers will probably become no more difficult to use than a car. Until then, however, for smooth encounters you need to know something about how systems software works. Today people communicate one way, computers another. People speak words and phrases; computers process bits and bytes (✓ p. 6). For us to communicate with these machines, we need an intermediary, an interpreter. This is the function of systems software. We interact mainly with the applications software, which interacts with the systems software, which controls the hardware.

Three Types of Systems Software

Preview & Review: Systems software is of three basic types: operating systems, utility programs, and language translators.

As we've said, **software, or programs, consists of the step-by-step instructions that tell the computer how to perform a task.** Software is of two types—*applications software* and *systems software.* As Chapter 2 described, **applications software is software that can perform useful work on general-purpose tasks,** such as word processing or spreadsheets. **Systems software enables the applications software to interact with the computer and helps the computer manage its internal and external resources.** Systems software is required to run applications software; however, the reverse is not true. Buyers of new computers will find the systems software has already been installed by the manufacturer.

There are three basic types of systems software—*operating systems, utility programs*, and *language translators.*

- **Operating systems:** An operating system is the principal piece of systems software in any computing system. We describe it at length in the next section.

- **Utility programs:** *Utility programs* **are generally used to support, enhance, or expand existing programs in a computer system.** Many operating systems have utility programs built in for common purposes such as merging two files into one file. Other external, or nonresident, utility programs (such as The Norton Utilities) are available separately to, for example, recover damaged files. We describe external utility programs later in this chapter.

- **Language translators:** *A language translator* **is software that translates a program written by a programmer in a language such as C—for example, a word processing applications program—into machine language (0s and 1s), which the computer can understand.**

The types of systems software are diagrammed below. (■ *See Panel 3.1.*)

The Operating System

Preview & Review: The operating system manages the basic operations of the computer. These operations include booting and housekeeping tasks. Other operations are managing computer resources and managing files. The operating system also manages tasks, through multitasking, multiprogramming, time-sharing, or multiprocessing. The user interface may be command-driven, menu-driven, or graphical.

The *operating system (OS)* **consists of the master system of programs that manage the basic operations of the computer.** These programs provide resource management services of many kinds, handling such matters as the control and use of hardware resources, including disk space, memory, CPU

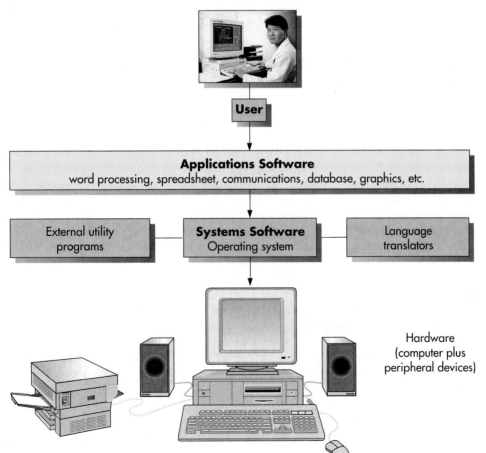

Applications Software
word processing, spreadsheet, communications, database, graphics, etc.

| External utility programs | **Systems Software** Operating system | Language translators |

Hardware (computer plus peripheral devices)

PANEL 3.1

The three types of systems software

An operating system is required for applications software to run on your computer. The user usually works with the applications software but can bypass it to work directly with the systems software for certain tasks.

time allocation, and peripheral devices. The operating system allows you to concentrate on your own tasks or applications rather than on the complexities of managing the computer.

Different sizes and makes of computers have their own operating systems. Software written for one operating system won't run on another operating system. Microcomputer users may readily experience the aggravation of such incompatibility when they buy a new microcomputer. Should they get an Apple Macintosh with Macintosh Systems Software, which won't run IBM-compatible programs? Or should they get an IBM or IBM-compatible (such as Compaq, Dell, or Zenith), which won't run Macintosh programs? Should they wait for a new operating system to be introduced that may resolve some of these differences?

Before we try to sort out these perplexities, we should see what operating systems do that deserves our attention. We consider:

- Booting
- Housekeeping tasks
- User interface
- Managing computer resources
- Managing files
- Managing tasks

Booting

The operating system begins to operate as soon as you turn on, or "boot," the computer. The term **booting refers to the process of loading an operating system into a computer's main memory from diskette or hard disk.** This loading is accomplished by a program (called the *bootstrap loader* or *boot routine*) that is stored permanently in the computer's electronic circuitry. When you turn on the machine, the program obtains the operating system from your diskette or hard disk and loads it into memory. Other programs called *diagnostic routines* also start up and test the main memory, the central processing unit (✓ p. 16), and other parts of the system to make sure they are running properly. Finally, other programs (indicated on your screen as "BIOS," for basic input-output system) will be stored in main memory to help the computer interpret keyboard characters or transmit characters to the display screen or to a diskette. The operating system remains in main memory until you turn the computer off. With newer operating systems, the booting process puts you into a graphically designed starting screen, from which you choose the applications programs you want to run.

Housekeeping Tasks

If you have not entered a command to start an applications program, what else can you do with the operating system? One important function is to perform common repetitious "housekeeping tasks."

One example of such a housekeeping task is formatting blank diskettes. Before you can use a new diskette that you've bought at a store, you may have to format it. **Formatting, or initializing, electronically prepares a diskette so it can store data or programs.** (Disks can also be purchased pre-formatted.)

User Interface

Many operating-system functions are never apparent on the computer's display screen. What you do see is the user interface. **The *user interface* is the part of the operating system that allows you to communicate, or interact, with it.**

There are three types of user interfaces, for both operating systems and applications software—*command-driven, menu-driven,* and *graphical. (■ See Panel 3.2.)* The latter two types of user interface are often called a *shell.*

- Command-driven: **A *command-driven interface* requires you to enter a command by typing in codes or words.** An example of such a command might be DIR (for "directory"). This command instructs the computer to display a directory list of all file names on a disk.

  You type a command at the point on the display screen where the cursor follows the prompt (such as following "C:\>"). Then you press the Enter key to execute the command. (A "prompt" is a symbol that shows you where to enter a command.)

  The command-driven interface is seen on IBM and IBM-compatible computers with the MS-DOS operating system (discussed shortly).

- Menu-driven: **A *menu-driven interface* allows you to choose a command from a menu.** Like a restaurant menu, **a software *menu* offers you options to choose from—in this case, commands available for manipulating data,** such as Print or Edit.

PANEL 3.2

Three types of user interfaces

(*Top*) A command-driven interface requires typing of codes or words. (*Middle*) A menu-driven interface contains menus offering displayed lists of options. (*Bottom*) A graphical user interface allows users to move an electronic mouse to select commands represented by pictorial figures (icons).

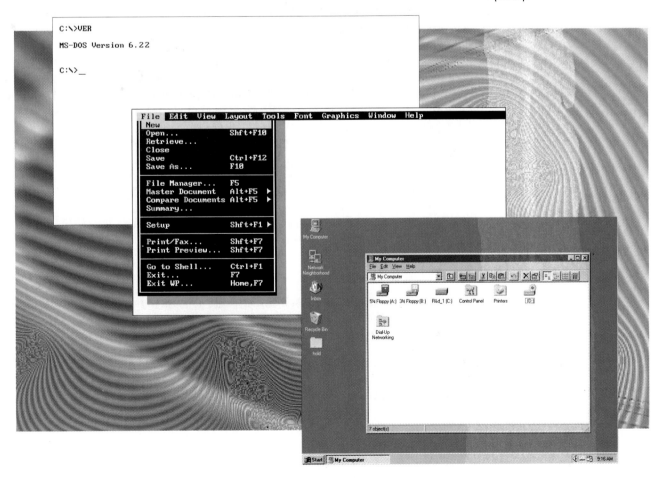

Menus are easier to use than command-driven interfaces, especially for beginners. Their disadvantage, however, is that they are slower to use. Thus, some software programs offer both features—menus for novice users and keyboard codes for experienced users.

- **Graphical:** The easiest interface to use, the **graphical user interface (GUI), uses images to represent options.** Some of these images take the form of icons. **Icons are small pictorial figures that represent tasks, functions, or programs**—for example, a trash can for a delete-file function.

Another feature of the GUI (pronounced "gooey") is the use of windows. **Windows divide the display screen into sections.** Each window may show a different display, such as a word processing document in one and a spreadsheet in another.

Finally, the GUI permits liberal use of the mouse. The mouse is used as a pointing device to move the cursor to a particular place on the display screen or to point to an icon or button. The function represented by the icon can be activated by pressing ("clicking") buttons on the mouse. Or, using the mouse, you can move ("drag") an image from one side of the screen to the other or change its size.

Microcomputer users first became aware of the graphical user interface in Apple Macintosh computers (although Apple got the idea from Xerox). Later Microsoft made a graphical user interface available for IBM and IBM-compatible computers through its Windows program. Now most operating systems on microcomputers feature a GUI.

Managing Computer Resources

Suppose you are writing a report using a word processing program and want to print out a portion of it while continuing to write. How does the computer manage both tasks?

Behind the user interface, the operating system acts like a police officer directing traffic. This activity is performed by the **supervisor, or kernel, the central component of the operating system. The supervisor, which manages the CPU, resides in main memory while the computer is on and directs other programs to perform tasks to support applications programs.**

The operating system also manages memory—it keeps track of the locations within main memory where the programs and data are stored. It can swap portions of data and programs between main memory and secondary storage, such as your computer's hard disk. This capability allows a computer to hold only the most immediately needed data and programs within main memory. Yet it has ready access to programs and data on the hard disk, thereby greatly expanding memory capacity.

Managing Files

Files of data and programs are located in many places on your hard disk and other secondary-storage devices. The operating system allows you to find them. If you move, rename, or delete a file, the operating system manages such changes and helps you locate and gain access to it. For example, you can *copy*, or duplicate, files and programs from one disk to another. You can *back up*, or make a duplicate copy of, the contents of a disk. You can *erase*, or remove, from a disk any files or programs that are no longer useful. You can *rename*, or give new filenames, to the files on a disk.

Managing Tasks

A computer is required to perform many different tasks at once. In word processing, for example, it accepts input data, stores the data on a disk, and prints out a document—seemingly simultaneously. Some computers' operating systems can also handle more than one program at the same time—word processing, spreadsheet, database searcher—displaying them in separate windows on the screen. Others can accommodate the needs of several different users at the same time. All these examples illustrate *task management*—a "task" being an operation such as storing, printing, or calculating.

Among the ways operating systems manage tasks in order to run more efficiently are *multitasking, multiprogramming, time-sharing,* and *multiprocessing.* Not all operating systems can do all these things.

- **Multitasking—executing more than one program concurrently:** **Multitasking is the execution of two or more programs by one user concurrently—not simultaneously—on the same computer with one central processor.** You may be writing a report on your computer with one program while another program searches an online database for research material. How does the computer handle both programs at once?

  The answer is that the operating system directs the processor (CPU) to spend a predetermined amount of time executing the instructions for each program, one at a time. In essence, a small amount of each program is processed, and then the processor moves to the remaining programs, one at a time, processing small parts of each. This cycle is repeated until processing is complete. The processor speed is usually so fast that it may seem as if all the programs are being executed at the same time. However, the processor is still executing only one instruction at a time, no matter how it may appear to the user.

- **Multiprogramming—concurrent execution of different users' programs:** **Multiprogramming is the execution of two or more programs on a *multiuser* operating system.** As with multitasking, the CPU spends a certain amount of time executing each user's program, but it works so quickly it seems as though all the programs are being run at the same time.

- **Time-sharing—round-robin processing of programs for several users:** **Time-sharing is a single computer's processing of the tasks of several users at different stations in round-robin fashion.** Time-sharing is used when several users are linked by a communications network to a single computer. The computer will first work on one user's task for a fraction of a second, then go on to the next user's task, and so on.

  How is this done? The answer is through *time slicing.* Computers operate so quickly that it is possible for them to alternately apportion slices of time (fractions of a second) to various tasks. Thus, the computer's operating system may rapidly switch back and forth among different tasks, just as a hairdresser or dentist works with several clients or patients concurrently. The users are generally unaware of the switching process.

  Multitasking and time-sharing differ slightly. With multitasking, the processor directs the programs to take turns accomplishing small tasks or events within the programs. These events may be making a calculation, searching for a record, printing out part of a document, and so on. Each event may take a different amount of time to accomplish. With time-sharing, the computer spends a *fixed amount* of time with each program before going on to the next one.

- **Multiprocessing—simultaneous processing of two or more programs by multiple computers:** **Multiprocessing is processing done by two or more computers or processors linked together to perform work simultaneously**—that is, at precisely the same time. This can entail processing instructions from different programs or different instructions from the same program.

Multiprocessing goes beyond multitasking, which works with only one microprocessor. In both cases, the processing should be so fast that, by spending a little bit of time working on each of several programs in turn, a number of programs can be run at the same time. With both multitasking and multiprocessing, the operating system keeps track of the status of each program so that it knows where it left off and where to continue processing. But the multiprocessing operating system is much more sophisticated than multitasking.

Operating system functions are summarized below. (■ *See Panel 3.3.*)

Microcomputer Operating Systems & Operating Environments

Preview & Review: The principal microcomputer operating systems and operating environments are DOS, Macintosh Operating System, Windows 3.X (for DOS), OS/2 Warp, Windows 95, Windows NT, Unix, and NetWare.

As a microcomputer user, you'll have to learn not only whatever applications software you want to use but also, to some degree, the operating system with which they work. Moreover, when you buy a PC, it comes with an operating system. You have to know which one you want.

In this section, we describe the following:

- DOS
- Macintosh Operating System
- Windows 3.X for DOS
- Windows 95
- Windows NT
- OS/2 Warp
- Unix
- NetWare

PANEL 3.3

Some operating system functions

Booting	House-keeping Tasks	User Interface	Managing Computer Resources	Managing Files	Managing Tasks
Loads operating system into computer's main memory	Formats diskettes	Provides a way for user to interact with the operating system—can be command-driven, menu-driven, or graphical	Via the supervisor, manages the CPU and directs other programs to perform tasks to support applications programs	Copies files/programs from one disk to another	May be able to perform multitasking, multiprogramming, time-sharing, or multiprocessing
Uses diagnostic routines to test system for equipment failure	Displays information about operating system version		Keeps track of locations in main memory where programs and data are stored	Backs up files/programs	
Stores BIOS programs in main memory	Displays disk space available		Moves data and programs back and forth between main memory and secondary storage (swapping)	Erases (deletes) files/programs	
				Renames files	

But, before we proceed, we need to briefly define what an operating *environment* is, since some people have trouble distinguishing it from an operating *system*. An *operating environment*—also known as a *windowing environment* or *shell*—adds a graphical user interface or a menu-driven interface as an outer layer to an operating system. The most well known operating environment is the Windows 3.X program sold by Microsoft, which adds a graphical user interface to DOS.

DOS

There are reportedly over 100 million users of DOS. This makes it the most popular software of any sort ever adopted, and certainly the most popular systems software.[3] **DOS—for *Disk Operating System*—runs primarily on IBM and IBM-compatible microcomputers,** such as Compaq, Zenith, AST, Dell, Tandy, and Gateway. (■ *See Panel 3.4.*)

There are now two main operating systems calling themselves DOS.

- **Microsoft's MS-DOS:** DOS is sold under the name MS-DOS by software maker Microsoft. The "MS" stands for Microsoft. Microsoft launched its original version, MS-DOS 1.0, in 1981, and there have been several upgrades since then.

- **IBM's PC-DOS:** Microsoft licenses a version to IBM called PC-DOS. The "PC" stands for "Personal Computer." The most recent version is PC-DOS 7, released March 1995.

What do the numbers in the names mean? The number before the period refers to a *version.* The number after the period refers to a *release,* which has fewer refinements than a version. The most recent versions are all backward compatible. For operating systems, *backward compatible* means that users can run the same applications on the later versions of the operating system that they could run on earlier versions.

Although DOS is still in wide use, its importance is declining in favor of graphical-user-interface types of operating systems, which we discuss next.

PANEL 3.4

The aging system of DOS

With DOS, you work with precise, typed commands.

"Like a turn-of-the-century building constructed according to venerable principles of design, DOS perpetuates principles designed for prior generations of computers. That's because DOS was developed before computers had their present speed and the ability to generate elaborate on-screen images.

The instructions you use to run a program, retrieve previous work, and the like are adaptations of the typewritten instructions used to communicate with early computers. To use DOS, you must type in an idiosyncratic string of words or letters or (in later versions) choose a written command from an on-screen menu."

—"The Basic Choice in Computers," *Consumer Reports*

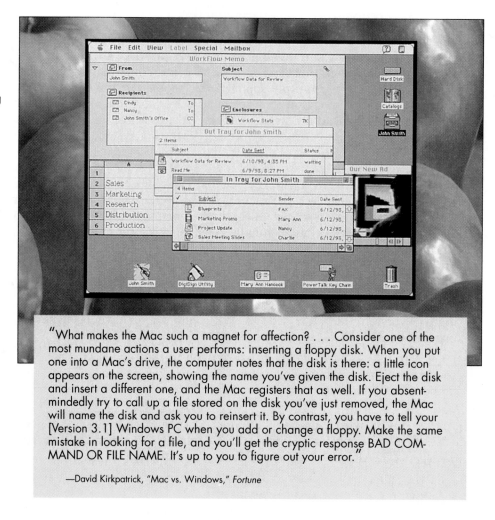

"What makes the Mac such a magnet for affection? . . . Consider one of the most mundane actions a user performs: inserting a floppy disk. When you put one into a Mac's drive, the computer notes that the disk is there: a little icon appears on the screen, showing the name you've given the disk. Eject the disk and insert a different one, and the Mac registers that as well. If you absent-mindedly try to call up a file stored on the disk you've just removed, the Mac will name the disk and ask you to reinsert it. By contrast, you have to tell your [Version 3.1] Windows PC when you add or change a floppy. Make the same mistake in looking for a file, and you'll get the cryptic response BAD COMMAND OR FILE NAME. It's up to you to figure out your error."

—David Kirkpatrick, "Mac vs. Windows," *Fortune*

Macintosh Operating System (Mac OS)

The Apple Macintosh has always had one outstanding feature: it is easy to use. The Macintosh was the first microcomputer to use a graphical user interface (1982), and the easy-to-use interface has generated a strong legion of fans. (■ *See Panel 3.5.*) The most recent version of Mac OS is System 7.5.

The DOS operating system was designed for microcomputers using Intel-type microprocessors (✓ p. 110), whereas Mac OS was designed to work with Motorola microprocessors. Without special hardware and software, applications software designed for one system will not run on the other.

Although the Macintosh is easy to use, not as many programs have been written for it as for DOS/Windows-based systems. Only about 6900 commercial applications packages have been written for Macs, according to BIS, a Norwell, Massachusetts, market research firm.[4] By contrast, some *29,400* applications packages are available for DOS computers. However, its graphics capabilities make the Macintosh a popular choice for people working in commercial art, desktop publishing, multimedia, and engineering design.

Windows for DOS (Windows 3.X)

As we introduce the Windows environment, it's worth expanding on what we said in Chapter 2 about how windows differ from *Windows* (✓ p. 51). A *window* (lowercase "w") is a portion of the video display area dedicated to some specified purpose. An operating system (or operating environment) can display several windows on a computer screen, each showing a different applications program, such as word processing and spreadsheets. However, *Windows* (capital "W") is something else. **Windows is an operating environment made by Microsoft that lays a graphical user interface shell around the MS-DOS or PC-DOS operating system.** Like Mac OS, Windows contains windows, which can display multiple applications. (■ *See Panel 3.6.*)

Note: It's important to realize that Windows 3.X ("3.X" represents versions 3.0, 3.1, and 3.11) is different from *Windows 95* (discussed next), which is not just an operating environment but a true operating system.

Microsoft's Windows 3.X is designed to run on IBM-style microcomputers with Intel microprocessors—the '386 and '486 chips. Although Windows is far easier to use than DOS, its earlier versions have not been as easy to use as the Mac operating system. This is because Windows sat atop the 11-year-old command-driven DOS operating system, which required certain compromises on ease of use. Windows 95 solved this problem.

Windows 95 & Later

"Thank God it's finally over," wrote a reporter. "Or thank Bill Gates."[5]

He was referring to Gates's and Microsoft's months of promotional buildup that ended in a spectacular display of excess when Windows 95 finally made its debut on August 24, 1995. **Windows 95, the successor to Windows 3.X for DOS, is a true operating system for IBM-style personal computers.**

Following are just some of the features of Windows 95:

● **Clean "Start":** Instead of encountering a confusing array of similar program groups (as with Windows 3.X), you'll first see a clean "desktop" with a "Taskbar" of important icons at the bottom of the screen and one button labeled START. (■ *See Panel 3.7 on the next page.*)

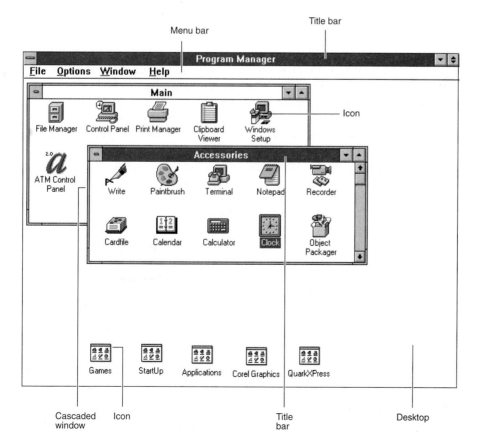

PANEL 3.6

Windows 3.X's graphical user interface

PANEL 3.7

Windows 95 screen

Start button: Click for an easy way to start using the computer.

Microsoft Network: Click here to connect to the Microsoft Network, the company's online service.

My Briefcase: Allows you to synchronize files in two computers—say, an office PC and a laptop.

Recycle Bin: Allows you to dispose of files—or retrieve them later.

Network Neighborhood: If your PC is linked to a network of PCs, click here to get a glimpse of everything available on the network.

My Computer: Gives you a quick overview of all the files and programs installed in your PC.

Document: New multitasking capabilities allow you to smoothly run more than one program at once.

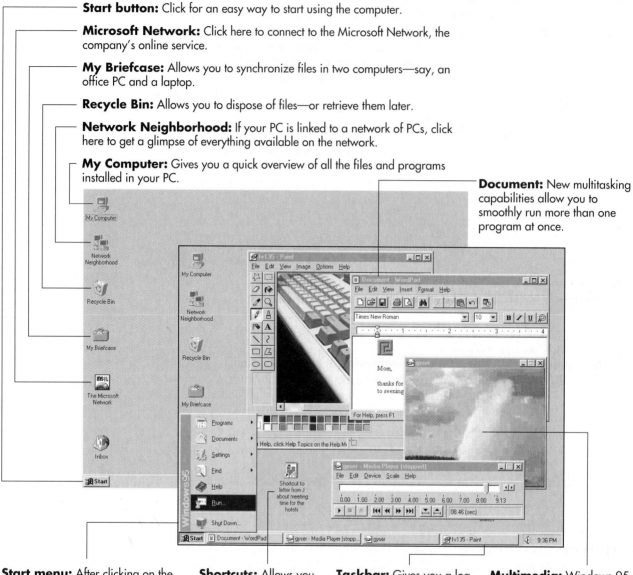

Start menu: After clicking on the start button, a menu appears, giving you a quick way to handle common tasks. You can launch programs, call up documents, change system settings, get help, and shut down your PC.

Shortcuts: Allows you to immediately launch often-used files and programs.

Taskbar: Gives you a log of all programs you have opened. To switch programs, click on the buttons that appear in the taskbar.

Multimedia: Windows 95 features sharper graphics and improved video capabilities.

"Windows has long since stretched the definition of *operating* system past the breaking point. The original DOS was little more than a thin (and clumsy) layer of hooks that applications could use for reading and writing data to memory, screen and disks. Windows 95 not only provides a rich environment for controlling many programs at once; it also offers, built in, a word processor, communications software, a fax program, an assortment of games, screen savers, a telephone dialer, a paint program, back-up software and a host of other housekeeping utilities and, of course, Internet software. By historical standards, you get a remarkable bargain."

—James Gleick, "Making Microsoft Safe for Capitalism," *New York Times Magazine*

- **Better menus:** Windows 3.X's quirky Program Manager and File Managers have been replaced by more accessible features called THE EXPLORER and MY COMPUTER, which let you quickly see what's stored on your disk drives and make tracking and moving files easier.
- **Long file names:** File names can now be up to 256 characters instead of the 8 characters (plus 3-character extension) of DOS and Windows 3.X. This means you can now have a file name for your resume, for example, of "Resume—January 15, 1997, version" instead of "RES11597." (Macintosh OS and OS/2 have always permitted long file names.)
- **The "Recycle Bin":** This feature allows you to delete complete files and then get them back if you change your mind.
- **32-bit instead of 16-bit:** The new software is a 32-bit program, whereas most Windows 3.X software is 16-bit. **Bit numbers refer to how many bits of data a computer chip, and software written for it, can process at one time.** Such numbers are important because they refer to the amount of information the hardware and software can use at any one time. This doesn't mean that 32-bit software will necessarily be twice as fast as 16-bit software, but it does promise that new 32-bit applications software will offer better speed and features once software developers take advantage of the design.[6]
- **Plug and play:** It has always been easy to add new hardware components to Macintoshes. It used to be extremely difficult with IBM-compatible PCs. *Plug and play* refers to the ability to add a new hardware component to a computer system and have it work without needing to perform complicated technical procedures.

More particularly, *Plug and Play* (abbreviated *PnP*) is a standard developed for IBM-style PCs by Microsoft and chip maker Intel and incorporated into Windows 95 to eliminate user frustration when one is adding new components. Now when you add a new printer or modem, your PC will recognize the model and set it up.

Windows 95 was expected to be superseded in 1996 by an upgrade called *Nashville,* or *Windows 96.*

Windows NT

Unveiled by Microsoft in May 1993, **Windows NT, for *New Technology,* is an operating system intended to support large networks of computers.** Examples of such networks are those used in airline reservations systems. Unlike the early Windows operating *environment* (which ran with DOS), Windows NT is a true operating *system,* interacting directly with the hardware. Although on a screen it looks identical to Windows 3.X and can run on the most modern PCs, it is primarily designed to run on workstations or other more powerful computers. Indeed, NT will run on the most powerful of microprocessors, both Intel's Pentium and Motorola's PowerPC.

OS/2 & OS/2 Warp

OS/2 (there is no OS/1) was initially released in April 1987 as IBM's contender for the next mainstream operating system. **OS/2—for *Operating System/2*—is designed to run on many recent IBM and compatible microcomputers.** Unlike Windows 3.X, OS/2 does not require DOS to run underneath it, so it generally processes more efficiently. Like Windows, it has a graphical user interface, called the Workplace Shell (WPS), which uses icons resembling documents, folders, printers, and the like. OS/2 can also run most DOS, Windows, and OS/2 applications simultaneously. This means that users don't have to throw out their old applications software to take advantage of new features. In addition, OS/2 is the first microcomputer operating system to take full advantage of the power of the newer Intel microprocessors, such as '486 and Pentium chips. Lastly, this operating system is designed to connect everything from small handheld personal computers to large mainframes.[7]

OS/2 can perform some advanced feats. It can, for example, receive a fax and run a video while at the same time recalculating a spreadsheet. This is the kind of multitasking (and even multimedia) activity that is increasingly important for networked computers. It is also the first operating system created just for today's new "workgroup" environments. In workgroups, individuals work in groups sharing electronic files and databases over communications lines. Yet, despite its seeming complexity, OS/2 has been considered reasonably easy for beginners to use. Indeed, its file management resembles Macintosh's System 7 (allowing long file names, for example), making it easier to use than the older versions of Windows.

In late 1994 IBM unveiled a souped-up version of OS/2 called *Warp*. Despite spending $2 billion on OS/2 in its long struggle against Windows for DOS the company failed to increase its market share. OS/2 remained hopelessly mired with only 4% of the market for desktop operating systems, versus roughly 80% for Windows 3.X—and Microsoft had not even released its much ballyhooed Windows 95 yet.

Unix

Unix was invented more than two decades ago by American Telephone & Telegraph, making it one of the oldest operating systems. **Unix is an operating system for multiple users and has built-in networking capability, the ability to run multiple tasks at one time, and versions that can run on all kinds of computers.** Because it can run with relatively simple modifications on different types of computers—from micros to minis to mainframes—Unix is called a "portable" operating system. The primary users of Unix are government agencies, large corporations, and banks, which use the software for everything from airplane-parts design to currency trading.

Unix is a popular operating system in Europe, where users have discovered that its applications can survive changes in hardware, so that business is not unduly disrupted when new hardware is introduced. Perhaps, with agreement on Unix standards, the same thing will be true in the United States.

NetWare

Novell, of Orem, Utah, is the maker of NetWare, the software Microsoft is trying to beat with its Windows NT. Developed during the 1980s, **NetWare has become the most popular operating system for coordinating microcomputer-based local area networks (LANs) throughout a company or campus.** (LANs allow PCs to share data files, printers, and other devices.)

Can you continue to use, say, MS-DOS on your office personal computer while it is hooked up to a LAN running NetWare? Indeed you can. NetWare provides a shell around your own operating system. If you want to work "off network," you respond to the usual prompt (for example, the DOS-based A:\>, B:\>, or C:\>) and run the PC's regular operating system. If you want to work "on network," you respond to another prompt (for instance, F:\>) and type in whatever password will admit you to the network.

Novell's vision is of a network that will extend beyond office networks of PCs, even beyond the global Internet.[8] It envisions wireless networks linking automobiles, appliances, vending machines, electronic cash registers, factory automation, security systems, and other nontraditional computing devices, as well as telephones, fax machines, and copiers. Novell has also developed ways to exchange information over ordinary electric power lines. With this technology, even standard electrical outlets could become NetWare connections.[9]

External Utility Programs

Preview & Review: External utility programs provide services not performed by other systems software. They often include screen savers, data recovery, backup, virus protection, file defragmentation, and data compression. Multiple-utility packages are available.

"You wouldn't take a cruise on a ship without life preservers, would you?" asks one writer. "Even though you probably wouldn't need them, the terrible *what if* is always there. Working on a computer without the help and assurance of utility software is almost as risky."[10]

The "what if" being referred to is an unlucky event, such as your hard-disk drive "crashing" (failing), risking loss of all your programs and data; or your computer system being invaded by someone or something (a virus) that disables it.

***External utility programs* are special programs that provide specific useful services not performed or performed less well by other systems software programs.** Examples of such services are backup of your files for storage, recovery of damaged files, virus protection, data compression, and memory management. Some of these features are essential to preventing or rescuing you from disaster. Others merely offer convenience.

Some Specific Utility Tasks

Some of the principal services offered by utilities are the following:

* **Screen saver:** A *screen saver* is a utility that supposedly prevents a monitor's display screen from being etched by an unchanging image ("burn-in"). Some people believe that if a computer is left turned on without keyboard or mouse activity, whatever static image is displayed may burn into the screen. Screen savers automatically put some moving patterns on the screen, supposedly to prevent burn-in. Actually, burn-in doesn't happen on today's monitors. Nevertheless, people continue to buy screen savers, often just to have a kind of "visual wallpaper." Some of these can be quite entertaining, such as flying toasters.

* **Data recovery:** A *data recovery utility* is used to *undelete* a file or information that has been accidentally deleted. *Undelete* means to undo the last delete operation that has taken place. The data or program you are trying to recover may be on a hard disk or a diskette.

* **Backup:** Suddenly your hard-disk drive fails, and you have no more programs or files. Fortunately, you have (we hope) used a utility to make a backup, or duplicate copy, of the information on your hard disk. Examples of backup utilities are Norton Backup from Symantec, Backup Exec from Arcada Software, Colorado Backup, and Fastback Plus from Fifth Generation Systems.

* **Virus protection:** Few things can make your heart sink faster than the sudden failure of your hard disk. The exception may be the realization that your computer system has been invaded by a virus. **A *virus* consists of hidden programming instructions that are buried within an applications or**

systems program. **They copy themselves to other programs, causing havoc.** Sometimes the virus is merely a simple prank that pops up a message. Sometimes, however, it can destroy programs and data. Viruses are spread when people exchange diskettes or download (make copies of) information from computer networks or the Internet.

Fortunately, antivirus software is available. ***Antivirus software* is a utility program that scans hard disks, diskettes, and the microcomputer's memory to detect viruses.** Some utilities destroy the virus on the spot. Others notify you of possible viral behavior, in case the virus originated after the antivirus software was released.

Examples of antivirus software are Anti-Virus from Central Point Software, Norton AntiVirus from Symantec, McAfee virus protection software, and ViruCide from Parsons Technology.

• **File defragmentation:** Over time, as you delete old files from your hard disk and add new ones, something happens: the files become *fragmented*. ***Fragmentation* is the scattering of portions of files about the disk in nonadjacent areas, thus greatly slowing access to the files.**

When a hard disk is new, the operating system puts files on the disk contiguously (next to one another). However, as you update a file over time, new data for that file is distributed to unused spaces. These spaces may not be contiguous to the older data in that file. It takes the operating system longer to read these fragmented files. By using a utility program, you can "defragment" the file and speed up the drive's operation.

An example of a program for unscrambling fragmented files is Norton SpeedDisk utility.

• **Data compression:** As you continue to store files on your hard disk, it will eventually fill up. You then have three choices: You can delete old files to make room for the new. You can buy a new hard disk with more capacity and transfer the old files and programs to it. Or you can buy a data compression utility.

***Data compression* removes redundant elements, gaps, and unnecessary data from a computer's storage space so less space is required to store or transmit data.** With a data compression utility, files can be made more compact for storage on your hard-disk drive. The files are then "stretched out" again when you need them.

Examples of data compression programs are Stacker from Stac Electronics, Double Disk from Verisoft Systems, and SuperStor Pro from AddStor.

Other examples of utilities are file conversion, file transfer, and security. A *file conversion utility* converts files between any two applications or systems formats—such as between WordPerfect and Word for Windows or between Windows and Mac OS. A *file transfer utility* allows files from a portable computer to be transferred to a desktop computer or a mainframe computer and vice versa. A *security utility* protects unauthorized people from gaining access to your computer without using a password, or correct code. Other utilities also exist.

Multiple-Utility Packages

Some utilities are available singly, but others are available as "multipacks." These multiple-utility packages provide several utility disks bundled in one box, affording considerable savings. Examples are Symantec's Norton Desktop (for DOS, Windows, or Macintosh), 911 Utilities from Microcom, and PC Tools from Central Point Software.

The Future: What's Coming?

Preview & Review: Some future computers might be "network PCs," without their own operating systems and dominated by Web browsers.

Almost without warning, the Internet and the World Wide Web have dramatically changed the picture of computing. One possible result, the "network PC," may eliminate the need for systems software, as we discuss next.

The Problem with Personal Computing Today

Today personal computing is complicated because of conflicting standards. Could it be different tomorrow as more and more people join the trend toward networked computers and access to the World Wide Web?

As we've seen, there are different hardware and software standards, or "platforms." *Platform* **means the particular hardware or software standard on which a computer system is based.** Examples are the Macintosh platform versus the IBM-compatible platform, or Unix versus Windows NT. Developers of applications software, such as word processors or database managers, need to make different versions to run on all the platforms.

Today microcomputer users who wish to access online data sources must provide not only their own computer, modem, and communications software but also their own operating system software and applications software. (■ *See Panel 3.8, top.*)

Could this change in the future?

Personal computing today

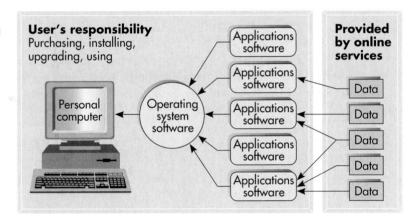

Personal computing tomorrow

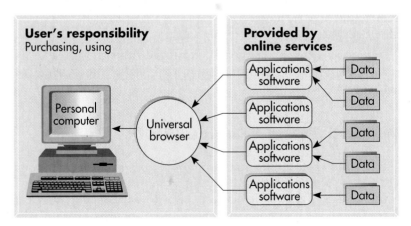

PANEL 3.8

Online personal computing—today and tomorrow

(*Top*) Today users provide their own operating system software and their own applications software and are usually responsible for installing them on their personal computers. They are also responsible for any upgrades of hardware and software. Data can be input or downloaded from online sources. (*Bottom*) Tomorrow, according to this model, users would not have to worry about operating systems or even about having to acquire and install (and upgrade) their own applications software. Using a universal Web browser, they could download not only data but also different kinds of applications software from an online source.

Personal Computing Tomorrow

Today you must take responsibility for making sure your computer system will be compatible with others you have to deal with. (For instance, if a Macintosh user sends you a file to run on your IBM PC, it's up to you to take the trouble to use special software that will translate the file so it will work on your system.) What if the responsibility for ensuring compatibility between different systems were left to online service providers?

In this future model, you would use your Web browser (✓ p. 64) to access the World Wide Web and take advantage of applications software anywhere on the network. (■ *Refer back to Panel 3.8, bottom.*) It would not matter what operating system you used. Applications software would become nearly disposable. You would download applications software and pay a small amount for each use. You would store frequently used software on your own computer. You would not need to worry about buying the right software, since it could be provided online whenever you needed to accomplish a task.[11]

The Network Computer

Engineers have proposed the idea of the "network computer" or "hollow PC." This view—which not everyone accepts—is that the expensive PCs with muscular microprocessors and operating systems would be replaced by network computers costing perhaps $500 or so.[12] Also known as the *Internet PC*, the *network computer (NC)* would theoretically be a "hollowed out" computer, perhaps without even a hard disk, serving as a mere terminal or entry point to the online universe. The computer thus becomes a peripheral to the Internet.

The concept of the "hollow PC" raises some questions: [13–15]

- **Would the browser really become the OS?** Would a Web browser become the operating system? Or will existing operating systems expand, as in the past, taking over browser functions?

- **Would communications functions really take over?** Would communications functions become the entire computer, as proponents of the network PC contend? Or would they simply become part of the personal computer's existing repertoire of skills?

- **Would an NC really be easy to use?** Would a network computer really be user friendly?

- **Aren't high-speed connections required?** Even users equipped with the fastest modems would find downloading even small programs ("applets") time-consuming. Doesn't the network computer ultimately depend on faster connections than are possible with the standard telephone lines and modems now in place?

- **Would users go for it?** Would computer users really prefer scaled-down generic software that must be retrieved from the Internet each time it is used? Would a pay-per-use system tied to the Internet really be cheaper in the long run?

Onward: Toward Compatibility

The push is on to make computing and communications products compatible. Customers are demanding that computer companies work together to create products that will make it easy to access and use great amounts of information. As technological capabilities increase, so will the demand for simplicity.

SUMMARY

What It Is / What It Does	Why It's Important
antivirus software *(p. 98, LO 6)* Software utility that scans hard disks, floppy disks, and microcomputer memory to detect viruses; some antivirus utilities also destroy viruses.	Computer users must find out what kind of antivirus software to install on their systems in order to protect them against damage or shut-down.
applications software *(p. 84, LO 1)* Software that can perform useful work on general-purpose tasks.	Applications software such as word processing, spreadsheet, database manager, graphics, and communications packages have become commonly used tools for increasing people's productivity.
bit number *(p. 94, LO 5)* Measure of how much data a computer chip, and software written for it, can process at one time.	Higher bit numbers indicate higher speed.
booting *(p. 86, LO 3)* Refers to the process of loading an operating system into a computer's main memory from floppy disk or hard disk.	When a computer is turned on, a program (called the *bootstrap loader* or *boot routine*) stored permanently in the computer's electronic circuitry obtains the operating system from the floppy disk or hard disk and loads it into main memory. Only after this process is completed can the user begin work.
command-driven interface *(p. 87, LO 4)* Type of user interface that requires users to enter a command by typing in codes or words.	The command-driven interface is used on IBM and IBM-compatible computers with the DOS operating system.
data compression *(p. 98, LO 6)* Software utility that removes redundant elements, gaps, and unnecessary data from computer files so less space is required to store or transmit data.	Many of today's files, with graphics, sound, and video, require too much storage space; data compression utilities allow users to reduce the space they take up.
disk operating system (DOS) *(p. 91, LO 5)* Microcomputer operating system that runs primarily on IBM and IBM-compatible microcomputers. DOS is sold under the names MS-DOS by Microsoft Corporation, PC-DOS by IBM, and, until recently, DOS 7 by Novell.	DOS is the most common microcomputer operating system.
external utility programs *(p. 97, LO 6)* Special programs that provide specific useful services not provided or performed less well by other system software programs.	Some of these programs are essential for helping rescue users from disaster—for example, recovery of damaged files, virus protection, data compression, and memory management.
formatting *(p. 86, LO 3)* Also called *initializing*; a computer process that electronically prepares a diskette so it can store data or programs.	Before you can use a new diskette, you usually have to format it.
fragmentation *(p. 98, LO 6)* Uneven distribution of data on a hard disk.	Fragmentation causes operating systems to run slower; to solve this problem, users can buy a file defragmentation software utility.

graphical user interface (GUI) *(p. 88, LO 4)*   User interface that uses images to represent options. Some of these images take the form of icons, small pictorial figures that represent tasks, functions, or programs.

GUIs are easier to use than command-driven interfaces and menu-driven interfaces; they permit liberal use of the electronic mouse as a pointing device to move the cursor to a particular icon or place on the display screen. The function represented by the icon can be activated by pressing ("clicking") buttons on the mouse.

icon *(p. 88, LO 4)*   Small pictorial figure that represents a task, function, or program.

The function represented by the icon can be activated by pointing at it with the mouse pointer and pressing ("clicking") on the mouse. The use of icons has simplified the use of computers.

language translator *(p. 85, LO 2)*   Systems software that translates a program written in a computer language by a computer programmer (such as BASIC) into the language (machine language) that the computer can understand.

Without language translators, software programmers would have to write all programs in machine language, which is difficult to work with.

Macintosh Operating System (Mac OS) *(p. 91, LO 5)* Operating system used on Apple Macintosh computers.

Although not used in as many offices as DOS and Windows, the Macintosh operating system is easier to use.

menu *(p. 87, LO 4)*   List of available commands displayed on the screen.

Menus are used in graphical user interface programs to make software easier to use.

menu-driven interface *(p. 87, LO 4)*   User interface that allows users to choose a command from a menu.

Like a restaurant menu, a software menu offers you options to choose from—in this case commands available for manipulating data. Two types of menus are available, menu bars and pull-down menus. Menu-driven interfaces are easier to use than command-driven interfaces.

multiprocessing *(p. 89, LO 3)*   Operating system software feature that allows two or more computers or processors linked together to perform work simultaneously. Whereas "concurrently" means at almost the same time, "simultaneously" means at precisely the same time.

Multiprocessing is faster than multitasking and time-sharing. Microcomputer users may encounter an example of multiprocessing in specialized microprocessors called *coprocessors*. Working simultaneously with a computer's CPU microprocessor, a coprocessor will handle such specialized tasks as display screen graphics and high-speed mathematical calculations.

See multitasking.

multiprogramming *(p. 89, LO 3)*   Operating system software feature that allows the execution of two or more programs on a multiuser system. Program execution occurs concurrently, not simultaneously.

multitasking *(p. 89, LO 3)*   Operating system software feature that allows the execution of two or more programs by one user concurrently on the same computer with one central processor.

Allows the computer to rapidly switch back and forth among different tasks. The user is generally unaware of the switching process and is able to use more than one applications program at the same time.

NetWare *(p. 96, LO 5)*   Most popular operating system, from Novell, for orchestrating microcomputer-based local area networks (LANs) throughout a company or campus.

NetWare allows PCs to share data files, printers, and file servers.

operating system (OS) *(p. 85, LO 2)*   Principal piece of systems software in any computer system; consists of the master set of programs that manage the basic operations of the computer. The operating system remains in main memory until the computer is turned off.

These programs act as an interface between the user and the computer, handling such matters as running and storing programs and storing and processing data. The operating system allows users to concentrate on their own tasks or applications rather than on the complexities of managing the computer.

OS/2 (Operating System/2) & OS/2 Warp *(p. 94, LO 5)* Microcomputer operating system designed to run on many recent IBM and compatible microcomputers.

Unlike traditional Windows, OS/2 and its most recent version, Warp, do not require DOS to run underneath and so generally process more efficiently. Like Windows, they have a graphical user interface, called the *Workplace Shell (WPS)*. OS/2 and Warp can also run most DOS, Windows, and OS/2 applications programs simultaneously, which means users don't have to throw out old applications to take advantage of new features.

What It Is / What It Does	Why It's Important

platform *(p. 99, LO 7)* Refers to the particular hardware or software standard on which a computer system is based—for example IBM platform or Macintosh platform.

Users need to be aware that, without special arrangements or software, different platforms are not compatible.

software *(p. 84, LO 1)* Also called *programs;* the step-by-step coded instructions that tell the computer how to perform a task. Software is of two types: applications software and systems software.

Without software, hardware would be useless.

supervisor *(p. 88, LO 3)* Also called *kernel;* central component of the operating system. It resides in main memory while the computer is on and directs other programs to perform tasks to support applications programs.

Were it not for the supervisor program, users would have to stop one task—for example, writing—and wait for another task to be completed—for example, printing out of a document.

systems software *(p. 84, LO 1)* Software that enables applications software to interact with the computer and helps the computer manage its internal resources.

Applications software cannot run without systems software.

time-sharing *(p. 89, LO 3)* Operating system software feature whereby a single computer processes the tasks of several users at different stations in round-robin fashion. Time-sharing and multitasking differ slightly. With time-sharing, the computer spends a fixed amount of time with each program before going on to the next one. With multitasking the computer works on each program until it encounters a logical stopping point, as in waiting for more data to be input.

Time sharing is used when several users are linked by a communications network to a single computer. The computer will work first on one user's task for a fraction of a second, then go on to the next user's task, and so on.

Unix *(p. 96, LO 5)* Operating system for multiple users, with built-in networking capability, the ability to run multiple tasks at one time, and versions that can run on all kinds of computers.

Because it can run with relatively simple modifications on many different kinds of computers, from micros to minis to mainframes, Unix is said to be a "portable" operating system. The main users of Unix are large corporations and banks that use the software for everything from designing airplane parts to currency trading.

user interface *(p. 87, LO 4)* Also called *shell;* part of the operating system that allows users to communicate, or interact, with it. There are three types of user interfaces, for both operating systems and applications software: command-driven, menu-driven, graphical user.

User interfaces are necessary for users to be able to use a computer system.

utility programs *(p. 84, LO 2)* Systems software generally used to support, enhance, or expand existing programs in a computer system.

Many operating systems have utility programs built in for common purposes such as copying the contents of one disk to another. Other external utility programs are available on separate diskettes to, for example, recover damaged files.

virus *(p. 97, LO 6)* Hidden programming instructions that are buried within an applications or systems program and that copy themselves to other programs, often causing damage.

Viruses can cause users to lose data or files or even shut down entire computer systems.

windows *(p. 88, LO 4)* Feature of graphical user interfaces; causes the display screen to divide into sections. Each window is dedicated to a specific purpose.

Using the windows feature, an operating system (or environment) can display several windows on a computer screen, each showing a different application program, such as word processing, spreadsheets, and graphics.

Windows *(p. 93, LO 5)* Operating environment made by Microsoft that places a graphical user interface shell around the MS-DOS and PC-DOS operating systems.

The Windows operating environment made DOS easier to use; far more applications have been written for Windows than for DOS alone.

Windows 95 *(p. 93, LO 5)* Successor to Windows 3.X for DOS; this is a true operating system for IBM-style computers, rather than just an operating environment.

Windows 95 and later versions may become the most common systems software used on microcomputers.

Windows NT (New Technology) *(p. 94, LO 5)* Operating system intended to support large networks of computers, such as those involved in airline reservations systems.

Unlike the traditional Windows operating environment, Windows NT is a true operating system, eliminating the need for DOS and interacting directly with the hardware. It is primarily designed to run on workstations or other more powerful computers.

(Selected answers appear at the back of the book.)

Short-Answer Questions

1. Why does a computer need systems software?
2. What are utility programs?
3. What does the term *booting* mean?
4. What does a language translator do? Why is such a program included in systems software?
5. Can an operating system designed for a mainframe run on a microcomputer?
6. Which is faster: multiprocessing, multitasking, or time-sharing?
7. Why have data compression utilities become necessary for some users?
8. Why do microcomputer users have to format their diskettes before using them?
9. What is NetWare? What does it do?
10. What is a computer virus?

Fill-in-the-Blank Questions

1. The three types of systems software are ___operating system___, ___utility program___, and ___language translators___.
2. The ___boot___ consists of the master programs that manage the basic operations of the computer.
3. Before you can use a new diskette on a microcomputer, sometimes you must ___format___ it.
4. A graphical user interface uses pictures, or ___images___, to represent processing functions.
5. Programs that are used to expand the existing capabilities of programs in a computer system are called _____.
6. In a command-driven user interface, the user types commands after the _____.
7. The main component of the operating system is the _____; it directs other programs to perform tasks.

8. Name three microcomputer operating systems besides DOS and the Macintosh Operating System:
   a. _____
   b. _____
   c. _____
9. _____ refers to the ability to add new hardware to a computer system and use it immediately without performing complicated installation procedures.
10. To scan a computer's hard disk, diskette, and memory for viruses, you should use _____ software.

Multiple-Choice Questions

1. Which of the following must you use to enable applications software to interact with the computer?
   a. software utilities
   b. systems software
   c. command-driven interface
   d. operating environment
   e. none of the above
2. Which of the following best describes the process of loading an operating system into a computer's memory?
   a. system prompt
   b. formatting
   c. backing up
   d. booting
   e. none of the above
3. Which of the following are you probably using if you're viewing windows and icons?
   a. command-driven interface
   b. menu-driven interface
   c. graphical user interface
   d. none of the above
4. Which of the following is the central component of an operating system?
   a. supervisor
   b. system prompt
   c. operating environment
   d. icons
   e. all of the above

5. _____ is the processing of tasks from several users at different locations in a round-robin fashion.
   a. multitasking
   b. multiprogramming
   c. time-sharing
   d. multiprocessing
   e. none of the above

6. _____ is the simultaneous processing of more than one program by multiple processors.
   a. multitasking
   b. multiprogramming
   c. time-sharing
   d. multiprocessing
   e. none of the above

7. Which of the following will you likely find in a company that uses networks of microcomputers?
   a. DOS
   b. Windows
   c. Unix
   d. NetWare
   e. all of the above

8. A popular service offered by an external utility program is:
   a. screen-saver assistance
   b. data recovery
   c. backup assistance
   d. virus protection
   e. all of the above

9. Which of the following should you use to optimize the amount of space that is used on a disk drive?
   a. antivirus software
   b. defragmentation software
   c. backup software
   d. memory-management utility
   e. none of the above

10. Which of the following do you need to transmit data between a computer and a peripheral device?
    a. file conversion utility
    b. driver
    c. memory-management software
    d. peripheral utility
    e. all of the above

True/False Questions

 T F 1. Applications software starts up the computer and functions as the principal coordinator of all hardware components.

 T F 2. A menu-driven user interface is the easiest type of operating system interface to use.

T F 3. Fragmentation is the uneven distribution of a file on the disk.

T F 4. Viruses can be found in both applications and systems programs.

T F 5. Screen burn-in is a bad problem with today's monitors.

T F 6. DOS is a 32-bit operating system.

T F 7. Multitasking is processing done by two or more processors at the same time.

T F 8. The central portion of the operating system is called the *supervisor.*

T F 9. All microcomputer diskettes must be formatted before you can use them.

T F 10. Applications software helps the computer manage its internal resources.

Projects/Critical-Thinking Questions

1. If you have been using a particular microcomputer for 2 years and are planning to upgrade the version of systems software you are using, what issues must you consider before you go ahead and buy the new version?

2. By the time this textbook goes to print, Microsoft will have released Windows 96. What is new in this latest release? What utilities are included in Windows 96? Which of these utilities were also included in Windows 95? Would you advise a Windows 95 user to upgrade to Windows 96? Why/why not?

3. PC-DOS 7 was released by IBM in March of 1995. Why would IBM spend money developing a new version of DOS if most experts believed that DOS would soon be obsolete? Who is using PC-DOS 7 now? Do you think IBM's decision to release a new version of DOS was a good one? Why/why not?

4. If you were in the market for a new microcomputer today, what software would you want to use on it? What systems software would you choose? Applications software? Why? How would you go about making your choices?

5. Pen-based computing uses its own particular type of systems software. Check some articles in computer magazines to find out what makes this type of systems software different from regular microcomputer systems software. Is pen-based systems software compatible with DOS? Windows? How would pen-based systems software limit a traditional microcomputer user?

Using the Internet

Objective: *In this exercise we describe how to navigate the Web using Netscape Navigator 2.0 or 2.01.*

Before you continue: *We assume you have access to the Internet through your university, business, or commercial service provider and to the Web browser tool named Netscape Navigator. Additionally, we assume you know how to connect to the Internet and then load Netscape Navigator. If necessary, ask your instructor or system administrator for assistance.*

1. The home page for Netscape Navigator should appear on your screen (see below).

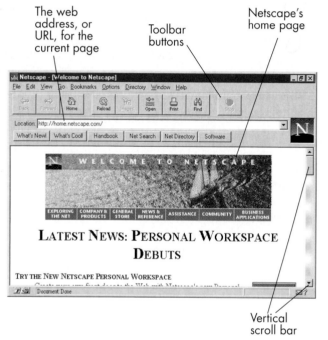

If at any time during the current work session you want to return to this home page, simply click the *Home icon* ( ) on the toolbar. The labeled toolbar appears below:

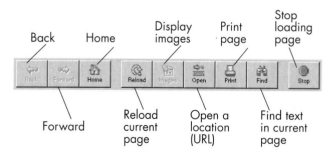

2. All underlined phrases are linked to other pages of information. When you click an underlined phrase, Netscape takes you to another page. For example:
   DRAG: the vertical scroll bar downward until you see the "Welcome to Netscape!" message and underlined phrases below
   CLICK: *an underlined phrase of your choice*
   A new page should appear.

3. Locate another underlined phrase of interest and then click. A new page should appear.

4. What if you want to go back to the previous page?
   CLICK: Back icon ( ) on the toolbar
   The page you displayed last should appear.

5. To go to the next page:
   CLICK: Forward icon ( ) on the toolbar

6. To return to the home page:
   CLICK: Home button ( )

7. To look at where you've been on the Web so far, you can use the Go menu. A list of the Web sites you've visited will appear at the bottom of the menu.
   CHOOSE: Go from the Menu bar
   Note the list of Web sites at the bottom of the pull-down menu. To go to one of the listed Web sites, simply click it with the mouse. To exit the pull-down menu:
   CHOOSE: Go from the Menu bar

8. To go directly to a specific page by typing in a web address, or URL (Uniform Resource Locator), you use the Open button ( ) on the toolbar.
   CLICK: Open button ( )
   TYPE: *a web address*
   For example, http://www.yahoo.com will take you to the home page for Yahoo, a popular Web library. Remember that web addresses often change, so any given address might no longer be in use.
   PRESS: Enter
   A new page should appear.

9. Now that you're familiar with some navigating basics, practice navigating the Web on your own. When you're finished, be sure to exit Netscape by choosing File, Exit from the Menu bar.

CHAPTER 4

Processors
Hardware for Power & Productivity

Concepts You Should Know

After reading this chapter, you should be able to:

1. Explain the miniaturization of processors and its link to micromachines and mobility.

2. Describe the function and operations of the CPU, including the machine cycle, and of main memory.

3. Define the different ways of measuring processing speeds.

4. Discuss how data and programs are represented in the computer.

5. Identify the components of the system unit and explain their uses.

6. Describe some adverse effects information technology has had on the environment.

Anticipointment" is a common experience for buyers of information technology.

Anticipointment, as Berkeley, California, editor Hank Roberts explained in an online computer conference, is a word coined "to describe always finding that, just as the techie-toy I've been dreaming about getting for six months has become affordable, there's something so much better on the horizon that I guess I have to wait just a bit longer."[1]

This feeling of anticipation-plus-disappointment could have been experienced by anyone observing trends in *portability* or *mobility* in electronic devices. For example, in 1955, Zenith ran ads showing a young woman holding a television set. The caption read: IT DOESN'T TAKE A MUSCLE MAN TO MOVE THIS LIGHTWEIGHT TV. That "lightweight" TV weighed a hefty 45 pounds. Today, by contrast, there is a handheld Casio color TV weighing a mere 6.2 ounces.

Similarly, tape recorders went from RCA's 35-pound machine in 1953 to today's Sony microcassette recorder of 3.5 ounces. Video cameras for consumers went from two components weighing 18.8 pounds in RCA's 1979 model to JVC's 18-ounce digital video camcorder today. Portable computers began in 1982 with Osborne's advertised "24 pounds of sophisticated computing power," a "luggable" size that most people would consider too unwieldy today. Since then portable computers have rapidly come down in weight and size. Now there are laptops (8–20 pounds), notebooks (4–7.5 pounds), subnotebooks (2.5–4 pounds), and pocket PCs (1 pound or less).

All this goes to show how relative the term *portability* is—and how much existing sizes are subject to obsolescence.[2]

Microchips, Miniaturization, & Mobility

Preview & Review: Computers used to be made from vacuum tubes. Then came the tiny switches called *transistors,* followed by integrated circuits made from the common mineral silicon. Integrated circuits called *microchips,* or *chips,* are printed and cut out of "wafers" of silicon. The microcomputer microprocessors, which process data, are made from microchips. They are also used in other instruments, such as phones and TVs.

Had the transistor not arrived, as it did in 1947, the Age of Portability and consequent mobility would never have happened. To us a "portable" telephone might have meant the 40-pound backpack radio-phones carried by some American GIs through World War II, rather than the 6-ounce shirt-pocket cellular models available today.

From Vacuum Tubes to Transistors to Microchips

Old-time radios used vacuum tubes—lightbulb-size electronic tubes with glowing filaments. The last computer to use these tubes, the ENIAC, which was turned on in 1946, employed 18,000 of them. Unfortunately, a tube failure occurred every 7 minutes, and it took more than 15 minutes to find and replace the faulty tube. Thus, it was difficult to get any useful computing work done. Moreover, the ENIAC was enormous, occupying 1500 square feet and weighing 30 tons.

The transistor changed all that. **A *transistor* is essentially a tiny electrically operated switch that can alternate between "on" and "off" many millions of times per second.** The first transistors were one-hundredth the size of a vacuum tube, needed no warmup time, consumed less energy, and were faster and more reliable. (■ *See Panel 4.1.*) Moreover, they marked the beginning of a process of miniaturization that has not ended yet.

In the old days, transistors were made individually and then formed into an electronic circuit with the use of wires and solder. Today transistors are part of an ***integrated circuit;* that is, an entire electronic circuit, including wires, is all formed together on a single chip of special material, silicon,** as part of a single manufacturing process. An integrated circuit embodies what is called *solid-state technology. Solid state* means that the electrons are traveling through solid material—in this case silicon. They do not travel through a vacuum, as was the case with the old radio vacuum tubes.

What is silicon, and why use it? *Silicon* is an element that is widely found in clay and sand. It is used not only because its abundance makes it cheap but also because it is a *semiconductor.* A semiconductor is material whose electrical properties are intermediate between a good conductor of electricity and a nonconductor of electricity. (An example of a good conductor of electricity is copper in household wiring; an example of a nonconductor is the plastic sheath around that wiring.) Because it is only a semiconductor, silicon has partial resistance to electricity. As a result, when good-conducting metals are overlaid on the silicon, the electronic circuitry of the integrated circuit can be created.

A computer's electronic circuitry is printed on a chip less than 1 centimeter square and about half a millimeter thick. (■ *See Panel 4.1 again.*) **A *chip*, or *microchip*, is a tiny piece of silicon that contains millions of microminiature electronic circuit components,** mainly transistors.

Miniaturization

There are different kinds of microchips—for example, microprocessor, memory, logic, communications, graphics, and math coprocessor chips. We discuss some of these later in this chapter. Perhaps the most important is the

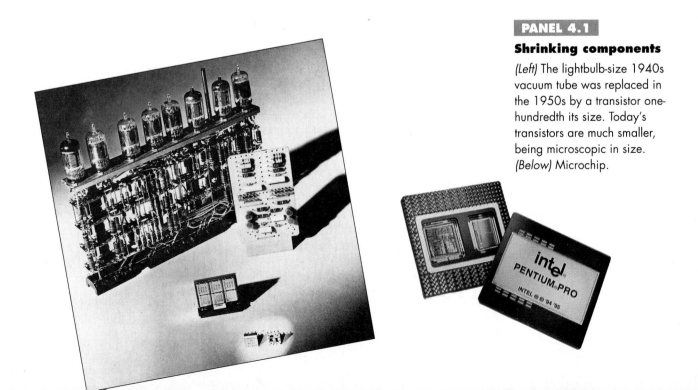

Shrinking components

(Left) The lightbulb-size 1940s vacuum tube was replaced in the 1950s by a transistor one-hundredth its size. Today's transistors are much smaller, being microscopic in size. *(Below)* Microchip.

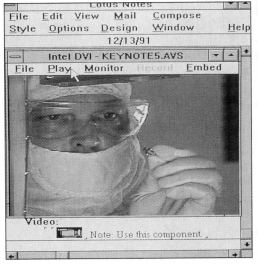

microprocessor chip. **A *microprocessor* ("microscopic processor" or "processor on a chip") is the miniaturized circuitry of a computer processor—the part that processes, or manipulates, data into information.** When modified for use in machines other than computers, microprocessors are called *microcontrollers*, or *embedded computers* (✓ p. 23).

The microprocessor, says Michael Malone, author of *The Microprocessor: A Biography*, "is the most important invention of the 20th century."[3] Quite a bold claim, considering the incredible products that have issued forth during the past nearly 100 years. Part of the reason, Malone argues, is the pervasiveness of the microprocessor in the important machines in our lives, from computers to transportation.

Mobility

Smallness—in TVs, phones, radios, camcorders, CD players, and computers—is now largely taken for granted. In the 1980s portability, or mobility, meant trading off computing power and convenience in return for smaller size and weight. Today, however, we are getting close to the point where we don't have to give up anything. As a result, experts predict that small, powerful, wireless personal electronic devices will soon transform our lives far more than the personal computer has done so far.[4]

Five Types of Computer Systems

Preview & Review: Computers are classified into microcontrollers, microcomputers, minicomputers, mainframe computers, and supercomputers.

Microcomputers may be personal computers (PCs) or workstations. PCs include desktop and tower units, laptops, notebooks, subnotebooks, pocket PCs, and pen computers. Workstations are sophisticated desktop microcomputers used for technical purposes.

Generally speaking, the larger the computer, the greater its processing power. As we mentioned in Chapter 1 (✓ p. 22), computers are often classified into five sizes—tiny, small, medium, large, and superlarge:

- Microcontrollers—embedded in "smart" appliances
- Microcomputers—both personal computers and workstations
- Minicomputers
- Mainframe computers
- Supercomputers

In this chapter, we pay particular attention to microcomputers.

Microcomputers: Personal Computers

***Microcomputers* are small computers that can fit on or beside a desk or are portable.** Microcomputers are considered to be of two types: personal computers and workstations.

***Personal computers (PCs)* are desktop, tower, or portable computers that can run easy-to-use programs such as word processing or spreadsheets.** PCs come in several sizes, as follows.

- **Desktop and tower units:** Even though many personal computers today are portable, buyers of new PCs often opt for nonportable systems, for reasons of price, power, or flexibility. For example, the television-tube-like (CRT, or cathode-ray tube) monitors that come with desktops have display screens that are easier to read than those of many portables. Moreover, you can stuff a desktop's roomy system cabinet with add-on circuit boards and other extras, which is not possible with portables.

  Desktop PCs are those in which the system cabinet sits on a desk, with keyboard in front and monitor often on top. A difficulty with this arrangement is that the system cabinet's "footprint" can deprive you of a fair amount of desk space. *Tower PCs* are those in which the system cabinet sits as a "tower" on the desk or on the floor next to the desk, giving you more usable desk space. LNSP

- **Laptops:** **A *laptop computer* is a portable computer equipped with a flat display screen and weighing 8–20 pounds.** The top of the computer opens up like a clamshell to reveal the screen.

- **Notebooks:** **A *notebook computer* is a portable computer that weighs 4–7.5 pounds and is roughly the size of a thick notebook,** perhaps 8½ by 11 inches. Notebook PCs can easily be tucked into a briefcase or backpack or simply under your arm.

  Notebook computers can be just as powerful as some desktop machines. However, because they are smaller, the keys on the keyboards are closer together and may be harder to use. Also, as with laptops, the display screens are more difficult to read.

- **Subnotebooks:** **A *subnotebook computer* weighs 2.5–4 pounds.** Clearly, subnotebooks have more of both the advantages and the disadvantages of notebooks.

- **Pocket PCs:** ***Pocket personal computers,* or *handhelds,* weigh about 1 pound or less.** These PCs are useful in specific situations, as when a driver of a package-delivery truck must feed hourly status reports to company headquarters. Another use allows police officers to check out suspicious car license numbers against a database in a central computer.

  Pocket PCs may be classified into three types:
  (1) *Electronic organizers* are specialized pocket computers that mainly store appointments, addresses, and "to do" lists. Recent versions feature wireless links to other computers for data transfer.
  (2) *Palmtop computers* are PCs that are small enough to hold in one hand and operate with the other.
  (3) *Pen computers* lack a keyboard or a mouse but allow you to input data by writing directly on the screen with a stylus, or pen. Pen computers are useful for inventory control, as when a store clerk has to count merchandise; for package-delivery drivers who must get electronic signatures as proof of delivery; and for more general purposes, like those of electronic organizers and PDAs.

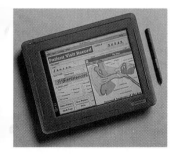

  ***Personal digital assistants (PDAs),* or *personal communicators,* are small, pen-controlled, handheld computers that, in their most developed form, can do two-way wireless messaging.**

Microcomputers: Workstations

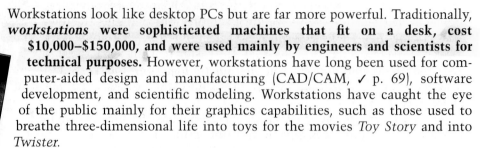

Workstations look like desktop PCs but are far more powerful. Traditionally, **_workstations_ were sophisticated machines that fit on a desk, cost $10,000–$150,000, and were used mainly by engineers and scientists for technical purposes.** However, workstations have long been used for computer-aided design and manufacturing (CAD/CAM, ✓ p. 69), software development, and scientific modeling. Workstations have caught the eye of the public mainly for their graphics capabilities, such as those used to breathe three-dimensional life into toys for the movies *Toy Story* and into *Twister.*

Now let's take a look at the inside of a microcomputer to see how it works.

The CPU & Main Memory

Preview & Review: The central processing unit (CPU) consists of the control unit and the arithmetic/logic unit (ALU). Main memory holds data in storage temporarily; its capacity varies in different computers. Registers in the CPU store data that is to be processed immediately.

The operations for executing a single program instruction are called the *machine cycle,* which has an instruction cycle and an execution cycle.

Processing speeds are expressed in several ways: fractions of a second, megahertz, MIPS, and flops.

How is the information in "information processing" in fact processed? As we indicated, this is the job of the circuitry known as the microprocessor. This device, the "processor-on-a-chip" found in a microcomputer, is also called the CPU. The CPU works hand in hand with other circuits known as main memory to carry out processing.

CPU

The *CPU,* for *central processing unit,* follows the instructions of the software to manipulate data into information. The CPU consists of two parts: (1) the control unit and (2) the arithmetic/logic unit (ALU). The two components are connected by a kind of electronic "roadway" called a *bus.* (■ *See Panel 4.2.*)

PANEL 4.2

The CPU and main memory

The two main CPU components (control unit and ALU), the registers, and main memory are connected by a kind of electronic "roadway" called a *bus.*

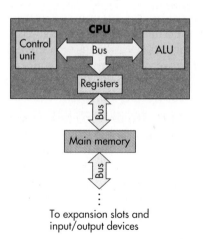

To expansion slots and input/output devices

- **The control unit:** **The *control unit* tells the rest of the computer system how to carry out a program's instructions.** It directs the movement of electronic signals between main memory and the arithmetic/logic unit. It also directs these electronic signals between main memory and the input and output devices.

- **The arithmetic/logic unit:** **The *arithmetic/logic unit*, or *ALU*, performs arithmetic operations and logical operations and controls the speed of those operations.**

    As you might guess, *arithmetic* operations are the fundamental math operations: addition, subtraction, multiplication, and division.

    Logical operations are comparisons. That is, the ALU compares two pieces of data to see whether one is equal to (=), greater than (>), or less than (<) the other. (The comparisons can also be combined, as in "greater than or equal to" and "less than or equal to.")

Main Memory

***Main memory*—variously known as *memory, primary storage, internal memory*, or *RAM* (for *random access memory*)—is working storage. (1) It holds data for processing. (2) It holds instructions (the programs) for processing the data. (3) It holds processed data (that is, information) waiting to be sent to an output or secondary-storage device.** Main memory is contained on special microchips called *RAM chips*, as we describe in a few pages. This memory is in effect the computer's short-term capacity. It determines the total size of the programs and data files it can work on at any given moment.

There are two important facts to know about main memory:

- **Its contents are temporary:** Once the power to the computer is turned off, all the data and programs within main memory simply vanish. This is why data must also be stored on disks and tapes—called "secondary storage" to distinguish them from main memory's "primary storage."

    Thus, main memory is said to be *volatile*. **Volatile storage is temporary storage; the contents are lost when the power is turned off.** Consequently, if you kick out the connecting power cord to your computer, whatever you are currently working on will immediately disappear. This impermanence is the reason you should *frequently save* your work in progress to a secondary-storage medium such as a diskette. By "frequently," we mean every 3–5 minutes.

- **Its capacity varies in different computers:** The size of main memory is important. It determines how much data can be processed at once and how big and complex a program may be used to process it. This capacity varies with different computers, with older machines holding less.

    For example, the original IBM PC, introduced in 1979, held only about 64,000 bytes (characters) of data or instructions. By contrast, new microcomputers can have 24 million bytes or more of memory.

Registers

The control unit and the ALU also use registers, or special areas that enhance the computer's performance. (■ *Refer back to Panel 4.2.*) **Registers are high-speed storage areas that temporarily store data during processing.** It could be said that main memory holds material that will be used "a little bit later." Registers hold material that is to be processed "immediately." The computer loads the program instructions and data from main memory into the registers just prior to processing, which helps the computer process faster.

Kilo —milli 1000
mega —micro 1 mill,
giab —nano 1 bill,
tera —pico 1 Trill,

Machine Cycle: How an Instruction Is Processed

How does the computer keep track of the data and instructions in main memory? Like a system of post-office mailboxes, it uses addresses. **An** *address* **is the location, designated by a unique number, in main memory in which a character of data or of an instruction is stored during processing.** To process each character, the control unit of the CPU retrieves that character from its address in main memory and places it into a register. This is the first step in what is called the *machine cycle.*

The *machine cycle* **is a series of operations performed to execute a single program instruction. The machine cycle consists of two parts: an instruction cycle, which fetches and decodes, and an execution cycle, which executes and stores.** (■ *See Panel 4.3.*)

Processing Speeds

With transistors switching off and on perhaps millions of times per second, the tedious repetition of the machine cycle occurs at blinding speeds.

There are several ways in which processing speeds are measured:

- **Time to complete one machine cycle, in fractions of a second:** The speeds for completing one machine cycle are measured in milliseconds for older and slower computers. They are measured in microseconds for most microcomputers and in nanoseconds for mainframes. Picosecond measurements occur only in some experimental machines.

  A *millisecond* is one-thousandth of a second. A *microsecond* is one-millionth of a second. A *nanosecond* is one-billionth of a second. A *picosecond* is one-trillionth of a second.

- **Time in millions of machine cycles per second (megahertz):** Microprocessor speeds are usually expressed in *megahertz (MHz),* **millions of machine cycles per second,** which, as we mentioned, is also the measure of a microcomputer's clock speed. For example, a 166-MHz Pentium-based microcomputer processes 166 million machine cycles per second.

- **Time to complete instructions, in millions of instructions per second (MIPS):** Another measurement is the number of instructions per second that a computer can process, which today is in the millions. *MIPS* **is a measure of a computer's processing speed; it stands for *m*illions of *i*nstructions *p*er *s*econd that the processor can perform.** A microcomputer (with an 80486 chip) might perform 54 MIPS, a mainframe 240 MIPS.

PANEL 4.3

The machine cycle

(Left) The machine cycle executes instructions one at a time during the instruction cycle and execution cycle. *(Right)* Example of how the addition of two numbers, 50 and 75, is processed and stored in a single cycle.

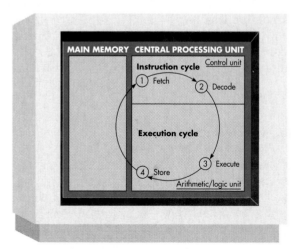
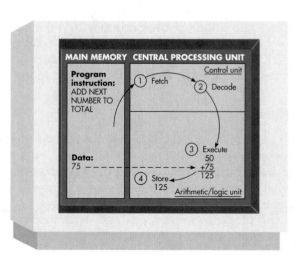

- **Time in floating-point operations per second (flops):** The abbreviation *flops* stands for *floating-point operations per second*, a floating-point operation being a special kind of mathematical calculation. This measure, usually expressed in *megaflops—millions of floating-point operations per second*—is used mainly with supercomputers.

Now that you know *where* data and instructions are processed, we need to review how those data and instructions are represented in the CPU, registers, buses, and RAM.

How Data & Programs Are Represented in the Computer

Preview & Review: Computers use the two-state 0/1 binary system to represent data.

A computer's capacity is expressed in bits, bytes, kilobytes, megabytes, gigabytes, or terabytes.

One common binary coding scheme is ASCII-8.

Parity-bit schemes are used to check for accuracy.

Human-language-like programming languages are processed as 0s and 1s by the computer in machine language.

As we explained in Chapter 1 (✓ p. 5), electricity is the basis for computers and communications because electricity can be either *on* or *off* (or low-voltage or high-voltage). This two-state situation allows computers to use the *binary system* to represent data and programs.

Binary System: Using Two States

The decimal system that we are accustomed to has 10 digits (0, 1, 2, 3, 4, 5, 6, 7, 8, 9). By contrast, **the *binary system* has only two digits: 0 and 1.** Thus, in the computer the 0 can be represented by the electrical current being off (or low voltage) and the 1 by the current being on (or high voltage). All data and programs that go into the computer are represented in terms of these binary numbers. (■ *See Panel 4.4, next page.*) For example, the letter *H* is a translation of the electronic signal 01001000, or off-on-off-off-on-off-off-off. When you press the key for *H* on the computer keyboard, the character is automatically converted into the series of electronic impulses that the computer can recognize.

How Capacity Is Expressed

How many 0s and 1s will a computer or a storage device such as a hard disk hold? To review what we covered in Chapter 1, the following terms are used to denote capacity.

- **Bit:** In the binary system, **each 0 or 1 is called a *bit*, which is short for "binary dig*it*."**
- **Byte:** To represent letters, numbers, or special characters (such as ! or *), bits are combined into groups. **A group of 8 bits is called a *byte*, and a byte represents one character, digit, or other value.** (As we mentioned, in one scheme, 01001000 represents the letter *H*.) The capacity of a computer's memory or a diskette is expressed in numbers of bytes or multiples such as kilobytes and megabytes.

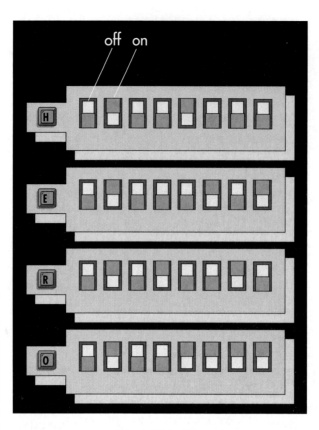

off   on

PANEL 4.4

Binary data representation

How the letters *H-E-R-O* are represented in one type of off/on, 0/1 binary code (ASCII-8).

PANEL 4.5

ASCII-8 binary code

There are many more characters than those shown here, such as punctuation marks, math symbols, Greek letters, and other foreign-language symbols.

Character	ASCII-8
A	0100 0001
B	0100 0010
C	0100 0011
D	0100 0100
E	0100 0101
F	0100 0110
G	0100 0111
H	0100 1000
I	0100 1001
J	0100 1010
K	0100 1011
L	0100 1100
M	0100 1101
N	0100 1110
O	0100 1111
P	0101 0000
Q	0101 0001
R	0101 0010
S	0101 0011
T	0101 0100
U	0101 0101
V	0101 0110
W	0101 0111
X	0101 1000
Y	0101 1001
Z	0101 1010
0	0011 0000
1	0011 0001
2	0011 0010
3	0011 0011
4	0011 0100
5	0011 0101
6	0011 0110
7	0011 0111
8	0011 1000
9	0011 1001
;	0011 1011
!	0010 0001

- **Kilobyte:** A *kilobyte (K, KB)* is about 1000 bytes. (Actually, it's precisely 1024 bytes, but the figure is commonly rounded.) The kilobyte was a common unit of measure for memory or secondary-storage capacity on older computers.
- **Megabyte:** A *megabyte (M, MB)* is about 1 million bytes (1,048,576 bytes). Many measures of microcomputer capacity today are expressed in megabytes.
- **Gigabyte:** A *gigabyte (G, GB)* is about 1 billion bytes (1,073,741,824 bytes). This measure is often used with "big iron" types of computers and now also with smaller systems.
- **Terabyte:** A *terabyte (T, TB)* represents about 1 trillion bytes (1,009,511,627,776 bytes).

Binary Coding Schemes

Letters, numbers, and special characters are represented within a computer system by means of *binary coding schemes.* That is, the off/on 0s and 1s are arranged in such a way that they can be made to represent characters, digits, or other values. One popular binary coding scheme, ASCII-8, uses 8 bits to form each byte. (■ *See Panel 4.5.*)

Pronounced "*as*-key," **ASCII stands for American Standard Code for Information Interchange and is the binary code most widely used with microcomputers.**

ASCII originally used seven bits, but a zero was added in the left position to provide an 8-bit code, which offers more possible combinations with which to form characters, such as math symbols and Greek letters.

When you type a word on the keyboard (for example, *HERO*), the letters are converted into bytes—eight 0s and 1s for each letter. The bytes are represented in the computer by a combination of eight transistors, some of which are closed (representing the 0s) and some of which are open (representing the 1s).

The Parity Bit

Dust, electrical disturbance, weather conditions, and other factors can cause interference in a circuit or communications line that is transmitting a byte. How does the computer know if an error has occurred? Detection is accomplished by use of a parity bit. **A *parity bit*, also called a *check bit*, is an extra bit attached to the end of a byte for purposes of checking for accuracy.**

Parity schemes may be *even parity* or *odd parity*. In an even-parity scheme, for example, the ASCII letter *H* (01001000) contains two 1s. Thus, the ninth bit, the parity bit, would be 0 in order to make the sum of the bits come out even. With the letter *O* (01001111), which has five 1s, the ninth bit would be 1 to make the byte come out even. (■ *See Panel 4.6.*) The systems software in the computer automatically and continually checks the parity scheme for accuracy. (If the message "Parity Error" appears on your screen, you need a technician to look at the computer to see what is causing the problem.)

Machine Language

Why won't word processing software that runs on an Apple Macintosh run (without special arrangements) on an IBM microcomputer? It's because each computer has its own machine language. **Machine language is a binary-type programming language that the computer can run directly.** To most people an instruction written in machine language is incomprehensible, consisting only of 0s and 1s. However, it is what the computer itself can understand, and the 0s and 1s represent precise storage locations and operations.

How do people-comprehensible program instructions become computer-comprehensible machine language? Special systems programs called *language translators* rapidly convert the instructions into machine language—language that computers can understand. This translating occurs virtually instantaneously, so that you are not aware it is happening.

Because the type of computer you will most likely be working with is the microcomputer, we'll now take a look at what's inside the microcomputer's system unit.

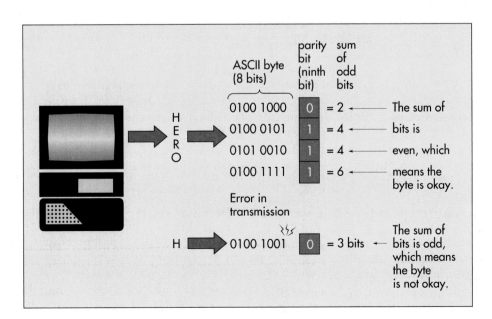

PANEL 4.6

Parity bit

This example uses an even-parity scheme.

ASCII byte (8 bits) | parity bit (ninth bit) | sum of odd bits
0100 1000 | 0 | = 2 ← The sum of
0100 0101 | 1 | = 4 ← bits is
0101 0010 | 1 | = 4 ← even, which
0100 1111 | 1 | = 6 ← means the byte is okay.

Error in transmission

H 0100 1001 | 0 | = 3 bits ← The sum of bits is odd, which means the byte is not okay.

The Microcomputer System Unit

Preview & Review: The system unit, or cabinet, contains the following electrical components: the power supply, the motherboard, the CPU chip, specialized processor chips, the system clock, RAM chips, ROM chips, other forms of memory (cache, video, flash), expansion slots and boards, bus lines, ports, and PC (PCMCIA) slots and cards.

What is inside the gray or beige box that we call "the computer"? **The box or cabinet is the *system unit*; it contains the electrical and hardware components that make the computer work.** These components actually do the processing in information processing.

The system unit of a desktop microcomputer does not include the keyboard or printer. Quite often it also does not include the monitor or display screen. It usually does include a hard-disk drive and one or two diskette drives, and sometimes a tape drive. We describe all these and other *peripheral devices—hardware that is outside the central processing unit—*in the chapters on input, output, and secondary-storage devices. Here we are concerned with 12 parts of the system unit, as follows:

- Power supply
- Motherboard
- CPU
- Specialized processor chips
- System clock
- RAM chips
- ROM chips
- Other forms of memory—cache, video, flash
- Expansion slots and boards
- Bus lines
- Ports
- PC (PCMCIA) slots and cards

These are terms that appear frequently in advertisements for microcomputers. After reading this section, you should be able to understand what these ads are talking about.

Power Supply

The electricity available from a standard wall outlet is alternating current (AC), but a microcomputer runs on direct current (DC). **The *power supply* is a device that converts AC to DC to run the computer.** (■ *See Panel 4.7.*) The on/off switch in your computer turns on or shuts off the electricity to the power supply. Because electricity can generate a lot of heat, a fan inside the computer keeps the power supply and other components from becoming too hot. (See page 177 for information about computers and power problems.)

Motherboard

The *motherboard*, or *system board*, is the main circuit board in the system unit. (■ *Refer to Panel 4.7 again.*)

The motherboard consists of a flat board that fills the bottom of the system unit. (It is accompanied by the power-supply unit and fan and probably one or more disk drives.) This board contains the "brain" of the computer, the CPU or microprocessor. It also contains electronic memory that assists the CPU, known as RAM, and some sockets, called *expansion slots*, where additional circuit boards, called *expansion boards*, may be plugged in. The processing is handled by the CPU and memory (RAM), as we explain next.

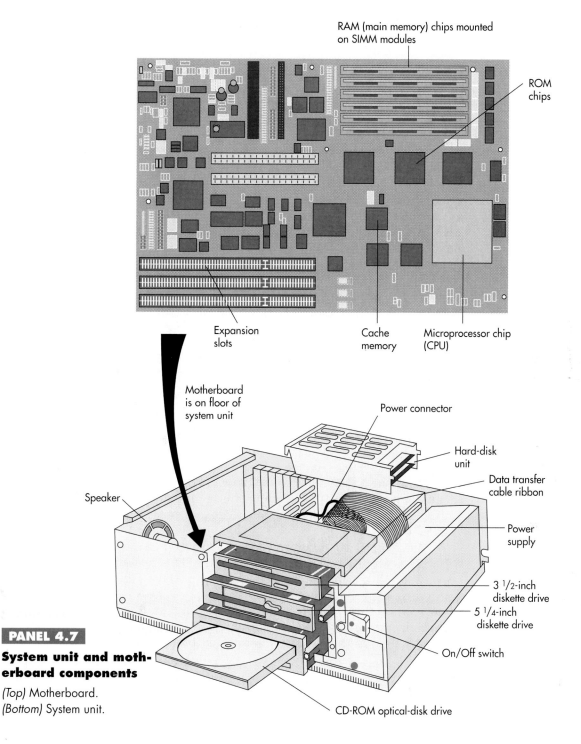

RAM (main memory) chips mounted on SIMM modules

ROM chips

Expansion slots

Cache memory

Microprocessor chip (CPU)

Motherboard is on floor of system unit

Power connector

Speaker

Hard-disk unit

Data transfer cable ribbon

Power supply

3 1/2-inch diskette drive

5 1/4-inch diskette drive

On/Off switch

CD-ROM optical-disk drive

PANEL 4.7

System unit and motherboard components

(Top) Motherboard.
(Bottom) System unit.

CPU Chip

Most personal computers today use CPU chips (microprocessors) of two kinds—those made by Intel and those by Motorola—although that situation may be changing. (■ *See Panel 4.8, next page.*) Workstations (✓ p. 23) generally use RISC chips.

- **Intel-type "86"-series chips:** Intel makes what are called "CISC" ("complex instruction set computer") chips for IBM and IBM-compatible computers such as Compaq, Dell, Gateway, Tandy, Toshiba, and Zenith. Variations of Intel chips are made by other companies—for example, Advanced Micro Devices (AMD), Cyrix, and Chips and Technologies.

Manufacturer and Chip	Date Introduced	Systems Chip	Clock Speed (MHz)	Bus Width
Intel 8088	1979	IBM PC, XT	4–8	8
Motorola 68000	1979	Macintosh Plus, SE; Commodore Amiga	8–16	16
Intel 80286	1981	IBM PC/AT, PS/2 Model 50/60; Compaq Deskpro 286	8–28	16
Motorola 68020	1984	Macintosh II	16–33	32
Sun Microsystems RISC	1985	Sun Sparcstation 1, 300	20–25	32
Intel 80386DX	1985	IBM PS/2; IBM-compatibles	16–33	32
Motorola 68030	1987	Macintosh IIx series, SE/30	16–50	32
Intel 80486DX	1989	IBM PS/2; IBM-compatibles	25-66	32
Motorola 68040	1989	Macintosh Quadras	25–40	32
IBM RISC 6000	1990	IBM RISC/6000 workstation	20–50	32
Sun Microsystems MicroSpar	1992	Sun Sparcstation LX	50	32
Intel Pentium	1993	Compaq Deskpro; IBM-compatibles	60–166	64
IBM/Motorola/Apple PowerPC RISC	1994	Power Macintoshes; Power Computing PowerWave	60–150	64
Intel Pentium Pro	1995	Compaq Proliant; Data General server	150–200	64

PANEL 4.8

Microcomputers and microprocessors

Some widely used microcomputer systems and their chips.

Intel has identified its chips by numbers—8086, 8088, 80286, 80386, 80486—and is now marketing its newest chips under the names Pentium and Pentium Pro. The higher the number, the newer and more powerful the chip and the faster the processing speed, which means that software runs more efficiently. The chips are commonly referred to by their last three digits, such as '386 and '486. (Most of today's software is written for CISC chips.)

Some chips have different versions—for example, "386SX" or "486DX." SX chips are usually less expensive than DX chips and run more slowly. (SX and DX are indicators of bus size [✓ p. 112]. There are several types of buses, as we will discuss shortly. Some of the SX chip's buses are narrower than the DX chip's.) SL chips are designed to reduce power consumption and so are used in portable computers. DX2 and DX4 chips are usually used for heavy-duty information processing.

• **Motorola-type "68000"-series chips:** Motorola makes chips for Apple Macintosh computers. These CISC chip numbers include the 68000, 68020, 68030, and 68040.

• **RISC chips:** Sun Microsystems, Hewlett-Packard, and Digital Equipment use RISC chips in their desktop workstations, although the technology is also showing up in some portables. *RISC* **stands for reduced instruction set computing.** A RISC computer system operates with fewer instructions than those required in conventional CISC-based computer systems. RISC-equipped workstations have been found to work 10 times faster than conventional computers. A problem, however, is that software has to be modified to work with them. (RISC computers can run CISC-based software, but they lose their speed advantage when they do.)

The capacities of CPUs are expressed in terms of *word size*. **A *word*, also called *bit number*, is the number of bits that may be manipulated or stored at one time by the CPU.** Often, the more bits in a word, the faster the computer. An 8-bit-word computer will transfer data within each CPU chip itself in 8-bit chunks. A 32-bit-word computer is faster, transferring data in 32-bit chunks.

Specialized Processor Chips

A motherboard usually has slots for plugging in specialized processor chips. Two in particular that you may encounter are math and graphics coprocessor chips. **A *math coprocessor chip* helps programs using lots of mathematical equations to run faster. A *graphics coprocessor chip* enhances the performance of programs with lots of graphics and helps create complex screen displays.** Specialized chips significantly increase the speed of a computer system.

System Clock

When people talk about a computer's "speed," they mean how fast it can do processing—turn data into information. Every microprocessor contains a system clock. **The *system clock* controls how fast all the operations within a computer take place.**

As we mentioned earlier in the chapter, processing speeds are expressed in megahertz (MHz), with 1 MHz equal to 1 million cycles per second. An old IBM PC had a clock speed of 4.77 MHz, whereas computers with '486 chips may run at 66 MHz. The high-end Macintosh-compatible PowerWave computer, from Power Computing, uses a PowerPC microprocessor running at 150 MHz. The most recent Intel Pentium Pro chip, used in workstations, runs at speeds up to 200 MHz.

RAM Chips

volatile

secondary storage is permanent

RAM, for *random access memory,* is memory that temporarily holds data and instructions that will be needed shortly by the CPU. RAM is what we have been calling *main memory, internal memory,* or *primary storage.* It operates like a chalkboard that is constantly being written on, then erased, then written on again. (The term *random access* comes from the fact that data can be stored and retrieved at random—from anywhere in the electronic RAM chips—in approximately equal amounts of time, no matter what the specific data locations are.)

Like the microprocessor, RAM consists of circuit-inscribed silicon chips attached to the motherboard. **RAM chips are often mounted on a small circuit board, such as a *SIMM* (for *single inline memory module*), which is plugged into the motherboard.** (■ *Refer back to Panel 4.7.*) The two principal types of RAM chips are *DRAM* (for *dynamic random access memory*) chips, used for most main memory, and *SRAM* (for *static random access memory*) chips, used for some specialized purposes within main memory.

Microcomputers come with different amounts of RAM. In many cases, additional RAM chips can be added by plugging a memory-expansion card into the motherboard, as we will explain. The more RAM you have, the faster the computer operates, and the better your software performs.

Having enough RAM has become a critical matter! Before you buy a software package, look at the outside of the box to see how much RAM is required. Windows 95 supposedly will run with 4 megabytes of RAM, but a realistic minimum is 8–12 megabytes, and 16 is preferable.

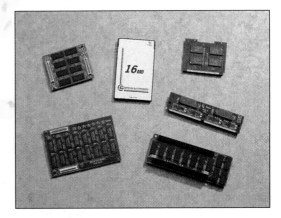

ROM Chips

Unlike RAM, which is constantly being written on and erased, **ROM, which stands for read-only memory and is also known as firmware, cannot be written on or erased by the computer user.** (■ *Refer back to Panel 4.7.*) ROM chips contain programs that are built in at the factory; these are special instructions for basic computer operations, such as those that start the computer or put characters on the screen.

There are variations of the ROM chip that allow programmers to vary information stored on the chip and also to erase it.

Other Forms of Memory

The performance of microcomputers can be enhanced further by adding other forms of memory, as follows.

- **Cache memory:** Pronounced "cash," **cache memory is a special high-speed memory area that the CPU can access quickly.** Cache memory can be located on the microprocessor chip or elsewhere on the motherboard. Cache memory is used in computers with very fast CPUs. The most frequently used instructions are kept in cache memory so the CPU can look there first. This allows the CPU to run faster because it doesn't have to take time to swap instructions in and out of main memory. Large, complex programs benefit the most from having a cache memory available.

- **Video memory:** **Video memory or video RAM (VRAM) chips are used to store display images for the monitor.** The amount of video memory determines how fast images appear and how many colors are available. Video memory chips are particularly desirable if you are running programs that display a lot of graphics.

- **Flash memory:** Used primarily in notebook and subnotebook computers, **flash memory, or flash RAM, cards consist of circuitry on credit-card-size cards that can be inserted into slots connected to the motherboard.** Unlike standard RAM chips, flash memory is *nonvolatile*. That is, it retains data even when the power is turned off. Flash memory can be used not only to simulate main memory but also to supplement or replace hard-disk drives for permanent storage.

Expansion Slots & Boards

Today all new microcomputer systems can be expanded. *Expandability* refers to a computer's capacity for adding more memory or peripheral devices. Having expandability means that when you buy a PC you can later add devices to enhance its computing power. This spares you from having to buy a completely new computer.

Expandability is made possible with expansion slots and expansion boards. **Expansion slots are sockets on the motherboard into which you can plug expansion cards. Expansion cards, or add-on boards, are circuit boards that provide more memory or control peripheral devices.** (■ *Refer back to Panel 4.7.*) The words *card* and *board* are used interchangeably. Some slots may be needed right away for ordinary functions, but if your system unit leaves enough slots open, you can use them for expansion later.

Among the types of expansion cards are the following.

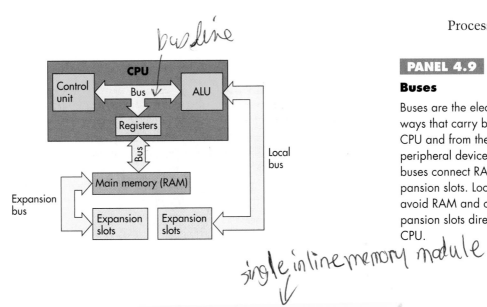

busline

single inline memory module

PANEL 4.9

Buses

Buses are the electrical pathways that carry bits within the CPU and from the CPU to peripheral devices. Expansion buses connect RAM with expansion slots. Local buses avoid RAM and connect expansion slots directly with the CPU.

- **Expanded memory:** Memory expansion cards (or SIMMs) allow you to add RAM chips, giving you more main memory.
- **Display adapter or graphics adapter cards:** These cards allow you to adapt different kinds of color video display monitors for your computer.
- **Other add-ons:** You can also add special circuit boards for modems, fax, sound, and networking, as well as math or graphics coprocessor chips.

Bus Lines

A *bus line*, **or simply** *bus,* **is an electrical pathway through which bits are transmitted within the CPU and between the CPU and other devices in the system unit.** There are different types of buses (address bus, control bus, data bus), but for our purposes the most important is the *expansion bus,* **which carries data between RAM and the expansion slots.** To obtain faster performance, some users will use a bus that avoids RAM altogether. **A bus that connects expansion slots directly to the CPU is called a _local bus._** (■ *See Panel 4.9.)*

A bus resembles a multilane highway: The more lanes it has, the faster the bits can be transferred. The old-fashioned 8-bit bus of early microprocessors had only eight pathways. It was therefore four times slower than the 32-bit bus of later microprocessors, which had 32 pathways. Intel's Pentium chip is a 64-bit processor. Some supercomputers contain buses that are 128 bits. Today there are several principal expansion bus standards, or "architectures," for microcomputers.

Ports

A *port* **is a socket on the outside of the system unit that is connected to an expansion board on the inside of the system unit.** A port allows you to plug in a cable to connect a peripheral device, such as a monitor, printer, or modem, so that it can communicate with the computer system.

Ports are of five types. (■ *See Panel 4.10, next page.)*

- **Parallel ports:** **A** *parallel port* **allows lines to be connected that will enable 8 bits to be transmitted simultaneously,** like cars on an eight-lane highway. Parallel lines move information faster than serial lines do, but they can transmit information efficiently only up to 15 feet. Thus, parallel ports are used principally for connecting printers.

Ports

Shown are the backs of an IBM-style computer and an Apple Macintosh.

IBM-compatible

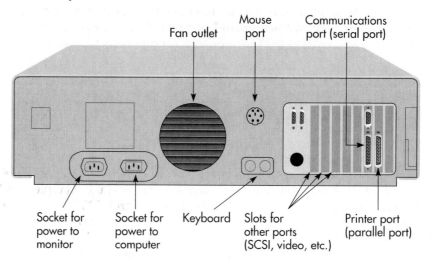

Fan outlet — Mouse port — Communications port (serial port)

Socket for power to monitor — Socket for power to computer — Keyboard — Slots for other ports (SCSI, video, etc.) — Printer port (parallel port)

Apple Macintosh

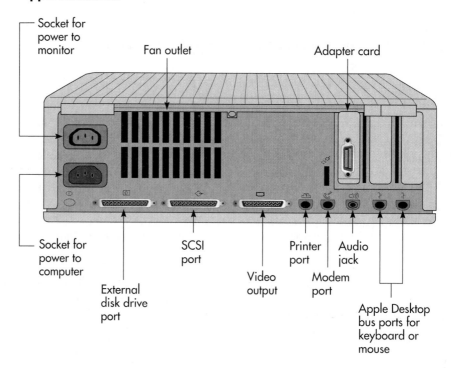

Socket for power to monitor — Fan outlet — Adapter card

Socket for power to computer — External disk drive port — SCSI port — Video output — Printer port — Modem port — Audio jack — Apple Desktop bus ports for keyboard or mouse

- **Serial ports:** A *serial port,* or *RS-232 port,* **enables a line to be connected that will send bits one after the other on a single line,** like cars on a one-lane highway. Serial ports are used principally for communications lines, modems, and mice. (They are frequently labeled "COM" for communications.)

- **Video adapter ports:** *Video adapter ports* **are used to connect the video display monitor outside the computer to the video adapter card inside the system unit.** Monitors may have either a 9-pin plug or a 15-pin plug. The plug must be compatible with the number of holes in the video adapter card.

- **SCSI ports:** Pronounced "scuzzy" (and **short for small computer system interface**), a *SCSI port* **provides an interface for transferring data at high speeds for up to eight SCSI-compatible devices.** These devices include external hard-disk drives, CD-ROM drives, and magnetic-tape backup units.

- **Game ports:** *Game ports* allow you to attach a joystick or similar game-playing device to the system unit.

Plug-In Cards: PC (PCMCIA) Slots & Cards

Although its name doesn't exactly roll off the tongue, PCMCIA is changing mobile computing more dramatically than any technology today.

Short for *Personal Computer Memory Card International Association*, *PCMCIA* is a relatively new bus standard for portable computers. *PC cards*—renamed because it's easier to say than PCMCIA—are 2-by-3-inch cards that may be used to plug peripherals into slots in portable computers. Slots are now being designed into desktop-size machines as well. The PC cards may be used to hold credit-card-size modems, sound boards, hard disks, extra memory, and even pagers and cellular communicators.

Computers & Environmental Questions

Preview & Review: Information technology has had some adverse effects on the environment, including energy consumption and environmental pollution.

The worldwide personal computer market is predicted to jump in size, with 100 million PCs expected to be shipped in 1999, twice the 1994 figure of 48 million units.[5]

Everyone hopes, of course, that the principal effects of this growth will be beneficial. But you need not be anti-technology to wonder just what negative impact computerization will have. How, for instance, will it affect the environment—energy consumption and environmental pollution, for example?

Energy Consumption & "Green PCs"

All the computers and communications devices discussed in this book run on electricity. Much of this is simply wasted. Computers themselves have in the past been built in ways that used power unnecessarily. An office full of computers also generates a lot of heat, so that additional power is required to run air-conditioning systems to keep people comfortable. Finally, people leave their computer systems on even when they're not sitting in front of them—not just during the day but overnight and weekends as well.

In recent years, the U.S. Environmental Protection Agency launched Energy Star, a voluntary program to encourage the use of computers that consume a minimum amount of power. The goal of Energy Star is to reduce the amount of electricity microcomputers and monitors use from the typical 150 watts of power to 60 watts or less. This goal is about the power requirement for a moderately bright lightbulb.[6] (Half the wattage would be for the system unit, the other half for the monitor.) As a result, manufacturers are now coming forth with Energy Star–compliant "green PCs."

"If you use your PC for 8 hours a day but always leave it on," says one writer, "a green PC could save about $70 a year. If everyone used only green PCs, $2 billion could be saved annually."[7]

Environmental Pollution

Communities like to see computer manufacturers move to their areas because they are viewed as being nonpolluting. Is this true? Actually, in the past, chemicals used to manufacture semiconductors polluted air, soil, and groundwater. Today, however, computer makers are literally cleaning up their act.

However, another problem is that the rush to obsolescence has produced numbers of computers, printers, monitors, fax machines, and so on, that have wound up as junk in landfills, although some are stripped by recyclers for valuable metals. More problematic is the disposal of batteries, as from portable computers. Nickel-cadmium batteries contain the toxic element cadmium, which, when buried in a landfill garbage dump, can leach into groundwater supplies. Disposal of such batteries should be through local toxic-waste disposal programs. Newer battery technology, such as nickel-hydride and lithium cells, may eventually replace nickel-cadmium.

If you have an old-fashioned computer system, consider donating it to an organization that can make use of it. Don't abandon it in a closet. Don't dump it in the trash. "Even if you have no further use for a machine that seems horribly antiquated," writes *San Jose Mercury News* computing editor Dan Gillmor, "someone else will be grateful for all it will do."[8]

Onward

New work habits have led to changes in how computers are used, and new computer uses have also changed work habits. For instance, employers have been seeking to trim costs and to respond to employees' demands for more flexibility about when and where they work—at home, on weekends, or out of the office. This situation has led to greater use of portable computers that can be taken anywhere. Conversely, distributing portables to employees has altered ways of doing business. For example, Wilsons The Leather Experts, a Minneapolis leather retailer, distributed notebook computers to its district sales managers, who formerly had used desktop computers in an office. Now the managers, keeping in touch with headquarters through modems on their notebooks, spend more time on the road. And their offices have been eliminated.[9] More and more, it is not so much computing as communications-linked *mobile* computing that is transforming our lives.

Using Software to Access the World Wide Web

What's the easiest way to use the *Internet* ("the Net"), that international conglomeration of thousands of smaller networks? Getting on that part known as the *World Wide Web* ("the Web") is no doubt the best choice. Increasingly, systems software is coming out with features for accessing and exploring the Net and the Web (as OS/2 already has). This Experience Box, however, describes ways to tour both the Net and the Web independent of whatever systems software you have.

The Web resembles a huge encyclopedia filled with thousands of general topics or so-called *Web sites* that have been created by computer users and businesses around the world. The entry point to each Web site is a *home page,* which may offer cross-reference links to other pages. Pages may theoretically be in *multimedia* form—meaning they can appear in text, graphical, sound, animation, and video form. At present, however, the Web is dominated by lots of pictures and text but little live, moving content, although that is changing.

To get on the Internet and its World Wide Web, you need a microcomputer, a modem, a telephone line, and communications software. (For details about the initial setup, see the Experience Box at the end of Chapter 2.) You then need to gain access to the Web and, finally, to get a browser. Some browsers—software programs that help you navigate the Web—come in kits that handle the setup for you, as we will explain.[10-14]

Gaining Access to the Web

There are three principal ways of getting connected to the Internet: (1) through school or work, (2) through commercial online services, or (3) through an Internet service provider.

Connecting Through School or Work The easiest access to the Internet is available to students and employees of universities and government agencies, most colleges, and certain large businesses. If you're involved with one of these, you can simply ask another student or co-worker with an Internet account how you can get one also. In the past, college students have often been able to get a free account through their institutions. However, students and faculty living off-campus may not be able to use the connections of campus computers.

Connections through universities and business sites are called *dedicated connections* and consist of phone lines (called *T1* or *T3 carrier lines*) that typically cost thousands of dollars to install and maintain every month. Their main advantage is their high speed, so that the graphic images and other content of the Web unfold more quickly.

Connecting Through Online Services The large commercial online services—such as America Online (AOL), CompuServe, Microsoft Network, or Prodigy—also offer access to the Internet. Some offer their own Web browsers, but some (such as AOL) offer Netscape Navigator and Internet Explorer. Commercial online services may also charge more than independent Internet service providers, although they are probably better organized and easier for beginners.

Web access through online services is usually called a *dial-up connection.* As long as you don't live in a rural area, there's no need to worry about long-distance telephone charges; you can generally sign on ("log on") by making a local call. When you receive membership information from the online service, it will tell you what to do.

Connecting Through Internet Service Providers Internet service providers (ISPs) are local or national companies that will provide public access to the Internet for a fee. Examples of national companies include PSI, UUNet, Netcom, and Internet MCI. Telephone companies such as AT&T and Pacific Bell have also jumped into the fray by offering Internet connections. Most ISPs offer a flat-rate monthly fee for a set number of hours of service. The connections offered by ISPs (called *SLIP/PPP connections*—discussed in Chapter 8) may offer faster access to the Internet than those of commercial online services. However, setting up a basic system on a microcomputer can require considerable fussing. Generally, ISPs are better for Internet experts than for beginners.

The whole industry of Internet connections is so new that many ISP users have had problems with uneven service (such as busy signals or severing of online "conversations"). Often ISPs signing up new subscribers aren't prepared to handle traffic jams caused by a great influx of newcomers. Some suggestions for choosing an ISP are given in the box. (■ *See Panel 4.11, next page.*)

You can also ask someone who is already on the Web to access for you the worldwide list of ISPs at *http://www.thelist.com.* Besides giving information about each provider in your area, "thelist" provides a rating (on a scale of 1 to 10) by users of different ISPs.

Accessing the Web: Browser Software

Once you're connected to the Internet, you then need a Web browser. This software program will help you to get whatever information you want on the Web by clicking your mouse pointer on words or pictures the browser displays on your screen.

Tips for choosing an Internet service provider (ISP)

ISPs may be less expensive and faster than online services for accessing the Internet, but they can also cause problems. Here are some questions to which you'll want answers before you sign up.

- *Is the ISP connection a local call?* Some ISPs are local, some are national. Be sure the ISP is in your local calling area, or the telephone company will charge you by the minute for your ISP connection. To find out if your ISP is local, call the phone company's directory assistance operator, and provide the prefix of your (modem's) phone number and the prefix of the ISP. The operator can tell you if the call to the ISP is free.

- *How much will it cost?* Ask about setup charges. Ask what the fee is per month for how many hours. Ask if any software (such as browsers) is included when you join.

- *How good is the service?* Ask how long the ISP has been in business, how fast it has been growing, what the peak use periods are, and how frequently busy signals occur. Ask if customer service (a help line) is available evenings and weekends as well as during business hours.

  Ask about the ratio of subscribers to ISP modems. If the service's modems are all in use, you will get a busy signal when you dial in. A ratio of 15 or 20 subscribers to every one modem probably means frequent busy signals (a 10-to-1 ratio is better).

There are all sorts of Web browsers, the best known being Netscape Navigator (the most popular), Mosaic, Microsoft's Internet Explorer, and Netcom's NetCruiser. (■ See Panel 4.12.)

Features of Browsers What kinds of things should you consider when selecting a browser? Here are some features:

- *Price:* Some browsers are free ("freeware," ✓ p. 46), such as the original Mosaic. Some come free with membership in an online service or ISP—America Online, for instance, offers browsers for both IBM-style computers and Macintoshes. Some may be acquired for a price separately from any online connection.

  You can get a kit that offers other features besides a browser. For instance, Macintosh users can buy the Apple Internet Connection Kit ($59), which contains the browser Netscape Navigator and several other programs. (They include Claris E-mailer Lite, News Watcher, Fetch, Alladin Stuffit Expander, NCSA Telnet, Adobe Acrobat Reader, Sparkle, Real Audio, MacTCP, MacPPP, and Apple Quicktime VR Player).[15] The kit comes with an Apple Internet Dialer application that helps you find an Internet service provider. Another kit, for Windows, is Internet In a Box ($149.95), which includes the browser Enhanced Mosaic and coupons for various ISPs and for CompuServe.

- *Ease of setup:* Especially for a beginner, the browser should be easy to set up. Ease of setup favors the university/business dedicated lines or commercial online services, of course, which already have browsers. If not provided by your online service or ISP, the browser should be compatible with it. Most online services allow you to use other browsers besides their own.

- *Ease of use:* If you have a multimedia PC, the browser should allow you to view and hear all of the Web's multimedia—not only text and images but also sounds and video. It should be easy to use for saving "hot lists" of frequently visited Web sites and for saving text and images to your hard disk. Finally, the browser should allow you to do "incremental" viewing of images, so that you can go on reading or browsing while a picture is slowly coming together on your screen, rather than having to wait with browser frozen until the image snaps into view.[16]

Surfing the Web

Once you are connected to the Internet and have used your browser to access the Web, you begin by clicking on the *Home* button found on most browsers. This will take you to a predefined home page, established by the software maker that developed the browser. (■ See Panel 4.13.)

Product	Price	Company & Contact
Mosaic	Free	NCSA, Champaign, IL; 217-244-0072, ftp.ncsa.uiuc.edu
Netscape Navigator 2.0 (Windows and Mac versions)	Free for education and nonprofit uses if downloaded; otherwise $49	Netscape Communications Corp., Mountain View, CA; 415-528-2555, http://www.mcom.com
NetCruiser (Windows only)	Free with subscription	Netcom, Inc., San Jose, CA; 800-501-8649
Internet Explorer	Free with Windows 95	Microsoft Corp., Bellevue, WA; 800-386-5550
Enhanced Mosaic	$29.95	Spyglass, Inc., Naperville, IL; 800-505-1010, info@spyglass.com, http://www.spyglass.com

Web Untanglers Where do you go from here? You'll find that, unlike a book, there is no page 1 where everyone is supposed to start reading and, unlike an encyclopedia, the entries are not in alphabetical order. Moreover, there is no definitive listing of everything available.

There are, however, a few search tools for helping you find your way around, which can be classified as directories and indexes. (■ See Panel 4.14, next page.)

- *Directories:* Directories are lists of Web sites classified by topic. Perhaps the best known example is the Yahoo directory, but there are others.
- *Indexes:* Indexes allow you to find specific documents through keyword searches or menu choices. Examples are WebCrawler, Lycos, and InfoSeek.

We describe directories and indexes in more detail in the Experience Box at the end of Chapter 9.

Home button

129

Web search tools

You may wish to use more than one search tool, since some data will be found on some sources but not others.

Directories

These are lists of Web sites classified by topic.

- **Yahoo** *(http://www.yahoo.com)* is supposedly the largest index of Web places, with more than 36,000 entries, and is used by more than 400,000 people a day. It is supported by on-screen advertising. The opening screen offers 14 general categories. You can "drill down" through layers of increasingly specialized lists to find what you're looking for or you can enter a keyword, which Yahoo will try to match.

- **World Wide Web Virtual Library** *(http://info.cern.ch/hypertext/ DataSources/by-Subject/Overview.html)* is a subject index of Web pages run by volunteers all over the world.

- **Yanoff's Special Internet Connections** *(http://www.uwm.edu/Mirror/ inet.services.html)* is an interesting though subjective list of selected Web locations, organized by topic.

Indexes

These search tools allow you to use keywords or menu choices to look for documents.

- **WebCrawler** *(http://webcrawler.com or http://webcrawler.cs.washington. edu/WebCrawler/Home.html)* is one of the best search tools and easier to use than Lycos or InfoSeek, though more limited.

- **Lycos** *(http://www.lycos.com or http://lycos2.cs.cmu.edu/)* is the Carnegie-Mellon search page, now a commercial product. Its searching mechanism takes some effort to learn.

- **InfoSeek** *(http://www.infoseek.com)* includes access to proprietary data, such as news from the Associated Press. Unlike other index-type search tools, it charges for its services (20 cents per transaction).

Web Addresses: URLs Getting to a directory, index, or any other Web location is easy if you know the address. Just click your mouse pointer on the *Open* button.   This will open the address, or *URL* (for *Uniform Resource Locator*). Web addresses usually start with *http* (for Hypertext Transfer Protocol) and are followed by a colon and double slash (://). For example, to reach the home page of Yahoo, you would type the address *http://www.yahoo.com.* Your browser uses the address to connect you to the computer of the Web site; then it downloads (transfers) the Web page information to display it on your screen.

If you get lost on the Web, you can return to your home page by clicking on the *Home* button.

Visiting Web Sites

Nowadays you see Web site addresses appearing everywhere, in all the mass media, and some of it's terrific and some of it's awful. For a sample of the best, try Yahoo's What's Cool list *(http://www.yahoo.com/Entertainment/ COOL_Links)*. For less-than-useful information, try Worst of the Web *(http://turnpike.net/metro/mirsky/Worst.html)*.

SUMMARY

address *(p. 114, LO 2)* The location in main memory, designated by a unique number, in which a character of data or of an instruction is stored during processing.

To process each character, the control unit of the CPU retrieves it from its address in main memory and places it into a register. This is the first step in what is called the *machine cycle*.

American Standard Code for Information Interchange (ASCII) *(p. 116, LO 4)* Binary code used in microcomputers; ASCII originally used seven bits to form a character, but a zero was added in the left position to provide an eight-bit code, providing more possible combinations with which to form other characters and marks.

ASCII is the binary code most widely used with microcomputers.

arithmetic/logic unit (ALU) *(p. 113, LO 2)* The part of the CPU that performs arithmetic operations and logical operations and that controls the speed of those operations.

Arithmetic operations are the fundamental math operations: addition, subtraction, multiplication, and division. Logical operations are comparisons, such as is equal to (=), greater than (>), or less than (<).

binary system *(p. 115, LO 4)* A two-state system.

Computer systems use a binary system for data representation; two digits, 0 and 1, to refer to the presence or absence of electrical current or a pulse of light.

bit *(p. 115, LO 4)* Short for *binary digit,* which is either a 1 or a 0 in the binary system of data representation in computer systems.

The bit is the fundamental element of all data and information stored in a computer system.

bus *(p. 123, LO 2)* Electrical pathway through which bits are transmitted within the CPU and between the CPU and other devices in the system unit. There are different types of buses (address bus, control bus, data bus, input/output bus).

The larger a computer's buses, the faster it operates.

byte *(p. 115, LO 4)* A group of 8 bits.

A byte holds the equivalent of a character—such as a letter or a number—in computer data-representation coding schemes. It is also the basic unit used to measure the storage capacity of main memory and secondary storage devices.

cache memory *(p. 122, LO 2)* Special high-speed memory area on a chip that the CPU can access quickly. A copy of the most frequently used instructions is kept in the cache memory so the CPU can look there first.

Cache memory allows the CPU to run faster because it doesn't have to take time to swap instructions in and out of main memory. Large, complex programs benefit the most from having a cache memory available.

central processing unit (CPU) *(p. 112, LO 2)* The processor; it controls and manipulates data to produce information. In a microcomputer the CPU is usually contained on a single integrated circuit or chip called a *microprocessor*. This chip and other components that make it work are mounted on a circuit board called a *system board*. In larger computers the CPU is contained on one or several circuit boards. The CPU consists of two parts: (1) the control unit and (2) the arithmetic/logic unit. The two components are connected by a bus.

The CPU is the "brain" of the computer.

What It Is / What It Does	**Why It's Important**
chip (microchip) *(p. 109, LO 1)* Microscopic piece of silicon that contains thousands of microminiature electronic circuit components, mainly transistors.	Chips have made possible the development of small computers.
control unit *(p. 113, LO 2)* The part of the CPU that tells the rest of the computer system how to carry out a program's instructions.	The control unit directs the movement of electronic signals between main memory and the arithmetic/logic unit. It also directs these electronic signals between the main memory and input and output devices.
expansion bus *(p. 123, LO 2)* Bus that carries data between RAM and the expansion slots.	Without buses, computing would not be possible.
expansion card *(p. 122, LO 5)* Add-on circuit board that provides more memory or a new peripheral-device capability. (The words *card* and *board* are used interchangeably.) Expansion cards are inserted into expansion slots inside the system unit.	Users can use expansion cards to upgrade their computers instead of having to buy entire new systems.
expansion slot *(p. 122, LO 5)* Socket on the motherboard into which users may plug an expansion card.	*See expansion card.*
flash memory *(p. 122, LO 2)* Used primarily in notebook and subnotebook computers; flash memory, or flash RAM cards, consist of circuitry on credit-card-size cards that can be inserted into slots connecting to the motherboard.	Unlike standard RAM chips, flash memory is nonvolatile—it retains data even when the power is turned off. Flash memory not only can be used to simulate main memory but also to supplement or replace hard-disk drives for permanent storage.
gigabyte (G, GB) *(p. 116, LO 4)* Approximately 1 billion bytes (1,073,741,824 bytes); a measure of storage capacity.	Gigabyte is used to express the storage capacity of large computers, such as mainframes, although it is also applied to some microcomputer secondary-storage devices.
graphics coprocessor chip *(p. 121, LO 2)* Secondary, "assistant" processor that enhances the performance of programs with lots of graphics and helps create complex screen displays.	Specialized chips such as these can significantly increase the speed of a computer system.
integrated circuit (IC) *(p. 109, LO 1)* Collection of electrical circuits, or pathways, etched on tiny squares, or chips, of silicon half the size of a person's thumbnail. In a computer, different types of ICs perform different types of operations. An integrated circuit embodies what is called *solid-state technology.*	The development of the IC enabled the manufacture of the small, powerful, and relatively inexpensive computers used today.
kilobyte (K, KB) *(p. 116, LO 4)* Unit for measuring storage capacity; equals 1024 bytes (usually rounded off to 1000 bytes).	The sizes of stored electronic files are often measured in kilobytes.
laptop computer *(p. 111, LO 1)* Portable computer equipped with a flat display screen and weighing 8–20 pounds. The top of the computer opens up like a clamshell to reveal the screen.	Laptop and other small computers have provided users with computing capabilities in the field and on the road.
local bus *(p. 123, LO 2)* Bus that connects expansion slots to the CPU, bypassing RAM.	A local bus is faster than an expansion bus.
machine cycle *(p. 114, LO 2)* Series of operations performed by the CPU to execute a single program instruction; it consists of two parts: an instruction cycle and an execution cycle.	The machine cycle is the essence of computer-based processing.
machine language *(p. 117, LO 4)* Binary code (language) that the computer uses directly. The 0s and 1s represent precise storage locations and operations.	For a program to run, it must be in the machine language of the computer that is executing it.

What It Is / What It Does	Why It's Important

main memory *(p. 113, LO 2)* Also known as *memory, primary storage, internal memory,* or *RAM* (for *random access memory*). Main memory is working storage that holds (1) data for processing, (2) the programs for processing the data, and (3) data after it is processed and is waiting to be sent to an output or secondary-storage device.

Main memory determines the total size of the programs and data files a computer can work on at any given moment.

math coprocessor chip *(p. 121, LO 2)* Specialized microprocessor that helps programs that use lots of mathematical equations to run faster.

Math coprocessors can significantly increase the speed of processing of a computer system.

megabyte (M, MB) *(p. 116, LO 4)* About 1 million bytes (1,048,576 bytes).

Most microcomputer main memory capacity is expressed in megabytes.

megahertz (MHz) *(p. 114, LO 3)* Measurement of transmission frequency; 1 MHz equals 1 million beats (cycles) per second.

Generally, the higher the megahertz rate, the faster a computer can process data.

microcomputer *(p. 110, LO 1)* Small computer that can fit on or beside a desktop or is portable; uses a single microprocessor for its CPU. A microcomputer may be a workstation, which is more powerful and is used for specialized purposes, or a personal computer (PC), which is used for general purposes.

The microcomputer has lessened the reliance on mainframes and has enabled more ordinary users to use computers.

microprocessor *(p. 110, LO 1)* A CPU (processor) consisting of miniaturized circuitry on a single chip; it controls all the processing in a computer.

Microprocessors enabled the development of microcomputers.

millions of instructions per second (MIPS) *(p. 114, LO 3)* Another measure of a computer's execution speed; for example, .5 MIPS is 500,000 instructions per second.

This measure is often used for large, relatively powerful computers and new sophisticated microcomputers.

motherboard *(p. 118, LO 2, 5)* Also called *system board;* the main circuit board in the system unit of a microcomputer.

It is the interconnecting assembly of important components, including CPU, main memory, other chips, and expansion slots.

notebook computer *(p. 111, LO 1)* Type of portable computer weighing 4–7.5 pounds and measuring about 8½ × 11 inches.

Notebooks have more features than many subnotebooks yet are lighter and more portable than laptops.

parallel port *(p. 123, LO 5)* Part of the computer through which a parallel device, which transmits 8 bits simultaneously, can be connected.

Enables microcomputer users to connect to a printer using a cable.

parity bit *(p. 117, LO 4)* Also called a *check bit;* an extra bit attached to the end of a byte.

Enables a computer system to check for errors during transmission (the check bits are organized according to a particular coding scheme designed into the computer).

peripheral devices *(p. 118, LO 5)* Hardware that is outside the central processing unit, such as input/output and secondary storage devices.

These devices are used to get data into and out of the CPU and to store large amounts of data that cannot be held in the CPU at one time.

personal computer (PC) *(p. 110, LO 1)* Type of microcomputer; desktop, floor-standing, or portable computer that can run easy-to-use programs, such as word processing or spreadsheets.

The PC is designed for one user at a time and so has boosted the popularity of computers.

What It Is / What It Does	Why It's Important

Personal Computer Memory Card International Association (PCMCIA) *(p. 125, LO 5)* Completely open, nonproprietary bus standard for portable computers.

This standard enables users of notebooks and subnotebooks to insert credit-card-size peripheral devices, called *PC cards,* such as modems and memory cards, into their computers.

personal digital assistant (PDA) *(p. 111, LO 1)* Also known as a *pocket communicator;* type of handheld pocket personal computer, weighing 1 pound or less, that is pen-controlled and in its most developed form can do two-way wireless messaging.

PDAs may supplant book-style personal organizers and calendars, as well as allow transmission of personal messages.

pocket personal computer *(p. 111, LO 1)* Also known as a *handheld computer;* a portable computer weighing 1 pound or less. Three types of pocket PCs are electronic organizers, palmtop computers, and personal digital assistants.

Pocket PCs are useful to help workers with specific jobs, such as delivery people and parking control officers.

port *(p. 123, LO 5)* Connecting socket on the outside of the computer system unit that is connected to an expansion board on the inside of the system unit. Ports are of five types: parallel, serial, video adapter, SCSI, and game ports.

A port enables users to connect by cable a peripheral device such as a monitor, printer, or modem so that it can communicate with the computer system.

power supply *(p. 118, LO 5)* Device in the computer that converts AC current from the wall outlet to the DC current the computer uses.

The power supply enables the computer (and peripheral devices) to operate.

random-access memory (RAM) *(p. 121, LO 2)* Also known as *main memory* or *primary storage;* type of memory that temporarily holds data and instructions needed shortly by the CPU. RAM is a volatile type of storage.

RAM is the working memory of the computer; it is the workspace into which applications programs and data are loaded and then retrieved for processing.

read-only memory (ROM) *(p. 122, LO 5)* Also known as *firmware;* a memory chip that permanently stores instructions and data that are programmed during the chip's manufacture. Three variations on the ROM chip are PROM, EPROM, and EEPROM. ROM is a nonvolatile form of storage.

ROM chips are used to store special basic instructions for computer operations such as those that start the computer or put characters on the screen.

reduced instruction set computing (RISC) *(p. 120, LO 2)* Type of design in which the complexity of a microprocessor is reduced by reducing the amount of superfluous or redundant instructions.

With RISC chips, a computer system gets along with fewer instructions than those required in conventional computer systems. RISC-equipped workstations work 10 times faster than conventional workstations.

register *(p. 113, LO 2)* High-speed circuit that is a staging area for temporarily storing data during processing.

The computer loads the program instructions and data from the main memory into the staging areas of the registers just prior to processing.

serial port *(p. 124, LO 5)* Also known as *RS-232 port;* a port to which a cable is connected that transmits 1 bit at a time.

Serial ports are used principally for connecting communications lines, modems, and mice to microcomputers.

single inline memory module (SIMM) *(p. 121, LO 2)* Small circuit board plugged into the motherboard.

A SIMM holds RAM chips and can be used to increase a computer's main memory capacity.

small computer system interface (SCSI) port *(p. 125, LO 5)* Pronounced "scuzzy"; an interface for transferring data at high speeds for up to eight SCSI-compatible devices.

SCSI ports are used to connect external hard-disk drives, magnetic-tape backup units, and CD-ROM drives to the computer system.

subnotebook computer *(p. 111, LO 1)* Type of portable computer, weighing 2.5–4 pounds.

Subnotebooks are lightweight and thus extremely portable; however, they may lack features found on notebooks and other larger portable computers.

system clock *(p. 121, LO 2,3)* Internal timing device that uses a quartz crystal to generate a uniform electrical frequency from which digital pulses are created.

The system clock controls the speed of all operations within a computer. The faster the clock, the faster the processing.

system unit *(p. 118, LO 5)* The box or cabinet that contains the electrical components that do the computer's processing; usually includes processing components, RAM chips (main memory), ROM chips (read-only memory), power supply, expansion slots, and disk drives but not keyboard, printer, or often even the display screen.

The system unit protects many important processing and storage components.

terabyte (T, TB) *(p. 116, LO 4)* Approximately 1 trillion bytes (1,009,511,627,776 bytes); a measure of capacity.

Some forms of mass storage, or secondary storage for mainframes and supercomputers, are expressed in terabytes.

transistor *(p. 109, LO 1)* Semiconducting device that acts as a tiny electrically operated switch, switching between "on" and "off" many millions of times per second.

Transistors act as electronic switches in computers. They are more reliable and consume less energy than their predecessors, electronic vacuum tubes.

video adapter port *(p. 124, LO 5)* Part of the computer used to connect the video display monitor outside the computer to the video adapter card inside the system unit.

The video adapter port enables users to have different kinds of monitors, some having higher resolution and more colors than others.

video memory *(p. 122, LO 2)* Video RAM (VRAM) chips are used to store display images for the monitor.

The amount of video memory determines how fast images appear and how many colors are available on the display screen. Video memory chips are useful for programs displaying lots of graphics.

volatile storage *(p. 113, LO 2)* Temporary storage, as in main memory (RAM).

The contents of volatile storage are lost when power to the computer is turned off.

word *(p. 120, LO 2)* Also called *bit number*; group of bits that may be manipulated or stored at one time by the CPU.

Often the more bits in a word, the faster the computer. An 8-bit word computer will transfer data within each CPU chip in 8-bit chunks. A 32-bit word computer is faster, transferring data in 32-bit chunks.

workstation *(p. 112, LO 1)* Type of microcomputer; desktop or floor-standing machine that costs $10,000–$150,000 and is used mainly for technical purposes.

Workstations are used for scientific and engineering purposes and also for their graphics capabilities.

(Selected answers appear at the back of the book.)

Short-Answer Questions

1. What is the main difference between a 66-MHz computer and a 133-MHz computer?
2. What are the two main parts of the CPU? What is each part responsible for?
3. Describe why having more main memory, or RAM, in your computer (as opposed to less) is useful.
4. What is the function of the ALU in a microcomputer system?
5. Name three of the five kinds of ports discussed in this chapter. What are they used for?

Fill-in-the-Blank Questions

1. The _____ links the CPU to every hardware device in the computer system.
2. List five of the components located in the system unit.
   a. _____
   b. _____
   c. _____
   d. _____
   e. _____
3. The *arithmetic / logic* _____ is located inside the system unit and controls how fast all the operations within the computer take place.
4. The *power supply* _____ is located inside the system unit and converts AC current to DC current to run the computer.
5. The binary coding scheme most commonly used in microcomputers is _____.
6. The main circuit board in a system unit is called the *motherboard* _____.
7. *silicon* _____ is the semiconductor material commonly used to make integrated circuits (chips).

Multiple-Choice Questions

1. A(n) ___ *byte* ___, or character, is composed of 8 ___ *bytes* ___.
   a. byte, bits
   b. bit, bytes
   c. megabyte, kilobytes
   d. kilobyte, megabytes
   e. none of the above

2. Which of the following memory types can be used in a computer system?
   a. main memory
   b. cache memory
   c. video memory
   d. flash memory
   e. all of the above

3. Main memory is also known as:
   a. primary storage
   b. RAM
   c. memory
   d. all of the above

4. Which of the following would you want inside your computer's system unit if you plan to perform lots of mathematical calculations?
   a. graphics coprocessor
   b. vector processing
   c. math coprocessor
   d. extended memory
   e. none of the above

5. Which of the following would your computer need to use a modem?
   a. parallel port
   b. serial port
   c. SCSI port
   d. video adapter port
   e. none of the above

6. A sophisticated microcomputer that fits on a desktop and is used mainly by engineers and scientists for technical purposes is called a(n)
   _____.
   a. workstation
   b. minicomputer
   c. mainframe computer
   d. supercomputer
   e. none of the above

True/False Questions

T (F) 1. The amount of main memory in your computer system is an important factor in determining what software the system can run.

T (F) 2. A printer is sometimes contained inside the system unit of a microcomputer.

(T) F 3. To connect a peripheral device, users plug a cable into an expansion slot on the outside of the system unit.

T (F) 4. RAM is volatile.

T (F) 5. Registers are used to transmit bits between the CPU and other devices in the system unit.

T F 6. Data and instructions in ROM are lost when the computer is turned off.

T **F** 7. A gigabyte is smaller than a terabyte.

T **F** 8. The CPU consists of the control unit and the ALU.

T **F** 9. In machine language, data and instructions are represented with 0s and 1s.

Projects/Critical-Thinking Questions

1. Computer magazines often sponsor tests to compare PCs based on their speed. For example, in March of 1995, PC World determined that the Micron Millennia was the fastest PC. By reviewing current computer magazines, identify the fastest PC today. What processing components make this the fastest PC? How much does the PC cost? What retail products would make this PC even faster?

2. Research is underway at Carnegie Mellon University to develop a PC that you can wear. The wearable PC might consist of a head band and visor, and a necklace that functions as a video input device. What is the rationale behind the wearable PC? What processing components are in the wearable PC? How long have researchers been working on this project? When do researchers predict that the wearable PC will be available for purchase? What do you think the future holds for the wearable PC?

3. Most industry observers agree that in the years ahead computer chip makers will continue to develop faster and faster microprocessors. What impact do you think this will have on the software industry? Will users have to change systems software and/or applications software in order to take advantage of more sophisticated microprocessors? What impact do you think faster microprocessors will have on the hardware industry? Will users have to scrap existing hardware?

Using the Internet

Objective: *In this exercise we describe how to use Netscape Navigator 2.0 or 2.01 to search the Web for general and specific topics.*

Before you continue: *We assume you have access to the Internet through your university, business, or commercial service provider and to the Web browser tool named Netscape Navigator. Additionally, we assume you know how to connect to the Internet and load Netscape Navigator. If necessary, ask your instructor or system administrator for assistance.*

1. Make sure you have started Netscape. The home page for Netscape Navigator should appear on your screen.

2. If you're interested in finding information on a general topic—such as processing hardware or portable computers—use a *directory*. Directories contain lists of Web sites that are organized by topic. For example, Yahoo (*http://www.yahoo.com*) contains a list of over 80,000 Web sites which are divided into 14 categories. After selecting a topic of interest, you're presented with a sub-list of topics. You continue selecting topics until the information you're looking for appears. If you're just becoming familiar with the Web, directories can provide a useful means of seeing what the Web has to offer. Other examples of directories are Galaxy (*http://galaxy.einet.net/*), InfoSeek (*http://www2.infoseek.com/*), Magellan (*http://www.mckinley.com/*), and Scott Yanoff's Internet Services Directory (*http://www.uwm.edu/Mirror/inet.services.html*).

To display Yahoo's home page:
CLICK: Open button (⬚)
TYPE: `http://www.yahoo.com`
PRESS: **Enter** or CLICK: Open

After a few moments, Yahoo's home page should appear. (*Note:* The progress indicator in the bottom-right corner of the screen indicates how much of a page has been loaded at a given time.) Your screen should appear similar to the following:

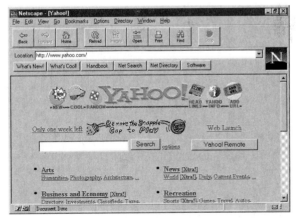

3. Point to some of the links (underlined words and phrases). Notice that the associated URL appears in the status area at the bottom of the screen.

4. Drag the vertical scroll bar downward and notice that additional topics appear.

5. In the next few steps you'll search for information on portable computing. The link that looks the most promising at this point is the CS link that

appears in the Science category. Drag the scroll bar until you see the Science category and then click the CS link.

6. Drag the vertical scroll bar downward to see a list of topics, as shown below.

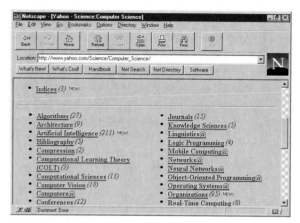

7. To continue with your search on portable computing, click the Mobile Computing@ link on the current page and then the Portable Computing link that is located on the next page.

8. As you can see, with each selection you make, the list of Web pages becomes more and more specific. The idea is that eventually the information you're looking for will appear on the screen. Now that you've had some practice using a directory, return to Netscape's home page and then practice using a search engine in the next few steps.

9. If you're interested in finding information on a specific topic, you may be better off using an *index*, or *search engine* rather than a directory. A search engine is a Web page that contains a form. You type a text string into the form that identifies the topic you want to search for. Whereas directories display lists of topics, search engines display a list of Web sites that match your search criteria. Some examples of general-interest search engines are Lycos (*http://lycos.cs.cmu.edu/*), Excite Netsearch (*http://www.excite.com/*),

Open Text (*http://www.opentext.com:8080/*), and Alta Vista (*http://www.altavista.digital.com/*).

To display Lycos' search engine:
CLICK: Open button (⊞)
TYPE: `http://lycos.cs.cmu.edu`
PRESS: **Enter** or CLICK: Open
The Lycos search form appears.

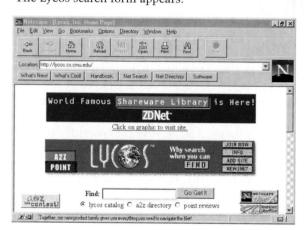

10. To display a list of Web sites that relate to the topic of "portable computing:"

CLICK: in the Find text box
TYPE: `portable computing`
CLICK: Go Get It button

After a few moments, a list of Web sites that relate to your search criteria will be loaded into the current page. (*Note:* You must drag the vertical scroll bar downward to see the list of Web sites.)

11. Continue dragging the vertical scroll bar downward until you see a list of related Web sites. Although Lycos found many Web sites that match your search criteria, Lycos only displays the first 10 Web sites. To see the next ten sites that match your criteria, click the "Next 10 hits" link on the bottom of the page.

12. Now that you've practiced using a directory and a search engine, display Netscape's home page and then exit Netscape.

CHAPTER 5

Input
Taking Charge of Computers & Communications

Concepts You Should Know

After reading this chapter, you should be able to:

1. Distinguish between the two types of input hardware.

2. List and explain the types of pointing devices, including operation of a mouse.

3. Identify and explain the different scanning devices.

4. Distinguish between magnetic-stripe, smart, and optical cards.

5. Discuss the basic operation and limitations of voice-recognition devices.

6. Briefly describe other advanced input devices: audio, video, cameras, sensors, and human-biology devices—biometric systems, line-of-sight systems, and cyber gloves and body suits.

7. Discuss some adverse effects of computers on health—repetitive strain injuries, eyestrain and headaches, and backstrain—and the significance of ergonomics.

For a sizable number of people, it's always "0:00" on the VCR clock.

About one in five owners of a videocassette recorder fails a basic test of the Information Age: setting the VCR's digital clock.[1] Even though 88% of Americans say their family owns a VCR, according to a Washington, D.C., polling firm, apparently 16% of them don't know how to set the time on it. In such households the VCR clock is always blinking "0:00" or "12:00."

Is this a sign that many people are being left behind in the Digital Revolution? If people can't set the time on the VCR, how will they be able to get the machine to automatically record programs from some of those hundreds of TV channels we are supposed to have some day? Or figure out how to download (transfer) crucial information into their "telecomputer" from some far-flung database?

Today, learning to benefit from information technology means becoming comfortable with the input and output devices that constitute its two principal interfaces with people. In this chapter we explain what input devices are.

Input Hardware: Keyboard Entry Versus Source Data Entry

Preview & Review: Input hardware is classified as keyboard entry or source data entry (automation).

Input hardware consists of devices that take data and programs that people can read or comprehend and convert them to a form the computer can process. The people-readable form may be words like the ones in these sentences, but the computer-readable form consists of 0s and 1s, or off and on electrical signals.

Input devices are categorized as *keyboard entry* and *source data entry* devices. (■ *See Panel 5.1, opposite page.*)

Keyboard Entry

In a computer, **a *keyboard* is a device that converts letters, numbers, and other characters into electrical signals that are machine-readable by the computer's processor.** The keyboard may look like a typewriter keyboard to which some special keys have been added. Or it may look like the keys on a bank's automated teller machine or the keypad of a pocket computer used by a bread-truck driver.

Source Data Entry

***Source data entry devices* refers to the many forms of data-entry devices that are not keyboards.** This type of data entry is also known as *source-data automation.*

Source data entry devices include the following:

- Pointing devices
- Scanning devices
- Magnetic-stripe, smart, and optical cards
- Voice-recognition devices

- Audio-input devices
- Video-input devices
- Digital cameras
- Sensors
- Human-biology input devices

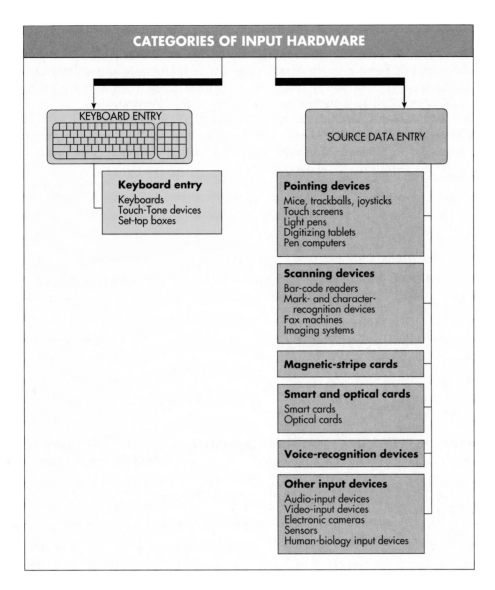

Often keyboard and source data entry devices are combined in a single computer system.

Keyboard Input

Preview & Review: Keyboard-type devices include computer keyboards, terminals, Touch-Tone devices, and set-top boxes.

Even if you aren't a ten-finger touch typist, you can use a keyboard. Yale University computer scientist David Gelernter, for instance, lost the use of his right hand and right eye in a mail bombing. However, he expressed not only gratitude at being alive but also recognition that he could continue to use a keyboard even with his limitations. "In the final analysis," he wrote in an online message to colleagues, "one decent typing hand and an intact head is all you really need. . . ."[2]

Here we describe the following keyboard-type devices:

- Computer keyboards
- Terminals
- Touch-Tone devices
- Set-top boxes

Computer Keyboards

People who always thought typing was an overrated skill will find themselves slightly behind in the Digital Age, since learning to use a keyboard is still probably the most important way of interacting with a computer. Of course, compared to a mouse or other source data entry devices, keyboards have one disadvantage: if you're to use them efficiently, you need training. Users who have to use the hunt-and-peck method waste a lot of time. Fortunately, there are software programs available that can help you learn typing skills or improve your existing ones.

You are probably already familiar with a computer keyboard. The illustration below provides a review of keyboard functions. (■ *See Panel 5.2.*)

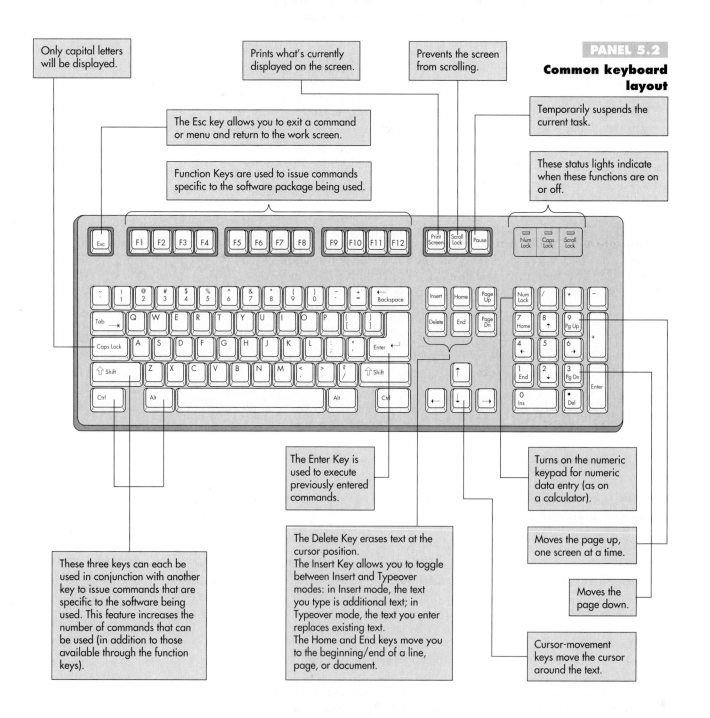

PANEL 5.2

Common keyboard layout

Only capital letters will be displayed.

Prints what's currently displayed on the screen.

Prevents the screen from scrolling.

Temporarily suspends the current task.

These status lights indicate when these functions are on or off.

The Esc key allows you to exit a command or menu and return to the work screen.

Function Keys are used to issue commands specific to the software package being used.

The Enter Key is used to execute previously entered commands.

Turns on the numeric keypad for numeric data entry (as on a calculator).

The Delete Key erases text at the cursor position. The Insert Key allows you to toggle between Insert and Typeover modes: in Insert mode, the text you type is additional text; in Typeover mode, the text you enter replaces existing text. The Home and End keys move you to the beginning/end of a line, page, or document.

Moves the page up, one screen at a time.

Moves the page down.

Cursor-movement keys move the cursor around the text.

These three keys can each be used in conjunction with another key to issue commands that are specific to the software being used. This feature increases the number of commands that can be used (in addition to those available through the function keys).

As the use of computer keyboards has become widespread, so has the incidence of various hand and wrist injuries. Accordingly, keyboard manufacturers have been giving a lot of attention to ergonomics. *Ergonomics* is the study of the physical relationships between people and their work environment. Various attempts are being made to make keyboards more ergonomically sound in order to prevent injuries. (We discuss ergonomics in the Experience Box at the end of Chapter 6.)

Terminals

A *terminal* is a device that consists of a keyboard, a video display screen, and a communications line to a large (usually mainframe) computer system. Terminals are generally used for input; they also display output. Terminals may be dumb or smart.

- **Dumb:** The most common type of terminal is dumb. A *dumb terminal* can be used only to input data to and receive information from a computer system. That is, it cannot do any processing on its own.

    An example of a dumb terminal is the type used by airline clerks at airport ticket and check-in counters.

- **Smart:** A *smart terminal*, also called an *X-terminal*, can do input and output and has some processing capability and RAM. However, a smart terminal is not designed to operate as a stand-alone computer. Thus it cannot be used to do programming—that is, create new instructions.

    One example of a smart terminal is the point-of-sale terminal. **A point-of-sale (POS) terminal is used much like a cash register. It records customer transactions at the point of sale but also stores data for billing and inventory purposes.** POS terminals are found in most department stores.

Touch-Tone Devices

The Touch-Tone, or push-button, telephone can be used like a dumb terminal to send data to a computer. FedEx customers, for example, can request pickup service for their packages by pushing buttons on their phones. Another common device is the *card dialer*, or *card reader*, used by merchants to verify credit cards over phone lines with a central computer.

Set-Top Boxes

If you receive television programs from a cable-TV service instead of free through the air, you may have what is called a *set-top box*. A *set-top box* works with a keypad to allow cable-TV viewers to change channels or, in the case of interactive systems, to exercise other commands. (■ *See Panel 5.3, next page.*) For example, in the 1970s the Warner-Amex interactive Qube system was instituted in Columbus, Ohio. The system's set-top box enabled users to request special programs, participate in quiz shows, order services or products, or respond to surveys.[3] Although Qube was discontinued because it was unprofitable, similar experiments are continuing, such as Time Warner's Full Service Network in Orlando, Florida.

We turn now from keyboard entry to various types of source data entry, in which data is converted into machine-readable form (✓ p. 117) as it is entered into the computer or other device. Source data entry is important because it cuts out the keyboarding step, thereby reducing mistakes.

Set-top box

Scientific-Atlanta's Model 8600^X Home Communications Terminal combines computing and cable-TV functions in a single device.

Pointing Devices

Preview & Review: Pointing devices include mice, trackballs, and joysticks; touch screens; light pens; digitizing tablets; and pen computers.

One of the most natural of all human gestures, the act of pointing, is incorporated in several kinds of input devices. The most prominent ones are the following:

- Mice, trackballs, and joysticks
- Touch screens
- Light pens
- Digitizing tablets
- Pen computers

Mice, Trackballs, & Joysticks

The principal pointing tools used with microcomputers are the mouse, the trackball, and the joystick, all of which have variations. (■ *See Panel 5.4, opposite.*)

- **Mouse:** **A *mouse* is a device that is rolled about on a desktop and directs a pointer on the computer's display screen.** The pointer may sometimes be, but is not necessarily the same as, the cursor. **The *mouse pointer* is the symbol that indicates the position of the mouse on the display screen.** It may be an arrow, a rectangle, or even a representation of a person's pointing finger. The pointer may change to the shape of an I-beam to indicate that it is a cursor and shows the place where text may be entered.

  On the bottom side of the mouse is a ball (trackball) that translates the mouse movement into digital signals. On the top side are one, two, or three buttons. Depending on the software, these buttons are used for such functions as *clicking, dropping,* and *dragging.*

- **Trackball:** Another form of pointing device, the trackball, is a variant of the mouse. **A *trackball* is a movable ball, on top of a stationary device, that is rotated with the fingers or palm of the hand.** In fact, the trackball looks like the mouse turned upside down. Instead of moving the mouse around on the desktop, you move the trackball with the tips of your fingers.

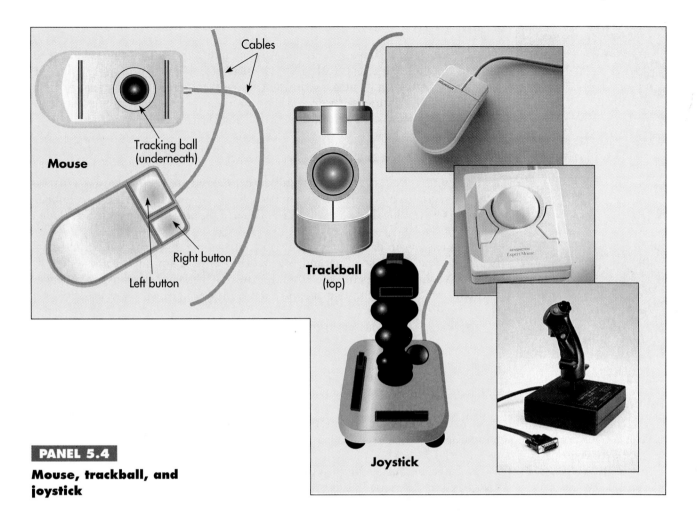

PANEL 5.4

Mouse, trackball, and joystick

- **Joystick:** A *joystick* is a pointing device that consists of a vertical handle like a gearshift lever mounted on a base with one or two buttons. Named for the control mechanism that directs an airplane's fore-and-aft and side-to-side movement, joysticks are used principally in videogames and in some computer-aided design systems.

Touch Screens

A *touch screen* is a video display screen that has been sensitized to receive input from the touch of a finger. (■ *See Panel 5.5.*) Because touch screens are easy to use, they can convey information quickly. You'll find touch screens in automated teller machines, tourist directories in airports, and campus information kiosks making available everything from lists of coming events to (with proper ID and personal code) student financial-aid records and grades.

PANEL 5.5

Touch screen

Making menu choices at a fast-food restaurant.

Light Pens

The *light pen* is a light-sensitive stylus, or pen-like device, connected by a wire to the computer terminal. The user brings the pen to a desired point on the display screen and presses the pen button, which identifies that screen location to the computer. *(■ See Panel 5.6.)* Light pens are used by engineers, graphic designers, and illustrators.

Digitizing Tablets

A *digitizing tablet* consists of a tablet connected by a wire to a stylus or puck. A *stylus* is a pen-like device with which the user "sketches" an image. A *puck* is a copying device with which the user copies an image. *(■ See Panel 5.7.)*

When used with drawing and painting software, a digitizing tablet and stylus allow you to do shading and many other effects similar to those artists achieve with pencil, pen, or charcoal. Alternatively, when you use a puck, you can trace a drawing laid on the tablet, and a digitized copy is stored in the computer.

Digitizing tablets are used primarily in design and engineering.

Pen Computers

In the next few years, students may be able to take notes in class without writing a word, if pen-based computer systems evolve as Depauw University computer science professor David Berque hopes they will. **Pen computers use a pen-like stylus to allow people to enter handwriting and marks onto a computer screen rather than typing on a keyboard.** Berque has developed a prototype for a system that would connect an instructor's electronic "whiteboard" on the classroom wall with students' pen com-

puters, so that students could receive notes directly, without having to copy information word for word. "The idea is that this might free the students up to allow them to think about what's going on," Berque says. "They wouldn't have to blindly copy things that maybe would distract them from what's going on."[4]

Several types of pen computer systems exist that will *store* handwriting as it is scrawled. What is more difficult is to *convert*—particularly without training—a person's distinctive script handwriting into typescript. After all, when you're trying to read someone else's notes, you may ask "Is that an *e* or an *a* or an *o*?"

This completes the section on the type of source data entry devices called *pointing devices*. Now we proceed to *scanning devices*.

Scanning Devices

Preview & Review: Scanning devices include bar-code readers, mark- and character-recognition devices, fax machines, and imaging systems.

Scanning devices **translate images of text, drawings, photos, and the like directly into digital form.** The images can then be processed by a computer, displayed on a monitor, stored on a storage device, or communicated to another computer. Scanning devices include:

- Bar-code readers
- Mark- and character-recognition devices
- Fax machines and fax modems
- Imaging systems

Bar-Code Readers

Bar codes **are the vertical zebra-striped marks you see on most manufactured retail products**—everything from candy to cosmetics to comic books. (■ *See Panel 5.8.*) In North America, supermarkets, food manufacturers, and others have agreed to use a bar-code system called the Universal Product Code. Other kinds of bar-code systems are used on everything from FedEx packages to railroad cars.

Bar codes are read by *bar-code readers,* **photoelectric scanners that translate the bar-code symbols into digital forms.** The price of a particular item

PANEL 5.8

Bar codes and bar-code readers

(Top left) Bar-coded groceries being scanned at the check-out counter. *(Bottom left)* Customer self-service scanning station. *(Bottom right)* Checking bar code inventory numbers.

is set within the store's computer and appears on the salesclerk's point-of-sale terminal and on your receipt. Records of sales are input to the store's computer and used for accounting, restocking store inventory, and weeding out products that don't sell well.

Mark-Recognition & Character-Recognition Devices

There are three types of scanning devices that sense marks or characters. They are usually referred to by their abbreviations MICR, OMR, and OCR.

- **Magnetic-ink character recognition:** *Magnetic-ink character recognition (MICR)* **reads the strange-looking numbers printed at the bottom of checks.** MICR characters, which are printed with magnetized ink, are read by MICR equipment, producing a digitized signal. This signal is used by a bank's reader/sorter machine to sort checks.

- **Optical-mark recognition:** *Optical-mark recognition (OMR)* **uses a device that reads pencil marks and converts them into computer-usable form.** The most well-known example is the OMR technology used to read the College Board Scholastic Aptitude Test (SAT) and the Graduate Record Examination (GRE).

- **Optical-character recognition:** *Optical-character recognition (OCR)* **uses a device that reads preprinted characters in a particular font (typeface design) and converts them to digital code.** Examples of the use of OCR characters are utility bills and price tags on department-store merchandise. The *wand reader* is a common OCR scanning device. (■ *See Panel 5.9.*)

Fax Machines

A *fax machine—or facsimile transmission machine—***scans an image and sends it as electronic signals over telephone lines to a receiving fax machine, which re-creates the image on paper.** Fax machines are useful to anyone who wants to send images rather than text. Many businesses have found them to also be a fast, reliable way to transmit text documents.

There are two types of fax machines—dedicated fax machines and fax modems:

- **Dedicated fax machines:** The type we generally call "fax machines," *dedicated fax machines* **are specialized devices that do nothing except send and receive fax documents.** They are found not only in offices and homes but also alongside regular phones in public places such as airports.

PANEL 5.9

Optical-character recognition

Various typefaces can be read by a special scanning device called a *wand reader.*

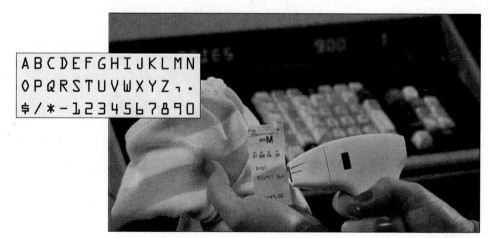

• **Fax modem:** **A *fax modem* is installed as a circuit board inside the computer's system cabinet. It is a modem with fax capability that enables you to send signals directly from your computer to someone else's fax machine or computer fax modem.** With this device, you don't have to print out the material from your printer and then turn around and run it through the scanner on a fax machine. The fax modem allows you to send information much more quickly than if you had to feed it page by page into a machine.

The main disadvantage of a fax modem is that you cannot scan in outside documents. Thus, if you have a photo or a drawing that you want to fax to someone, you need an image scanner, as we describe next.

By now, communication by fax has become cheaper than first-class mail for many purposes. (Fax machines are also covered in Chapter 8.)

Imaging Systems

An *imaging system*—or *image scanner* or *graphics scanner*—converts text, drawings, and photographs into digital form that can be stored in a computer system and then manipulated using different software programs. (■ *See Panel 5.10.)* The system scans each image with light and breaks the image into light and dark dots, which are then converted to digital code.

An example of an imaging system is the type used in desktop publishing (✓ p. 65). This device scans in artwork or photos that can then be positioned within a page of text. Other systems are available for turning paper documents into electronic files so that people can reduce their paperwork.

The next category of source data entry we'll cover are devices the size of credit cards.

Magnetic-Stripe, Smart, & Optical Cards

Preview & Review: Magnetic-stripe cards are encoded with data specific to a particular use. A smart card contains a microprocessor and a memory chip. An optical card is a plastic, laser-recordable card used with an optical card reader.

Magnetic-Stripe Cards

ATM cards and credit cards are ***magnetic-stripe cards,* with stripes of magnetically encoded data on their backs.** These cards are used only for specific purposes, such as getting cash or charging a purchase, so the encoded data might include your name, account number, and PIN number (personal identification number).

Smart Cards

The next level of data entry via card is represented by the smart card. **A *smart card* looks like a credit card but contains a microprocessor and memory chip.** Smart cards can be used as telephone debit cards. You insert the card into a slot in the phone, wait for a tone, and dial the number. The time of your call is automatically calculated on the chip inside the card and deducted from the balance. Recently, some credit-card companies and banks have begun to promote smart cards, combining ATM, credit, and debit cards in one instrument.

Optical Cards

The conventional magnetic-stripe credit card holds the equivalent of a half page of data. The smart card with a microprocessor and memory chip holds the equivalent of 30 pages. The optical card presently holds about 2000 pages of data. Optical cards use the same type of technology as music compact disks (or CD-ROMs) but look like silvery credit cards. **Optical cards are plastic, laser-recordable, wallet-type cards used with an optical-card reader.** Because they can cram so much data (6.6 megabytes) into so little space, they may become popular in the future. With an optical card, for instance, there's enough room for a person's health card to hold not only his or her medical history and health-insurance information but also digital images.

Voice-Recognition Input Devices

Preview & Review: Voice-recognition systems, which convert human speech into digital code, still have several limitations.

London native Caroline Goldie now lives in Washington, D.C., but she still retains her British accent. While in Vermont on a ski trip she tried to place an operator-assisted long-distance call. When she dialed the telephone company's long-distance access number, a tinny recorded voice asked her to speak a word from a number of choices, including the word "operator." Every time Ms. Goldie said "operator," the uncomprehending voice-recognition system said "Sorry, please repeat." In frustration, she finally handed the phone to an American friend, whose flat *"ah-per-aid-er"* finally got her connected to the real thing. Since then Ms. Goldie has worked out an American accent that she uses for the phone company. (Actually, if a caller waits long enough, an operator will come on the line.)[5]

Such are the difficulties with which telephone-company linguists and computer experts must deal in improving voice-recognition technology. **A voice-recognition system converts a person's speech into digital code by comparing the electrical patterns produced by the speaker's voice with a set of prerecorded patterns stored in the computer.**

Voice-recognition systems are finding many uses. Among the more commonplace uses, warehouse workers are able to speed inventory-taking by recording inventory counts verbally and traders on stock exchanges can communicate their trades by speaking to computers. In addition, blind and other disabled people can give verbal commands to their PCs rather than use a keyboard, and nurses can fill out patient charts by talking to a computer.

So far, most voice-recognition technology has been hindered by three basic limitations:

- **Speaker dependence:** Most systems need to be "trained" by the speaker to recognize his or her distinctive speech patterns and even variations in the way a particular word is said.

- **Single words versus continuous speech:** Most systems can handle only single words and have vocabularies of 1000 words or less. However, some technologies, such as DragonDictate, now offer continuous-speech recognition—that is, one need not artificially pause between words when speaking—and 30,000-word dictionaries (vocabularies).

- **Lack of comprehension:** Most systems merely translate sounds into characters. A more useful technology would be one that actually comprehends the *meaning* of spoken words. You could then ask a question, and the system could check a database and formulate a meaningful answer.

Other Input Devices: Audio, Video, Cameras, Sensors, & Human-Biology Devices

Preview & Review: Five other types of input devices are audio-input devices, video-input devices, digital cameras, sensors, and human-biology input devices.

Audio-Input Devices

Voice-recognition devices are only one kind of *audio input*, which can include music and other sounds. **An *audio-input device* records or plays analog sound and translates it for digital storage and processing.** You'll recall that an *analog sound signal* represents a continuously variable wave within a certain frequency range (✓ p. 6). Such continuous fluctuations are usually represented with an analog device such as an audiocassette player. For the computer to process them, these variable waves must be converted to digital 0s and 1s.

There are two ways by which audio is digitized:

- **Audio board:** Analog sound from, say, a cassette player goes through a special circuit board called an audio board. An *audio board* is an add-on circuit board in a computer that converts analog sound to digital sound and stores it for further processing.

- **MIDI board:** A *MIDI board*—MIDI stands for *Musical Instrument Digital Interface*—is an add-on board that creates digital music. That is, the music is created in digital form as the musician performs, as on a special MIDI keyboard.

The principal use of audio-input devices such as these is to provide digital input for multimedia PCs. A *multimedia system* is a computer that incorporates text, graphics, sound, video, and animation in a single digital presentation. Video input is also often used for this purpose, as we describe next.

Video-Input Devices

As with sound, most film and videotape is in analog form, with the signal a continuously variable wave. Thus, to be used by a computer, the signals that come from a VCR, a videodisk or laserdisk, or a camcorder must be converted to digital form through a special *video card* installed in the computer.

Two types of video cards are frame-grabber video and full-motion video:

- **Frame-grabber video card:** Some video cards, called *frame grabbers*, can capture and digitize only a single frame at a time.

- **Full-motion video card:** Other video cards, called *full-motion video cards*, can convert analog to digital signals at the rate of 30 frames per second, giving the effect of a continuously flowing motion picture.

Digital Cameras

The digital camera is a particularly interesting piece of hardware because it foreshadows major change for the entire industry of photography. Instead of using traditional (chemical) film, **a *digital camera* captures images in digital (electronic) form for immediate viewing on a television or computer display screen or for direct storage on diskette or compact disk, or for printing out. No film is used.**

PANEL 5.11

Earthquake sensor

Sensor instruments of a tele-metered weak motion seismic station (earthquake motion detection)

Sensors

A *sensor* is a type of input device that collects specific kinds of data directly from the environment and transmits it to a computer. Although you are unlikely to see such input devices connected to a PC in an office, they exist all around us, often in invisible form. Sensors can be used for detecting all kinds of things: speed, movement, weight, pressure, temperature, humidity, wind, current, fog, gas, smoke, light, shapes, images, and so on.

Beneath the pavement, for example, are sensors that detect the speed and volume of traffic. These sensors send data to computers that can adjust traffic lights to keep cars and trucks away from gridlocked areas. In California, sensors have been planted along major earthquake fault lines in an experiment to see whether scientists can predict major earth movements. (■ *See Panel 5.11.*) In aviation, sensors are used to detect ice buildup on airplane wings or to alert pilots to sudden changes in wind direction.

Human-Biology Input Devices

Characteristics and movements of the human body, when interpreted by sensors, optical scanners, voice recognition, and other technologies, can become forms of input. Some examples:

• **Biometric systems:** *Biometrics* **is the science of measuring individual body characteristics.** *Biometric security devices* identify a person through a fingerprint, voice intonation, or other biological characteristic. For example, retinal-identification devices use a ray of light to identify the distinctive network of blood vessels at the back of one's eyeball.

• **Line-of-sight systems:** *Line-of-sight systems* enable a person to use his or her eyes to "point" at the screen, a technology that allows physically handicapped users to direct a computer. For example, the Eyegaze System from LC Technologies allows you to operate a computer by focusing on particular areas of a display screen. A camera mounted on the computer analyzes the point of focus of the eye to determine where you are looking. You operate the computer by looking at icons on the screen and "press a key" by looking for a specified period.

• **Cyber gloves and body suits:** Special gloves and body suits—often used in conjunction with "virtual reality" games (described in Chapter 6)—use sensors to detect body movements. The data for these movements is sent to a computer system. Similar technology is being used for human-controlled robot hands, which are used in nuclear power plants and hazardous-waste sites.

Perhaps the ultimate input device is that using brainwaves to direct the computer. We describe that shortly.

Input Technology & Quality of Life: Health & Ergonomics

Preview & Review: The use of computers and communications technology can have important effects on our health. Some of these are repetitive strain injuries such as carpal tunnel syndrome, eyestrain and headaches, and backstrain.

Negative health effects have increased interest in the field of ergonomics, the study of the relationship of people to a work environment.

Susan Harrigan, a financial reporter for *Newsday*, a daily newspaper based on New York's Long Island, now writes her stories using a voice-activated computer. She does not do so by choice, nor is she as efficient as she used to be. She does it because she is too disabled to type at all. After 20 years of writing articles with deadline-driven flying fingers at the keyboard, she developed a crippling hand disorder. At first the pain was so severe she couldn't even hold a subway token. "Also, I couldn't open doors," she says, "so I'd have to stand in front of doors and ask someone to open them for me."[6]

Health Matters

Harrigan suffers from one of the computer-induced disorders classified as repetitive strain injuries (RSIs). The computer is supposed to make us efficient. Unfortunately, it has made some users—journalists, postal workers, data-entry clerks—anything but. The reasons are repetitive strain injuries, eyestrain and headache, and back and neck pains.[7] In this section we consider these health matters along with the effects of electromagnetic fields and noise.

- **Repetitive strain injuries:** *Repetitive strain injuries (RSIs)* **are several wrist, hand, arm, and neck injuries resulting when muscle groups are forced through fast, repetitive motions.** Most victims of RSI are in meat-packing, automobile manufacturing, poultry slaughtering, and clothing manufacturing. Musicians, too, are often troubled by RSI (because of long hours of practice).

  People who use computer keyboards—some of whom make as many as 21,600 keystrokes *an hour*—account for only 12% of RSI cases that result in lost work time. However, the number of cases is rising.[8] Keyboard users must devise their own mini-breaks to prevent excessive use of hands and wrists.

  RSIs cover a number of disorders. Some, such as muscle strain and tendinitis, are painful but usually not crippling. These injuries, often caused by hitting the keys too hard, may be cured by rest, anti-inflammatory medication, and change in typing technique. However, carpal tunnel syndrome is disabling and often requires surgery. *Carpal tunnel syndrome (CTS)* **consists of a debilitating condition caused by pressure on the median nerve in the wrist, producing damage and pain to nerves and tendons in the hands.**

  It's important to point out, however, that scientists still don't know what causes RSIs. They don't know why some people operating keyboards develop upper body and wrist pains and others don't. The working list of possible explanations for RSI includes wrist size, stress level, relationship with supervisors, job pace, posture, length of workday, exercise routine, workplace furniture, and job security.

- **Eyestrain and headaches:** Computers compel people to use their eyes at close range for a long time. However, our eyes were made to see most efficiently at a distance. It's not surprising, then, that people develop what's called computer vision syndrome.

   Computer vision syndrome (CVS) consists of eyestrain, headaches, double vision, and other problems caused by improper use of computer display screens. By "improper use," we mean not only staring at the screen for too long but also not employing the technology as it should be employed. This includes allowing faulty lighting and screen glare, and using screens with poor resolution.

- **Back and neck pains:** Many people use improper chairs or position keyboards and display screens in improper ways, leading to back and neck pains. All kinds of adjustable, special-purpose furniture and equipment is available to avoid or diminish such maladies.

- **Electromagnetic fields:** Like kitchen appliances, hairdryers, and television sets, many devices related to computers and communications generate low-level electromagnetic field emissions. **Electromagnetic fields (EMFs) are waves of electrical energy and magnetic energy.**

   In recent years, stories have appeared in the mass media reflecting concerns that high-voltage power lines, cellular phones, and CRT-type (✓ p. 111) computer monitors might be harmful. There have been worries that monitors might be linked, for example, to miscarriages and birth defects, and that cellular phones and power lines might lead to some types of cancers.

   Is there anything to this? The answer is: so far no one is sure. The evidence seems scant that weak electromagnetic fields, such as those used for cellular phones and found near high-voltage lines, cause cancer. Still, unlike car phones or older, luggable cellular phones, the handheld cellular phones do put the radio transmitter next to the user's head. This causes some health professionals concern about the effects of radio waves entering the brain as they seek out the nearest cellular transmitter. Customers are usually warned not to let the antenna touch them while they are talking.[9]

   CRT monitors made since the early 1980s produce very low emissions. Even so, users are advised to not work closer than arm's length to a CRT monitor. The strongest fields are emitted from the sides and backs of terminals. Alternatively, you can use laptop computers, because their liquid-crystal-display (LCD) screens emit negligible radiation.[10]

- **Noise:** The chatter of certain types of printers or hum of fans in computer power units can be psychologically stressful to many people. Sound-muffling covers are available for printers. Some system units may be placed on the floor under the desk to minimize noise from fans.

Ergonomics: Design with People in Mind

Previously workers had to fit themselves to the job environment. However, health and productivity issues have spurred the development of a relatively new field, called ergonomics, that is concerned with fitting the job environment to the worker.

Ergonomics is the study of the physical relationships between people and their work environment. It is concerned with designing hardware and software that is less stressful and more comfortable to use, that blends more smoothly with a person's body or actions. Examples of ergonomic hardware are tilting display screens, detachable keyboards, and keyboards hinged in the middle so that the user's wrists are presumably in a more natural position. We address some further ergonomic issues in the Experience Box at the end of Chapter 6.

Onward

When Stanley Adelman, late of New York City, died at age 72 in November 1995, he took with him some skills that some of his customers will find extremely hard to replace. Adelman was a typewriter repairer considered indispensable by many literary stars—many of whom could not manage the transition to word processors—from novelist Isaac Bashevis Singer to playwright David Mamet. He was able to fix all kinds of typewriters, including even those for languages he could not read, such as Arabic. Now his talent is no more.[11]

Adelman's demise followed by only a few months the near-death of Smith Corona Corporation, which filed for bankruptcy-court protection from creditors after losing the struggle to sell its typewriters and personal word processors in a world of microcomputers.[12] Clearly, the world is changing, and changing fast. The standard interfaces are obsolete. We look forward to interfaces that are more intuitive—computers that respond to voice commands, facial expressions, and thoughts, computers with "digital personalities" that understand what we are trying to do and offer assistance when we need it.[13,14]

R E A D M E

No Good Alternative to Typing

Q: Now that I've finally up and bought a computer, there's just one problem: I can't type! I mean, I really can't type. I tried the two-fingered method, but it just takes so long. A friend—the friend who is typing this for me now—told me there is some software I can use that will let me talk to the computer and have it automatically turn it into text. What is it?—Name withheld

A: Tell your friend to stick to typing. And you'd better learn, too. I know it sounds harsh. And yes, there is such a thing as the technology your friend is referring to—it is called speech-recognition technology—but for what you're looking for, it's not a good idea.

Basically, there are two types of speech-recognition software. The kind that lets you dictate letters and such requires a pricey, sophisticated system. It requires hours of

training to learn and lots of pauses between words when you actually do the dictating. And even then, it isn't always flawless. If you insist on going this route anyway, Dragon Systems' newest Dragon Dictate for Windows is the way to go. It costs under $700.

The other kind is cheaper and easier to set up, but it only lets you speak simple commmands, like "open window" or "print." My advice: Pay $60 for a program like Mavis Beacon Teaches Typing and hunt and peck your way to typing pleasure. Just do it. You'll find it well worth it in the end.

—Gina Smith, "No Good Alternative to Typing," *San Francisco Chronicle*

[Note: Other typing tutors include Typing Tutor, Typing Instructor, Expert Typing for Windows, UltraKey, and Mavis Teaches Typing.]

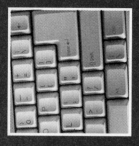

SUMMARY

What It Is / What It Does	Why It's Important
audio-input device *(p. 151, LO 6)* Device that records or plays analog sound and translates it for digital storage and processing.	Audio-input devices, such as audio boards and MIDI boards, are important for multimedia computing.
bar code *(p. 147, LO 3)* Vertical striped marks of varying widths that are imprinted on retail products and other items; when scanned by a bar-code reader, the code is converted into computer-acceptable digital input.	Bar codes may be used to input data from many items, from food products to overnight packages to railroad cars, for tracking and data manipulation.
bar-code reader *(p. 147, LO 3)* Photoelectric scanner, found in many supermarkets, that translates bar code symbols on products into digital code.	With bar-code readers and the appropriate computer system, retail clerks can total purchases and produce invoices with increased speed and accuracy; and stores can monitor inventory with greater efficiency.
biometrics *(p. 152, LO 6)* Science of measuring individual body characteristics.	Biometric systems are used in lieu of typed passwords to identify people authorized to use a computer system.
carpal tunnel syndrome (CTS) *(p. 153, LO 7)* Type of repetitive strain injury; condition caused by pressure on the median nerve in the wrist, producing damage and pain to nerves and tendons in the hands.	CTS is a debilitating, possibly disabling, condition brought about by overuse of computer keyboards; it may require surgery.
computer vision syndrome (CVS) *(p. 154, LO 7)* Computer-related disability; consists of eyestrain, headaches, double vision, and other problems caused by improper use of computer display screens.	Contributors to CVS include faulty lighting, screen glare, and screens with poor resolution.
dedicated fax machine *(p. 148, LO 3)* Specialized machine for scanning images on paper documents and sending them as electronic signals over telephone lines to receiving fax machines or fax-equipped computers; a dedicated fax machine will also received faxed documents.	Unlike fax modems installed inside computers, dedicated fax machines can scan paper documents.
digital camera *(p. 152, LO 6)* Type of electronic camera that uses a light-sensitive silicon chip to capture photographic images in digital form.	Digital cameras can produce images in digital form that can be transmitted directly to a computer's hard disk for manipulation, storage, and/or printing out. No film is used.
digitizing tablet *(p. 146, LO 2)* Tablet connected by a wire to a pen-like stylus, with which the user sketches an image, or a puck, with which the user copies an image.	A digitizing tablet can be used to achieve shading and other artistic effects or to "trace" a drawing, which can be stored in digitized form.
electromagnetic fields (EMFs) *(p. 154, LO 7)* Waves of electrical energy and magnetic energy.	Some users of cellular phones and CRT monitors have expressed concerns over possible health (cancer) effects of EMFs, but the evidence is weak.
ergonomics *(p. 154, LO 7)* Study of the physical and psychological relationships between people and their work environment.	Ergonomic principles are used in designing ways to use computers to further productivity while avoiding stress, illness, and injuries.
fax machine *(p. 148, LO 3)* Short for *facsimile transmission machine;* input device for scanning an image and sending it as electronic signals over telephone lines to a receiving fax machine, which re-creates the image on paper. Fax machines may be dedicated fax machines or fax modems.	Fax machines enable the transmission of text and graphic data over telephone lines quickly and inexpensively.

What It Is / What It Does	Why It's Important
fax modem *(p. 149, LO 3)* Modem with fax capability installed as a circuit board inside a computer; it can send and receive electronic signals via telephone lines directly to/from a computer similarly equipped or to/from a dedicated fax machine.	With a fax modem, users can send information much more quickly than they would if they had to feed it page by page through a dedicated fax machine. However, fax modems cannot scan paper documents for faxing.
imaging system *(p. 149, LO 3)* Also known as *image scanner,* or *graphics scanner;* input device that converts text, drawings, and photographs into digital form that can be stored in a computer system.	Image scanners have enabled users with desktop-publishing software to readily input images into computer systems for manipulation, storage, and output.
input hardware *(p. 140, LO 1)* Devices that take data and programs that people can read or comprehend and convert them to a form the computer can process. Devices are of two types: keyboard entry and direct entry.	Input hardware enables data to be put into computer-processable form.
joystick *(p. 145, LO 2)* Pointing device that consists of a vertical handle like a gearshift lever mounted on a base with one or two buttons; it directs a cursor or pointer on the display screen.	Joysticks are used principally in videogames and in some computer-aided design systems.
keyboard *(p. 140, LO 1)* Typewriter-like input device that converts letters, numbers, and other characters into electrical signals that the computer's processor can "read."	Keyboards are the most popular kind of input device.
light pen *(p. 146, LO 2)* Light-sensitive pen-like device connected by a wire to a computer terminal; the user brings the pen to a desired point on the display screen and presses the pen button, which identifies that screen location to the computer.	Light pens are used by engineers, graphic designers, and illustrators for making drawings.
magnetic-ink character recognition (MICR) *(p.147, LO 3)* Type of scanning technology that reads magnetized-ink characters printed at the bottom of checks and converts them to computer-acceptable digital form.	MICR technology is used by banks to sort checks.
magnetic-stripe cards *(p. 149, LO 4)* Credit-type cards with encoded magnetic stripes on their backs.	Magnetic-stripe cards are used for specific purposes, such as ATM cards and charge cards.
mouse *(p. 144, LO 2)* Direct-entry input device that is rolled about on a desktop to position a cursor or pointer on the computer's display screen, which indicates the area where data may be entered or a command executed.	For many purposes, a mouse is easier to use than a keyboard for communicating commands to a computer. With microcomputers, a mouse is needed to use most graphical user interface programs and to draw illustrations.
mouse pointer *(p. 144, LO 2)* Symbol on the display screen whose movement is directed by movement of a mouse on a flat surface, such as a table top.	The position of the mouse pointer indicates where information may be entered or a command (such as clicking, dragging, or dropping) may be executed. Also, the shape of the pointer may change, indicating a particular function that may be performed at that point.
optical card *(p. 150, LO 4)* Plastic, wallet-type card using laser technology like music compact disks, which can be used to input data.	Because they hold so much data, optical cards have considerable uses, as for a health card holding a person's medical history, including digital images such as X-rays.
optical-character recognition (OCR) *(p. 147, LO 3)* Type of scanning technology that reads special preprinted characters and converts them to computer-usable form. A common OCR scanning device is the wand reader.	OCR technology is frequently used with utility bills and price tags on department-store merchandise.
optical-mark recognition (OMR) *(p. 147, LO 3)* Type of scanning technology that reads pencil marks and converts them into computer-usable form.	OMR technology is frequently used for grading multiple-choice and true/false tests, such as parts of the College Board Scholastic Aptitude Test.

What It Is / What It Does	Why It's Important
pen computer *(p. 146, LO 2)* Input system that uses a pen-like stylus to enter handwriting and marks into a computer. The four types of systems are gesture recognition, handwriting stored as scribbling, personal handwriting stored as typed text with training, and standard handwriting "typeface" stored as typed text without training.	Pen-based computer systems benefit people who don't know how to or who don't want to type or need to make routinized kinds of inputs such as checkmarks.
point-of-sale (POS) terminal *(p. 143, LO 1)* Smart terminal used much like a cash register.	POS terminals record customer transactions at the point of sale but also store data for billing and inventory purposes.
puck *(p. 146, LO 2)* Copying device with which the user of a digitizing tablet may copy an image.	With a puck, users may "trace" (copy) a drawing and store it in digitized form.
repetitive strain injuries (RSI) *(p. 153, LO 7)* Several wrist, hand, arm, and neck injuries resulting when muscle groups are forced through fast, repetitive motions.	Computer users may suffer RSIs such as muscle strain and tendinitis, which are not disabling, or carpal tunnel syndrome, which is.
scanning devices *(p. 147, LO 3)* Input devices that translate images such as text, drawings, and photos into digital form.	Scanning devices—bar-code readers, fax machines, imaging systems—simplify the input of complex data.
sensor *(p. 152, LO 6)* Type of input device that collects specific kinds of data directly from the environment and transmits it to a computer.	Sensors can be used for detecting speed, movement, weight, pressure, temperature, humidity, wind, current, fog, gas, smoke, light, shapes, images, and so on.
smart card *(p. 149, LO 4)* Wallet-type card containing a microprocessor and memory chip that can be used to input data.	Telephone users may buy a smart card that lets them make telephone calls until the total cost limit programmed into the card has been reached.
source data entry device *(p. 140, LO 1)* Also called *source-data automation;* non-keyboard data-entry device. The category includes pointing devices; scanning devices; magnetic-stripe, smart, and optical cards; voice-recognition devices; audio-input devices; video-input devices; electronic cameras; sensors; and human-biology input devices.	Source data entry devices lessen reliance on keyboards for data entry and can make data entry more accurate. Some also enable users to draw graphics on screen and create other effects not possible with a keyboard.
stylus *(p. 146, LO 2)* Pen-like device with which the user of a digitizing tablet "sketches" an image.	With a stylus, users can achieve artistic effects similar to those achieved with pen or pencil.
terminal *(p. 143, LO 1)* Input device that consists of a keyboard, a video display screen, and a communications line to a main computer system.	A terminal is generally used to input data to, and receive visual data from, a mainframe computer system.
touch screen *(p. 145, LO 2)* Video display screen that has been sensitized to receive input from the touch of a finger. It is often used in automated teller machines and in directories conveying tourist information.	Because touch screens are easy to use, they can convey information quickly and can be used by people with no computer training; however, the amount of information offered is usually limited.
trackball *(p. 144, LO 2)* Movable ball, on top of a stationary device, that is rotated with the fingers or palm of the hand; it directs a cursor or pointer on the computer's display screen, which indicates the area where data may be entered or a command executed.	Unlike a mouse, a trackball is especially suited to portable computers, which are often used in confined places.
voice-recognition system *(p. 150, LO 5)* Input system that converts a person's speech into digital code; the system compares the electrical patterns produced by the speaker's voice with a set of prerecorded patterns stored in the computer.	Voice-recognition technology is useful for inputting data in situations in which people are unable to use their hands or need their hands free for other purposes.

(Selected answers appear at the back of the book.)

Short-Answer Questions

1. What are the two main categories of input hardware?

2. What is *ergonomics*?

3. What is the difference between dumb terminals and smart terminals?

4. What is the main difference between a mouse and a trackball?

5. What is the main disadvantage of using a fax modem instead of a dedicated fax machine?

6. What are the three main limitations of voice-recognition technology?

7. What is a point-of-sale terminal used for?

8. What is a digitizer and how is it used?

9. What determines what the function keys on a keyboard do?

Fill-in-the-Blank Questions

1. _____ describes the process in which data created while an event is taking place is entered directly into the computer system in a machine-processable form.

2. A(n) _____ converts a person's speech into digital code that your computer can understand.

3. A(n) _____ can collect information directly from the environment and transmit it to a computer.

4. A(n) _____ scans an image and sends it as electronic signals over telephone lines to a receiving machine, which prints the transmission on paper.

5. A(n) _____ captures pictures in digital form for immediate viewing on a television or computer screen.

6. A(n) _____ is rolled about on the desktop to direct a pointer on the computer's display screen.

7. A(n) _____ terminal can be used only to input and receive data; it cannot do any independent processing.

8. List three types of computer-induced health disorders.
   a. neck aches
   b. wrist
   c. eye strain/headaches

Multiple-Choice Questions

1. Which of the following doesn't use keyboard entry?
   a. optical cards
   b. terminals
   c. Touch-Tone devices
   d. microcomputers
   e. set-top boxes

2. Which of the following is often used to verify credit cards?
   a. optical card
   b. pen-based system
   c. Touch-Tone device
   d. microcomputer
   e. set-top box

3. Which of the following uses a stylus?
   a. light pen
   b. digitizing tablet
   c. touch screen
   d. joystick
   e. none of the above

4. Which of the following do banks use to sort checks?
   a. bar-code reader
   b. magnetic-ink character recognition
   c. fax machine
   d. imaging system
   e. all of the above

5. Which of the following is a type of human-biology input device?
   a. biometric system
   b. bar code
   c. magnetic stripe
   d. smart card
   e. all of the above

6. Which of the following would you use to enter handwriting and marks directly onto the display screen?
   a. electronic book
   b. digitizing tablet
   c. pen-based system
   d. joystick
   e. none of the above

7. Which of the following enables a person to use his or her eyes to "point" at the screen?
   a. biometric system
   b. line-of-sight system
   c. cyber gloves
   d. joystick
   e. none of the above

8. Which of the following can convert music into a computer-usable form?
   a. audio-input device
   b. musical sensor
   c. MTV-input device
   d. musical camera
   e. none of the above

True/False Questions

T F 1. Cursor-movement keys are used to execute commands.

T F 2. Many commands that you execute with a mouse can also be performed with a keyboard.

T F 3. A smart card contains a microprocessor and a memory chip.

T F 4. Function keys are used the same way with every software application.

T F 5. Digital cameras can capture pictures for direct storage on diskette or CD.

T F 6. Electronic imaging has become an important part of multimedia.

T F 7. A dedicated fax machine does nothing but send and receive faxes.

Projects/Critical-Thinking Questions

1. During the next week, make a list of all the input devices you notice—in stores, on campus, at the bank, in the library, on the job, at the doctor's, and so on. At the end of the week, report on the devices you have listed and name some of the advantages you think are associated with each device.

2. Research image and graphics scanning technology. What differentiates one scanner from another? Is it the clarity of the scanned image? Price? Software? If you were going to buy a scanner, which do you think you would buy? Why? What do you need to run a color scanner?

3. Given the many ways that data can be input to a computer system and that new input technologies will surely become available, do you think that keyboards might become obsolete one day? Why/why not?

4. Research how the legal system is dealing with ergonomic issues in the workplace. Are any ergonomic laws or standards in place in either the private or public sector? Do you think reform is necessary?

5. Identify an input task that you think would greatly benefit from an improved input technology. How is the task currently performed? By whom? In what industry? Is the input technology you describe currently available in some form?

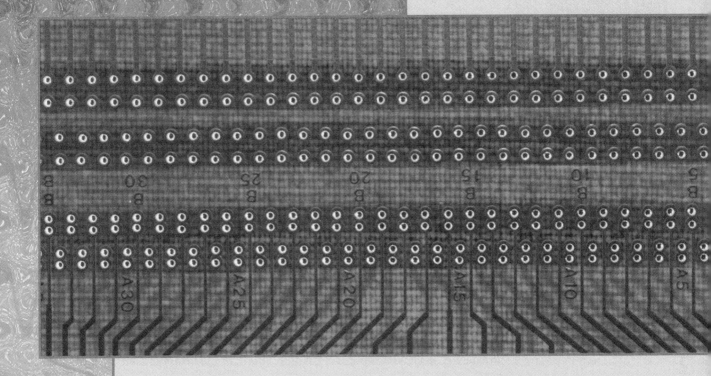

The Many Uses of Computers & Communications

Concepts You Should Know

After reading this chapter, you should be able to:

1. Distinguish between softcopy and hardcopy output.
2. Describe the different types of display screens.
3. List and explain three types of video display adapters.
4. Explain the operations of the different types of printers and plotters.
5. Briefly describe other forms of output: audio, video, virtual reality, and robots.

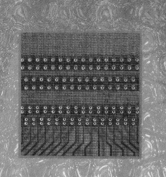

How long until speech robots call you during the dinner hour and use charming voices to get you to part with a charitable donation through your credit card?

Perhaps sooner than you think. Already University of Iowa scientists have created a computer program and audio output that simulates 90% of the acoustic properties made by one man's voice. "What we have now are extremely human-like speech sounds," says Brad H. Story, the lead researcher and also the volunteer whose vocal tract was analyzed and simulated. "We can produce vowels and consonants in isolation." A harder task, still to be accomplished, is to re-create the transitions between key linguistic sounds—what researchers call "running speech."

In the meantime, Mr. Story and a colleague have performed around the country with an electronic device they call "Pavarobotti." The operatic "singing" (such as Puccini arias) is not borrowed from recordings of humans but is created electronically. Singing, however, is easier to simulate than speech. Even singing that would bring an audience to its feet, such as holding a high pitch for a long time, is easy to perform with microchips and speakers.[1]

Output Hardware: Softcopy Versus Hardcopy

Preview & Review: Output devices translate information processed by the computer into a form that humans can understand.

The two principal kinds of output are hardcopy and softcopy.

Output devices include display screens; printers and plotters; audio-output devices; video-output devices; virtual reality; and robots.

If the foregoing example is any guide, the closing days of the 20th century represent exciting times in the development of computer output. In the following pages we describe present and future types of output devices, ranging from printers to high-definition television.

The principal kinds of output are *hardcopy* and *softcopy*. (■ *See Panel 6.1.*)

- **Hardcopy:** *Hardcopy* **refers to printed output.** The principal examples are printouts, whether text or graphics, from printers. Film, including microfilm and microfiche, is also considered hardcopy output.
- **Softcopy:** *Softcopy* **refers to data that is shown on a display screen or is in audio or voice form.** This kind of output is not tangible; it cannot be touched. Virtual reality and robots might also be considered softcopy devices.

There are several types of output devices. We will discuss the following ones.

- Display screens
- Printers and plotters
- Audio-output devices
- Video-output devices
- Virtual-reality devices
- Robots

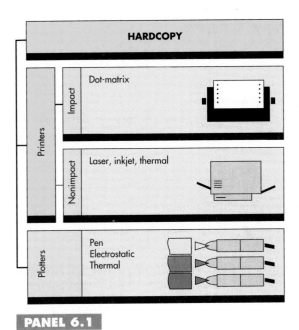

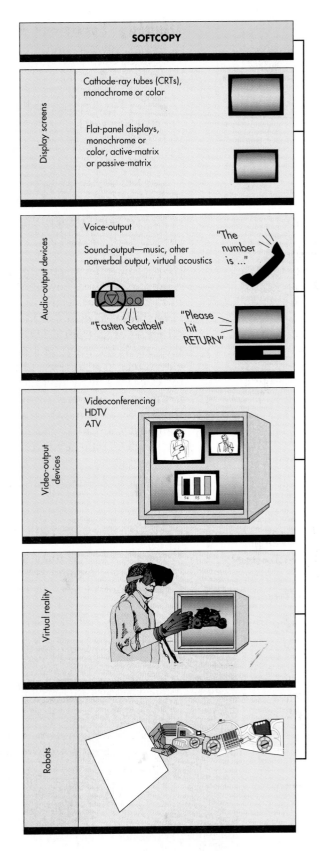

PANEL 6.1

Summary of output devices

Display Screens: Softcopy Output

Preview & Review: Display screens are either CRT (cathode-ray tube) or flat-panel display.

CRTs use a vacuum tube like that in a TV set.

Flat-panel displays are thinner, weigh less, and consume less power than CRTs but are not as clear. Principal flat-panel displays are liquid-crystal display (LCD) and gas-plasma display.

Users must decide about screen clarity, monochrome versus color, and text versus graphics (character-mapped versus bitmapped).

Various video display adapters (such as VGA, SVGA, and XGA) allow various kinds of resolution and colors.

Display screens are among the principal windows of information technology. As more and more refinements are made, we can expect to see them adapted to many innovative uses.

*Display screens—***also variously called *monitors, CRTs,* or simply *screens*—are output devices that show programming instructions and data as they are being input and information after it is processed.** Sometimes a display screen is also referred to as a *VDT,* for *video display terminal,* although technically a VDT includes both screen and keyboard. The size of a screen is measured diagonally from corner to corner in inches, just like television screens. For terminals on large computer systems and for desktop microcomputers, 15- to 17-inch screens are a common size. Portable computers of the notebook and subnotebook size may have screens ranging from 7.4 inches to 10.4 inches.

Display screens are of two types: *cathode-ray tubes* and *flat-panel displays.*

PANEL 6.2

CRT and flat-panel monitors

(Left) CRT monitor. *(Right)* LCD display.

Cathode-Ray Tubes (CRTs)

The most common form of display screen is the CRT. **A *CRT,* for *cathode-ray tube,* is a vacuum tube used as a display screen in a computer or video display terminal.** (■ *See Panel 6.2.*) This same kind of technology is found

not only in the screens of desktop computers but also in television sets and flight-information monitors in airports.

Images are represented on the screen (whether CRT or flat-panel display) by individual dots or "picture elements" called *pixels*. **A *pixel* is the smallest unit on the screen that can be turned on and off by the computer or made different shades.**

Flat-Panel Displays

If CRTs were the only existing technology for computer screens, we would still be carrying around 25-pound "luggables" instead of lightweight notebooks, subnotebooks, and pocket PCs. CRTs provide bright, clear images, but they add weight and consume space and power.

Compared to CRTs, ***flat-panel displays* are much thinner, weigh less, and consume less power. Thus, they are better for portable computers.** Flat-panel displays are made up of two plates of glass with a substance in between them, which is activated in different ways. One common type of flat-panel display is LCD. (■ *See Panel 6.2.*)

***Liquid-crystal display (LCD)* consists of a substance called *liquid crystal*, the molecules of which line up in a way that alters their optical properties. As a result, light—usually backlighting behind the screen—is blocked or allowed through to create an image.**

Flat-panel screens are either active-matrix or passive-matrix displays.

In an *active-matrix display*, each pixel on the screen is controlled by its own transistor. Active-matrix screens are much brighter and sharper than passive-matrix screens, but they are more complicated and thus more expensive.

In a *passive-matrix display*, a transistor controls a whole row or column of pixels. The advantage is that passive-matrix displays are less expensive and use less power than active-matrix screens.

Screen Clarity

Whether CRT or flat-panel display, screen clarity depends on three qualities: *resolution, dot pitch,* and *refresh rate.*

- **Resolution:** **The clarity or sharpness of a display screen is called its *resolution*; the more pixels there are per square inch, the better the resolution.** Resolution is expressed in terms of the formula *horizontal pixels × vertical pixels.* Each pixel can be assigned a color or a particular shade of gray. A screen with 640 × 480 pixels multiplied together equals 307,200 pixels. This screen will be less clear and sharp than a screen with 800 × 600 (equals 480,000) or 1024 × 768 (equals 786,432) pixels.

- **Dot pitch:** ***Dot pitch* is the amount of space between the centers of adjacent pixels; the closer the dots, the crisper the image.** For crisp images, dot pitches should be less than .31 millimeter.

- **Refresh rate:** *Refresh rate* is the number of times per second that the pixels are recharged so that their glow remains bright. In dual-scan screens, the tops and bottoms of the screens are refreshed independently at twice the rate of single-scan screens, producing more clarity and richer colors. In general, displays are refreshed 45 to 100 times per second.

Monochrome Versus Color Screens

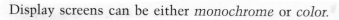

Display screens can be either *monochrome* or *color.*

- **Monochrome:** **Monochrome display screens display only two colors**—usually black and white, amber and black, or green and black.

- **Color:** **Color display screens can display between 16 and 16.7 million colors,** depending on their type. Most software today is developed for color, and—except for some pocket PCs—most microcomputers today are sold with color display screens.

Text Versus Graphics: Character-Mapped Versus Bitmapped Display

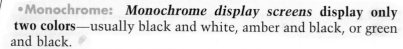

Another distinction in display screens relates to their capacity to display graphics. A screen lacking this capacity is referred to as *character-mapped;* a *bitmapped* screen can display graphics.

- **Character-mapped:** **Character-mapped display screens display only text**—letters, numbers, and special characters. They cannot display graphics unless a video display adapter card is installed, as explained next.

- **Bitmapped:** **Bitmapped display screens permit the computer to manipulate pixels on the screen individually rather than as blocks,** enabling software to create a greater variety of images. Today most screens can display text and graphics—icons, charts, graphs, and drawings.

Video Display Adapters

To display graphics, a display screen must have a video display adapter. **A *video display adapter,* also called a *graphics adapter card,* is a circuit board that determines the resolution, number of colors, and how fast images appear on the display screen.** Video display adapters come with their own memory chips, which determine how fast the card processes images and how many colors it can display. A video display adapter with 256 kilobytes of memory will provide 16 colors; one with 1 megabyte will support 16.7 million colors.

The video display adapter is often built into the motherboard (✓ p. 118), although it may also be an expansion card that plugs into an expansion slot (✓ p. 122). Video display adapters embody certain standards. Today's microcomputer monitors commonly use VGA and SVGA standards.

- **VGA:** Perhaps the most common video standard today, **VGA, for *Video Graphics Array,* will support 16 to 256 colors, depending on resolution.** At 320 × 200 pixels it will support 256 colors; at the sharper resolution of 640 × 480 pixels it will support 16 colors.

- **SVGA:** **SVGA, for *Super Video Graphics Array,* will support 256 colors at higher resolution than VGA.** SVGA has two graphics modes: 800 × 600 pixels and 1024 × 768 pixels.

- **XGA:** Also referred to as *high resolution,* **XGA, for *Extended Graphics Array,* supports up to 16.7 million colors at a resolution of 1024 × 768 pixels.** Depending on the video display adapter memory chip, XGA will support 256, 65,536, or 16,777,216 colors.

For any of these displays to work, video display adapters and monitors must be compatible. Your computer's software and the video display adapter must also be compatible. Thus, if you are changing your monitor or your video display adapter, be sure the new one will still work with the old.

Printers & Plotters: Hardcopy Output

Preview & Review: Printers and plotters produce printed text or images on paper.

Printers may be desktop or portable, impact or nonimpact. The most common impact printers are dot-matrix printers. Nonimpact printers include laser, inkjet, and thermal printers.

Plotters are pen, electrostatic, and thermal.

A *printer* **is an output device that prints characters, symbols, and perhaps graphics on paper.** Printers are categorized according to whether or not the image produced is formed by physical contact of the print mechanism with the paper. *Impact printers* do have contact; *nonimpact printers* do not. Some printers are portable. Some are even combined with a PC. (■ *See Panel 6.3.*)

Impact Printers

An impact printer has mechanisms resembling those of a typewriter. That is, **an** *impact printer* **forms characters or images by striking a mechanism such as a print hammer or wheel against an inked ribbon, leaving an image on paper.**

For microcomputer users, the most common type of impact printer is a dot-matrix printer. **A** *dot-matrix printer* **contains a print head of small pins, which strike an inked ribbon against paper, forming characters or images.** Print heads are available with 9, 18, or 24 pins, with the 24-pin head offering the best quality. Dot-matrix printers can print *draft quality,* a coarser-looking 72 dots per inch vertically, or *near-letter-quality (NLQ),* a crisper-looking 144 dots per inch vertically. (■ *See Panel 6.4.*)

Dot-matrix printers print 150–300 characters per second and print graphics as well as text. However, they are noisy.

Nonimpact Printers

Nonimpact printers are faster and quieter than impact printers because they have fewer moving parts. *Nonimpact printers* **form characters and images without making direct physical contact between printing mechanism and paper.**

Two types of nonimpact printers often used with microcomputers are *laser printers* and *inkjet printers.* A third kind, the *thermal printer,* is seen less frequently, except with ultraportable printers.

- **Laser printer:** Similar to a photocopying machine, **a** *laser printer* **uses the principle of dot-matrix printers in creating dot-like images. However, these images are created on a drum, treated with a magnetically charged ink-like toner (powder), and then transferred from drum to paper.** (■ *See Panel 6.5, next page.*)

  There are good reasons why laser printers are the most common type of nonimpact printer. They produce sharp, crisp images of both text and graphics, and they are quieter and faster than dot-matrix printers. Whereas most impact printers are serial printers, printing one character at a time,

PANEL 6.3

PC with printer inside

The Canon NoteJet is a notebook computer that combines computer and printer. It weighs 8.6 pounds, is 2.5 inches thick, and has a 9.5-inch screen. Paper is fed into the computer from a slot in front that can hold up to 10 pages at a time; printed documents exit from the rear.

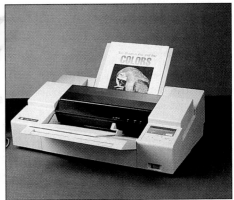

```
This is a sample of
draft quality.
```
**This is a sample of
near-letter-quality.**

PANEL 6.4

Dot-matrix printer

Under the printer are examples of draft-quality and near-letter-quality printing. The near-letter-quality image is printed twice. On the second pass the print head is positioned slightly to the right of the original image.

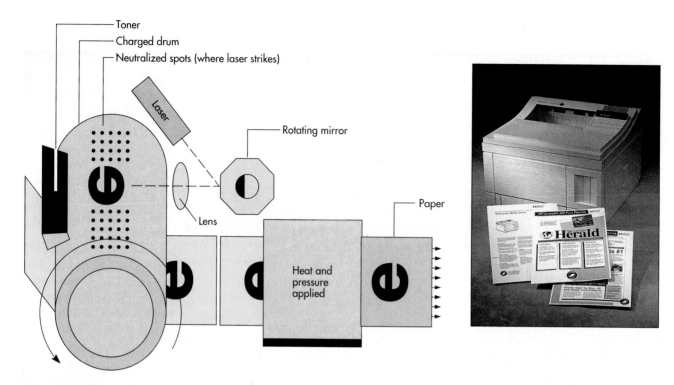

Toner
Charged drum
Neutralized spots (where laser strikes)
Laser
Rotating mirror
Lens
Paper
Heat and pressure applied

PANEL 6.5

Laser printer

A small laser beam is bounced off a mirror millions of times per second onto a positively charged drum. The spots where the laser beam hits become neutralized, enabling a special toner (powder) to stick to them and then print out on paper. The drum is then recharged for the next cycle.

laser printers print whole pages at a time. They can print 8–20 pages per minute for individual microcomputers (and up to 200 pages per minute for mainframes). They can print in different *fonts—that is, type styles and sizes.* The more expensive models can print in different colors.

One particular group of laser printers has become known as PostScript printers. *PostScript* **is a printer language, or page description language, that has become a standard for printing graphics with laser printers. A *page description language* (software) describes the shape and position of letters and graphics to the printer.** PostScript printers are essential if you are printing a lot of graphics or want to generate fonts in various sizes. Another page description language used with laser printers is *Printer Control Language (PCL),* which has resolutions and speeds similar to those of PostScript.

● **Inkjet printer:** Like laser and dot-matrix printers, inkjet printers also form images with little dots. *Inkjet printers* **spray small, electrically charged droplets of ink from four nozzles through holes in a matrix at high speed onto paper.**

Most color printing is done on inkjets because the nozzles can hold four different colors. Moreover, inkjet printers can match the speed of dot-matrix printers. They are even quieter than laser printers and produce an equally high-quality image.

● **Thermal printer:** For people who want the highest-quality color printing available with a desktop printer, thermal printers are the answer. However, they are expensive, and they require expensive paper. Thus, they are not generally used for jobs requiring a high volume of output.

Thermal printers **use colored waxes and heat to produce images by burning dots onto special paper.** The colored wax sheets are not required for black-and-white output.

Printers are compared on the opposite page. (■ *See Panel 6.6.)*

PANEL 6.6 **Printer comparisons**

Type	Technology	Advantages	Disadvantages	Typical Speed	Approximate Cost
Dot-matrix	Print head with small pins strikes an inked ribbon against paper	Inexpensive; produces draft quality and near-letter-quality; can output some graphics; can print multipart forms; low cost per page	Noisy; cannot produce high-quality output of text and graphics; limited fonts	30 to 500+ cps*	$100–$2000
Laser	Laser beam directed onto a drum, "etching" spots that attract toner, which is then transferred to paper	Quiet; excellent quality; output of text and graphics; very high speed	High cost, especially for color	8–200 ppm*	$400–$20,000
Inkjet	Electrostatically charged drops hit paper	Quiet; prints color, text, and graphics; less expensive; fast	Relatively slow; clogged jets; fewer dots per inch	35–400+ cps	$200–$2000
Thermal	Temperature-sensitive paper changes color when treated; characters are formed by selectively heating print head	Quiet; high-quality color output of text and graphics; can also produce transparencies	Special paper required; expensive; slow	11–80 cps	$2000–$22,000

*cps = characters per second; ppm = pages per minute.

Black & White Versus Color Printers

Today prices have plummeted to $400 or less for laser and inkjet printers, so that the cheap, noisy dot-matrix printer may well be going the way of the black-and-white TV set. (Impact printers are still needed, however, for any activity that involves printing multipart forms.) "The choice between a laser printer and an inkjet comes down to how much you print and whether you need color," says one analysis of microcomputer printers.[2]

Lasers, which print a page at a time, can handle thousands of black-and-white pages a month. Moreover, compared to inkjets, laser printers are faster and crisper (though not by much) at printing black-and-white and a cent or two cheaper per page.[3] Finally, a freshly printed page from a laser won't smear, as one from an inkjet might.

However, inkjets, which spray ink onto the page a line at a time, can give you both high-quality black-and-white text and high-quality color graphics. Moreover, color inkjets start at about $400 (black-and-white inkjets at about $200), whereas color lasers cost thousands. Thus, if you think you might have occasion to print multicolor charts for reports or presentations or do colorful children's school projects, a color inkjet would seem to be most desirable.

Plotters

A *plotter* is a specialized output device designed to produce high-quality graphics in a variety of colors. Plotters are especially useful for creating maps and architectural drawings, although they may also produce less complicated charts and graphs.

The three principal kinds of plotters are *pen, electrostatic,* and *thermal.* (■ *See Panel 6.7.*)

The Paper Glut: Whither the "Paperless Office"?

In producing hardcopy, printers and plotters produce *paper*—great quantities of it. In the past decade the United States' annual rate of paper consumption has nearly doubled. Supposedly, computers were going to make the use of paper obsolete, providing us with "the paperless office," but in fact the opposite has happened. As one writer has observed:

> What computers have done, in essence, is bestow independence on thousands of office workers who once needed a secretary to type and produce documents. With their own PCs, most of them can draft memos and reports. They can enlarge charts and make color reproductions on ultra-equipped copier machines. They can use the fax machine to disperse copies or program it to shoot batches of paper all over the country.[4]

Why don't we simply leave information in electronic form, as on a 3½-inch diskette, holding the equivalent of 240 sheets of paper? "People like to hold paper; they like printouts," says Stanford University professor Clifford Nass, who studies social patterns and technology. "If the information exists only in the computer, where is it? You can't tell people it's being stored as a bunch of ones and zeros. They want to touch it."[5]

We make printouts because we are afraid "anything can happen to a computer." Moreover, we get physical comfort from handling paper, from marking it up with colored pens, and from filing it away. As a result, however, there has been a paper explosion such as the world has never seen.

The Theater of Output: Audio, Video, Virtual Reality, & Robots

Preview & Review: Other output hardware includes audio-output devices, video-output devices, virtual-reality devices, and robots.

Audio output includes voice-output technology and sound-output technology.

Video output includes videoconferencing and digitized television.

Virtual reality is a kind of computer-generated artificial reality.

Robots perform functions ordinarily ascribed to human beings.

PANEL 6.7

Plotters

(Left) Pen plotter. *(Right)* Electrostatic plotter.

Many of the other forms of output technology that remain to be discussed relate—although by no means exclusively—to entertainment and recreation. They include:

- Audio-output devices
- Video-output devices
- Virtual-reality devices
- Robots

Audio-Output Devices

Audio-output devices **include those devices that output voice or voice-like sounds and those that output music and other sounds.**

- **Voice-output devices:** *Voice-output devices* **convert digital data into speech-like sounds.** By now these devices are no longer very unusual. You hear such forms of voice output on telephones ("Please hang up and dial your call again"), in soft-drink machines, in cars, in toys and games, and recently in vehicle-navigation devices.

- **Sound-output devices:** *Sound-output devices* **produce digitized sounds, ranging from beeps and chirps to music.** All these sounds are nonverbal. For example, PC owners can customize their machines to greet each new program with the sound of breaking glass or to moo like a cow every hour. To exercise these possibilities, you need both the necessary software and the sound card, or digital audio circuit board (such as the popular Sound Blaster and Sound Blaster Pro cards). The sound card plugs into an expansion slot in your computer. A sound card is also required in making computerized music.

The various kinds of audio outputs and their devices are important components of multimedia systems (✓ p. 27).

Video-Output Devices

Innovations in video technology are being seen in a number of areas, including the following:

- **Videoconferencing and video editing:** Want to have a meeting with someone across the country and go over some documents—without having to go there? The answer is videoconferencing. *Videoconferencing* **is a method whereby people in different geographical locations can have a meeting—and see and hear one another—using computers and communications.**

Videoconferencing was, and still is, a service offered by long-distance telephone companies to large businesses. Now, however, you can do it yourself. Say you're on the West Coast and want to go over a draft of a client proposal with your boss on the East Coast. The first thing you need for such a meeting is a high-capacity telephone line. To this you link your IBM-compatible PC running Windows to which you have added a hardware/software package called ProShare Video System from Intel. ProShare consists of a small video camera that sits atop your display monitor, a circuit board, and software that turns your microcomputer into a personal conferencing system.

Your boss's image appears in one window on your computer's display screen, and the document you're working on together is in another window. (An optional window shows your own image.) Although the display screen images are choppier than those on a standard TV set, they're clear enough to enable both of you to observe facial expressions and most body language. The software includes drawing tools and text tools for adding comments. Thus, you can go through and edit paragraphs and draw crude sketches on the proposal draft.[6]

- **HDTV or ATV?** *High-definition television (HDTV)* is a television system that features enhanced video and crisp, clear pictures. These pictures are far superior to any seen on television today, appearing within a frame that is twice as wide. They also have twice the present number of scan lines (1100 versus 525 for an average TV screen), thus offering twice the resolution, or sharpness. In addition, HDTV offers better color and better audio, with sound being of compact-disk quality.

  In the United States, a competition was organized to invent an American HDTV standard. The competition was won by a lab belonging to General Instrument, which abandoned the older *analog* (✓ p. 6) transmission standard (that using radio waves analogous to sound and light waves). Instead, it proposed to create better pictures using *digital* transmission, the 1s and 0s of computer code.

  To exploit the new technology, a consortium of seven companies and labs (including AT&T, the Massachusetts Institute of Technology, and Philips) known as "the Grand Alliance" was formed. But when the alliance unveiled its new technology in December 1994 it was discovered it could do more than show "stunningly realistic images of mountain vistas, springtime flowers, and sun-sparkled lakes," as one article described it.[7] It could also be used to transmit huge amounts of *digital data*—computer files, e-mail, paging messages, and similar services.

  Because the new system surpassed all expectations, the Federal Communications Commission also gave it a new name—advanced television. *Advanced television (ATV)* is television that, in place of one analog channel, can transmit not just one high-definition image but five or six digital channels as good as current ones. These channels could be used for non-TV-related services, such as paging signals and e-mail.

Virtual-Reality & Simulation Devices

Virtual reality (VR) **is a kind of computer-generated artificial reality that projects a person into a sensation of three-dimensional space.** To achieve this effect, you need the following interactive sensory equipment:

- **Headgear:** The headgear—which is called *head-mounted display (HMD)*—has two small video display screens, one for each eye, that create the sense of three-dimensionality. Headphones pipe in stereophonic sound or even "3-D" sound. Three-dimensional sound makes you think you are hearing sounds not only near each ear but also in various places all around you.

- **Glove:** The glove has sensors that collect data about your hand movements.

- **Software:** Software gives the wearer of this special headgear and glove the interactive sensory experience that feels like an alternative to the realities of the physical world.

You may have seen virtual reality used in arcade-type games, such as Atlantis, a computer simulation of The Lost Continent. You may even have tried to tee off on a virtual golf range. There are also a few virtual-reality home videogames, such as the 7th Sense. However, there are far more important uses, one of them being in simulators for training.

Simulators are devices that represent the behavior of physical or abstract systems. Virtual-reality simulation technologies are applied a great deal in training. For instance, they have been used to create lifelike bus control panels and various scenarios such as icy road conditions to train bus drivers.[8] They are used to train pilots on various aircraft and to prepare air-traffic controllers for equipment failures.[9] They also help children who prefer hands-on learning to explore subjects such as chemistry.[10]

Of particular value are the uses of virtual reality in health and medicine. For instance, surgeons-in-training can rehearse their craft through simulation on "digital patients."[11] Virtual-reality therapy has been used for autistic children and in the treatment of phobias, such as extreme fear of public speaking or of being in public places or high places.[12,13] It has also been used to rally the spirits of quadriplegics and paraplegics by engaging them in plays and song-and-dance routines. (■ *See Panel 6.8.*)

Robots

The first Robot Olympics was held in Toronto in November 1991. "Robots competed for honors in 15 events—jumping, rolling, fighting, climbing, walking, racing against each other, and solving problems," reported writer John Malyon.[14] For instance, in the Micromouse race, robots had to negotiate a standardized maze in the shortest possible time.

To get to definitions, **a robot is an automatic device that performs functions ordinarily ascribed to human beings or that operates with what appears to be almost human intelligence.** Forty years ago, in *Forbidden Planet*, Robby the Robot could sew, distill bourbon, and speak 187 languages.[15] We haven't caught up with science-fiction movies, but we may get there yet. Scrub-Mate—a robot equipped with computerized controls, ultrasonic "eyes," sensors, batteries, three different cleaning and scrubbing tools, and a self-squeezing mop—can clean bathrooms.[16] Rosie the HelpMate delivers special-order meals from the kitchen to nursing stations in hospitals.[17] Robodoc—notice how all these robots have names—is used in surgery to bore the thighbone

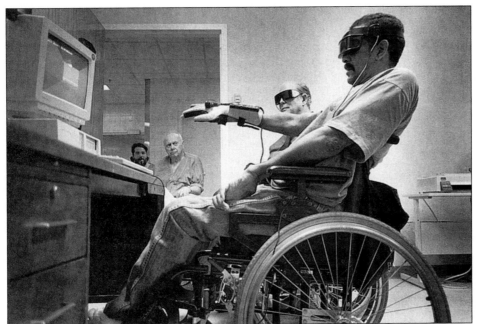

PANEL 6.8

Virtual-reality therapy

Miguel Rivera uses a glove and goggles with virtual-reality computer in the spinal-injury ward of the Veterans Affairs Medical Center, the Bronx, New York. The technology allows the disabled to "move" freely in an electronic world.

so that a hip implant can be attached.[18] Remote Mobile Investigator 9 is used by Maryland police to flush out barricaded gunmen and negotiate with terrorists.[19] A driverless harvester, guided by satellite signals and artificial vision system, is used to harvest alfalfa and other crops.[20]

E Creative Output: The Manipulation of Truth in Art & Journalism

Preview & Review: Users of information technology must weigh the effects of the digital manipulation of sound, photos, and video in art and journalism.

The ability to manipulate digitized output—images and sounds—has brought a wonderful new tool to art. However, it has created some big new problems in the area of credibility, especially for journalism. How can we now know that what we're seeing or hearing is the truth? Consider the following.

Manipulation of Sound

Frank Sinatra's 1994 album *Duets* paired him through technological tricks with singers like Barbra Streisand, Liza Minnelli, and Bono of U2. Sinatra recorded solos in a recording studio. His singing partners, while listening to his taped performance on earphones, dubbed in their own voices. This was done not only at different times but often, through distortion-free phone lines, from different places. The illusion in the final recording is that the two singers are standing shoulder to shoulder.

Newspaper columnist William Safire loves the way "digitally remastered" recordings recapture great singing he enjoyed in the past. However, he called *Duets* "a series of artistic frauds." Said Safire, "The question raised is this: When a performer's voice and image can not only be edited, echoed, refined, spliced, corrected, and enhanced—but can be transported and combined with others not physically present—what is performance? . . . Enough of additives, plasticity, virtual venality; give me organic entertainment."[21] Another critic said that to call the disk *Duets* seemed a misnomer. "Sonic collage would be a more truthful description."[22]

Some listeners feel that the technology changes the character of a performance for the better—that the sour notes and clinkers can be edited out. Others, however, think the practice of assembling bits and pieces in a studio drains the music of its essential flow and unity.

Whatever the problems of misrepresentation in art, however, they pale beside those in journalism. Could not a radio station, for instance, edit a stream of digitized sound to achieve an entirely different effect from what actually happened?

Manipulation of Photos

When O. J. Simpson was arrested in 1994 on suspicion of murder, the two principal American newsmagazines both ran pictures of him on their covers.[23,24] *Newsweek* ran the mug shot unmodified, as taken by the Los Angeles Police Department. *Time*, however, had the shot redone with special effects as a "photo-illustration" by an artist working with a computer. Simpson's image was darkened so that it still looked like a photo but, some critics said, with a more sinister cast to it.

Should a magazine that reports the news be taking such artistic license? Should *National Geographic* in 1982 have moved two Egyptian pyramids closer together so that they would fit on a vertical cover? Was it even right for *TV Guide* in 1989 to run a cover showing Oprah Winfrey's head placed on Ann-Margret's body?[25] In another case, to show what can be done, a photographer digitally manipulated the famous 1945 photo showing the meeting of the leaders of the wartime Allied powers at Yalta. Joining Stalin, Churchill, and Roosevelt are some startling newcomers: Sylvester Stallone and Groucho Marx. The additions are done so seamlessly it is impossible to tell the photo has been altered. (■ *See Panel 6.9.*)

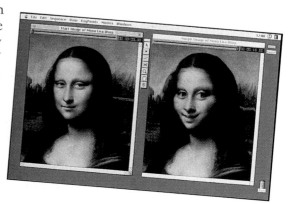

The potential for abuse is clear. "For 150 years, the photographic image has been viewed as more persuasive than written accounts as a form of 'evidence,'" says one writer. "Now this authenticity is breaking down under the assault of technology."[26] Asks a former photo editor of the *New York Times Magazine*, "What would happen if the photograph appeared to be a straightforward recording of physical reality, but could no longer be relied upon to depict actual people and events?"[27]

Many editors try to distinguish between photos used for commercialism (advertising) versus for journalism, or for feature stories versus for news stories. However, this distinction implies that the integrity of photos applies only to some narrow definition of news. In the end, it can be argued, altered photographs pollute the credibility of all of journalism.

PANEL 6.9

Photo manipulation

In this 1945 photo, World War II Allied leaders Joseph Stalin, Winston Churchill, and Franklin D. Roosevelt are shown from left to right. Digital manipulation has added Sylvester Stallone standing behind the bench and Groucho Marx seated at right.

Manipulation of Video

The technique of morphing, used in still photos, takes a quantum jump when used in movies, videos, and television commercials. In *morphing*, a film or video image is displayed on a computer screen and altered pixel by pixel, or dot by dot. The result is that the image metamorphoses into something else—a pair of lips into the front of a Toyota, for example.

Morphing and other techniques of digital image manipulation have had a tremendous impact on filmmaking. Director and digital pioneer Robert Zemeckis (*Death Becomes Her*) compares the new technology to the advent of sound in Hollywood.[28] It can be used to erase jet contrails from the sky in a western and to make digital planes do impossible stunts. It can even be used to add and erase actors. In *Forrest Gump*, many scenes involved old film and TV footage that had been altered so that the Tom Hanks character was interacting with historical figures.

Films and videotapes are widely thought to accurately represent real scenes (as evidenced by the reaction to the amateur videotape of the Rodney King beating by police in Los Angeles). Thus, the possibility of digital alterations raises some real problems. One is the possibility of doctoring videotapes supposed to represent actual events. Another concern is for film archives: Because digital videotapes suffer no loss in resolution when copied, there are no "generations." Thus, it will be impossible for historians and archivists to tell whether the videotape they're viewing is the real thing or not.[29]

Onward

The advances in digitization and refinements in output devices are producing more and more materials in *polymedia* form. That is, someone's intellectual or creative work, whether words, pictures, or animation, may appear in more than one form or medium. For instance, you could be reading this chapter printed in a traditional bound book. Or you might be reading it in a "course pack," printed on paper through some sort of electronic delivery system. Or it might appear on a computer display screen. Or it could be in multimedia form, adding sounds, pictures, and video to text. Thus, information technology changes the nature of how ideas are communicated.

If materials can be input and output in new ways, where will they be stored? That is the subject of the next chapter.

Good Habits: Protecting Your Computer System, Your Data, & Your Health

Whether you set up a desktop computer and never move it or tote a portable PC from place to place, you need to be concerned about protection. You don't want your computer to get stolen or zapped by a power surge. You don't want to lose your data. And you certainly don't want to lose your health for computer-related reasons. Here are some tips for taking care of these vital areas.

Protecting Your Computer System

Computers are easily stolen, particularly portables. They also don't take kindly to fire, flood, or being dropped. Finally, a power surge through the power line can wreck the insides.

Guarding Against Hardware Theft & Loss

Portable computers—laptops, notebooks, and subnotebooks—are easy targets for thieves. Obviously, anything conveniently small enough to be slipped into your briefcase or backpack can be slipped into someone else's. *Never* leave a portable computer unattended in a public place.

Desktop computers are also easily stolen. However, for under $25, you can buy a cable and lock, like those used for bicycles, that secure the computer, monitor, and printer to a work area. For instance, you can drill a quarter-inch hole in your equipment and desk, then use a product called LEASH-IT (from Z-Lock, Redondo Beach, California) to connect them together. LEASH-IT consists of two tubular locks and a quarter-inch aircraft-grade stainless steel cable.[30]

If your hardware does get stolen, its recovery may be helped if you have inscribed your driver's license number, Social Security number, or home address on each piece. Some campus and city police departments lend inscribing tools for such purposes. (And the tools can be used to mark some of your other possessions.)

Finally, insurance to cover computer theft or damage is surprisingly cheap. Look for advertisements in computer magazines. (If you have standard tenants' or homeowners' insurance, it may not cover your computer. Ask your insurance agent.)

Guarding Against Heat, Cold, Spills, & Drops

"We dropped 'em, baked 'em, we even froze 'em. Which ones survived?" proclaimed the *PC Computing* cover, ballyhooing a story about its third annual notebook "torture test."[31]

The magazine put eight notebook computers through durability trials. One approximated putting these machines in a car trunk in the desert heat, another leaving them outdoors in a Buffalo, New York, winter. A third test simulated sloshing coffee on a keyboard, and a fourth dropped them the equivalent of from desktop height to a carpeted floor. All passed the bake test, but one failed the freeze test. Three completely flunked the coffee-spill test, one other revived, and the rest passed. One that was dropped lost the right side of its display; the others were unharmed. Of the eight, half passed all tests unscathed.

This gives you an idea of how durable *some* computers are. Designed for portability, notebooks may be hardier than desktop machines. Pushing your computer off your desk and onto the floor is surely tempting fate, as it might cause your hard-disk drive to fail.

Guarding Against Power Fluctuations

Electricity is supposed to flow to an outlet at a steady voltage level. No doubt, however, you've noticed instances when the lights in your house suddenly brighten or, because a household appliance kicks in, dim momentarily. Such power fluctuations can cause havoc with your computer system, although most computers have some built-in protection. An increase in voltage may be a *spike*, lasting only a fraction of a second, or a *surge*, lasting longer. A surge can burn out the power supply circuitry in the system unit. A decrease may be a momentary voltage *sag*, a longer *brownout*, or a complete failure or *blackout*. Sags and brownouts can produce a slowdown of the hard-disk drive or a system shutdown.[32]

Extreme spikes and surges can be handled by plugging your computer, monitor, and other devices into a *surge protector* or *surge suppressor*. This device, which is plugged into the wall outlet and usually has six or more sockets, will diffuse excess voltage before it reaches your hardware.

Voltage brownouts and blackouts can be dealt with by buying a *UPS*, or *uninterruptible power supply* unit. A UPS is essentially a standby battery, which is charged by keeping it constantly plugged into a wall outlet. The UPS contains three or so sockets into which you may plug your computer hardware. The UPS filters incoming power, smoothing out power sags. In the event of a brownout or blackout, the UPS will keep your computer system running a few minutes, giving you time to save whatever work you are currently engaged in and shut your system down.

Guarding Against Damage to Software Systems software and applications software generally come on diskettes. The unbreakable rule is simply this: *Copy the original disk,* either onto your hard-disk drive or onto another diskette. Then store the original disk in a safe place. If your computer gets stolen or your software destroyed, you can retrieve the original and make another copy.

Protecting Your Data

Computer hardware and commercial software are nearly always replaceable, although perhaps with some expense and difficulty. Data, however, may be major trouble to replace or even be irreplaceable. (A report of an eyewitness account, say, or a complex spreadsheet project might not come out the same way when you try to reconstruct it.) The following are some precautions to take to protect your data.

Backup Backup Backup If your hard-disk drive fails ("crashes"), do you have the same data on a diskette? Or, if you're using just a diskette (no hard disk), do you have a duplicate of the information on it on another diskette?

Almost every microcomputer user sooner or later has the experience of accidentally wiping out or losing material and having no copy. This is what makes people true believers in *backing up* their data—making a duplicate in some form. If you're working on a research paper, for example, it's fairly easy to copy your work onto a second diskette at the end of your work session. You can then store that disk in another location. If your computer is destroyed by fire, at least you'll still have the data (unless you stored your disk right next to the computer).

If you do lose data because your disk has been physically damaged, you may still be able to recover it, using special software.

Treating Diskettes with Care Diskettes can be harmed by any number of enemies. Here are some diskette-maintenance tips:

• Insert the diskette *carefully* into the disk drive.

• Don't manipulate the metal "shutter" on the diskette; it protects the surface of the magnetic material inside.

• Do not expose the diskette to excessive heat or light.

• Do not use or place diskettes near a magnetic field, such as a telephone or paper clips stored in magnetic holders. Data can be lost if exposed.

• Do not use alcohol, thinners, or freon to clean the diskette.

Instead of leaving disks scattered on your desk, where they can be harmed by dust or beverage spills, it's best to store them in their boxes.

From time to time it's also best to clean the diskette drive, because dirt can get into the drive and cause data loss. You can buy an inexpensive drive-cleaning kit, which includes a disk that looks like a diskette and which cleans the drive's read/write heads.

Note: No disk lasts forever. Experts suggest that a diskette that is used properly might last 10 years. However, if you're storing data for the long term, you should copy the data onto new disks every 2 years (or use tape for backup).[33]

Guarding Against Viruses Computer *viruses* are programs—"deviant" programs—that can cause destruction to computers that contract them. They are spread from computer to computer in two ways: (1) They may be passed by way of an "infected" diskette, such as one a friend gives you containing a copy of a game you want. (2) They may be passed over a network or an online service. They may then attach themselves to software on your hard disk, adding garbage to or erasing your files or wreaking havoc with the systems software. They may display messages on your screen or they may evade detection and spread their influence elsewhere.

Each day, viruses are getting more sophisticated and harder to detect. There are several types of viruses. *Boot sector viruses* attach themselves to the part of a diskette or hard disk called the *boot sector*. *File viruses* attach themselves to your software files. *Polymorphic viruses* change their binary patterns each time they infect a new file in order to keep from being identified. There are several routine activities you can practice to minimize the possibility of viruses, as shown in the accompanying box. (■ *See Panel 6.10.*) Beyond these measures, the best protection is to install antivirus software. Some programs prevent viruses from infecting your system, others detect the viruses that have slipped through, and still others remove viruses or institute damage control. (Some of the major antivirus programs are Central Point Anti-Virus from Central Point Software, Norton AntiVirus from Symantec, Pro-Scan from McAfee Associates, and VirusCare from IMSI.)

Protecting Your Health

More important than any computer system and (probably) any data is your health. What adverse effects might computers cause? The most serious are painful hand and wrist injuries, eyestrain and headache, and back and neck pains. Some experts also worry about the long-range effects of exposure to electromagnetic fields and noise. All these matters can be addressed by *ergonomics,* the study of the physical relationships between people and their work environment. Let's see what you can do to avoid these problems.

Practical tips for avoiding viruses

Be careful about . . .

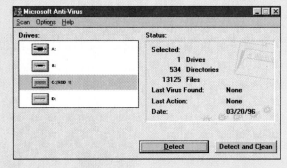

- *Sharing diskettes.* There are two kinds of disks—data disks and program disks. You're probably safe exchanging data files with friends, such as word processing files, because viruses can't be passed along via data files. However, *do not* borrow a friend's *program* disk, which may be infected.

    Don't share your software with someone else, since it might be returned to you with a virus on it.

    Don't let anyone else use your computer to run their program disks, which might be inadvertently infected.

- *Downloading programs from online services.* If your computer has a modem that lets you communicate over telephones lines with *online services or the Internet,* you can download (transfer to your computer) all kinds of wonderful programs, often for free. To ensure that a program is virus-free, use a commercial virus-protection program to scan the program yourself before using it.

- *Inspecting and write-protecting.* Don't buy new commercial software unless it's still in its shrink-wrapped package. An open package may have been returned by a previous customer or used by a dealer for demonstration purposes—both opportunities for infection.

    Before you use a new program, write-protect the disks. This will prevent them from becoming infected in case your own computer is infected.

- *Backing up your hard disk—often.* If a virus ever erases your hard disk, you may be able to kill the virus with a commercial antivirus program. Then, if you've been conscientious about regularly making backup copies of your programs and data on diskettes, you can restore them to your hard disk.

- *Booting.* When booting from your hard disk, make sure you have removed any diskettes from the diskette drives. Thus you will not inadvertently attempt to boot from a possibly infected diskette. (This helps protect against common boot sector viruses.)

Protecting Your Hands & Wrists To avoid difficulties, consider employing the following:

- *Hand exercises:* You should warm up for the keyboard just as athletes warm up before doing a sport in order to prevent injury. There are several types of warm-up exercises. You can gently massage the hands, press the palm down to stretch the underside of the forearms, or press the fist down to stretch the top side of the forearm. Experts advise taking frequent breaks, during which time you should rotate and massage your hands.

- *Work-area setup:* With a computer, it's important to set up your work area so that you sit with both feet on the floor, thighs at right angles to your body. The chair should be adjustable and support your lower back. Your forearms should be parallel to the floor. You should look down slightly at the screen. (See Panel 6.11, next page.) This setup is particularly important if you are going to be sitting at a computer for hours at a stretch.

- *Wrist position:* To avoid wrist and forearm injuries, you should keep your wrists straight and hands relaxed as you type. Instead of putting the keyboard on top of a desk, therefore, you should put it on a low table or in an underdesk keyboard drawer. Otherwise the nerves in

your wrists will rub against the sheaths surrounding them, possibly leading to RSI pains.[34] Some experts also suggest using a padded, adjustable wrist rest, which attaches to the keyboard.

Various kinds of ergonomic keyboards are also available, such as those that are hinged in the middle.

Guarding Against Eyestrain, Headaches, & Back & Neck Pains Eyestrain and headaches usually arise because of improper lighting, screen glare, and long shifts staring at the screen. Make sure your windows and lights don't throw a glare on the screen, and that your computer is not framed by an uncovered window. Headaches may also result from too much noise, such as listening for hours to an impact printer printing out.

Back and neck pains occur because furniture is not adjusted correctly or because of heavy computer use. Adjustable furniture and frequent breaks should provide relief here.

Some people worry about emissions of electromagnetic waves and whether they could cause problems in pregnancy or even cause cancer. The best approach is to simply work at an arm's length from computers with CRT-type monitors.

How to set up your computer work area

HEAD Directly over shoulders, without straining forward or backward, about an arm's length from screen.

NECK Elongated and relaxed.

SHOULDERS Kept down, with the chest open and wide.

BACK Upright or inclined slightly forward from the hips. Maintain the slight natural curve of the lower back.

ELBOWS Relaxed, at about a right angle.

WRISTS Relaxed, and in a neutral position, without flexing up or down.

KNEES Slightly lower than the hips.

CHAIR Sloped slightly forward to facilitate proper knee position.

LIGHT SOURCE Should come from behind the head.

SCREEN At eye level or slightly lower. Use an antiglare screen.

FINGERS Gently curved.

KEYBOARD Best when kept flat (for proper wrist positioning) and at or just below elbow level. Computer keys that are far away should be reached by moving the entire arm, starting from the shoulders, rather than by twisting the wrists or straining the fingers. Take frequent rest breaks.

FEET Firmly planted on the floor. Shorter people may need a footrest.

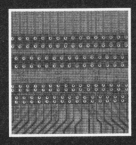

SUMMARY

What It Is / What It Does	Why It's Important
active-matrix display *(p. 165, LO 2)* Type of flat-panel display in which each pixel on the screen is controlled by its own transistor.	Active-matrix screens are much brighter and sharper than passive-matrix screens, but they are more complicated and thus more expensive.
audio-output device *(p. 171, LO 5)* Device that outputs voice or voice-like sounds (voice-output technology) or that outputs music and other sounds (sound-output technology).	Audio-output devices are important in speech synthesis and in multimedia computing.
bitmap *(p. 166, LO 2)* In computer graphics, an area in memory that represents an image. Depending on the screen, 1 or several bits represent 1 pixel or several pixels of the image.	Bitmapped display screens permit the computer to manipulate pixels on the screen individually rather than as blocks (character-map), enabling software to create a greater variety of images. However, bitmaps take up a lot of storage space.
cathode-ray tube (CRT) *(p. 164, LO 2)* Vacuum tube used as a display screen in a computer or video display terminal. Images are represented on the screen by individual dots or "picture elements" called *pixels.*	This technology is found not only in the screens of desktop computers but also in television sets and flight-information monitors in airports.
character-map *(p. 166, LO 2)* Fixed location on a video display screen where a predetermined character can be placed. Character-mapped display screens display only text—letters, numbers, and special characters (as opposed to bitmapped display screens).	Character-mapped display screens cannot display graphics unless a video adapter card is installed.
color display screen *(p. 166, LO 2)* Display screens that can display between 16 and 16.7 million colors, depending on their type (see VGA, SVGA, XGA).	Most software today is developed for color, and—except for some pocket PCs—most microcomputers today are sold with color display screens.
display screen *(p. 164, LO 2)* Also variously called *monitor, CRT,* or simply *screen;* softcopy output device that shows programming instructions and data as they are being input and information after it is processed. Sometimes a display screen is also referred to as a VDT, for video display terminal, although technically a VDT includes both screen and keyboard. The size of a screen is measured diagonally form corner to corner in inches, just like television screens.	Display screens enable users to immediately view the results of input and processing.
dot-matrix printer *(p. 167, LO 4)* Printer that contains a print head of small pins that strike an inked ribbon, forming characters or images. Print heads are available with 9, 18, or 24 pins, with the 24-pin head offering the best quality.	Dot-matrix printers can print draft quality, a coarser-looking 72 dots per inch vertically, or near-letter-quality (NLQ), a crisper-looking 144 dots per inch vertically. They can also print graphics.
dot pitch *(p. 165, LO 2)* Amount of space between pixels (dots); the closer the dots, the crisper the image.	Dot pitch is one of the measures of display screen capability.
Extended Graphics Array (XGA) *(p. 166, LO 3)* Graphics board display standard, also referred to as *high resolution;* supports up to 16.7 million colors at a resolution of 1024 × 768 pixels. Depending on the video display adapter memory chip, XGA will support 256, 65,536, or 16,777,216 colors. For any of these displays to work, video display adapters and monitors must be compatible. The computer's software and the video display adapter must also be compatible.	Extended Graphics Array offers the most sophisticated standard for color and resolution.

What It Is / What It Does	Why It's Important

flat-panel display *(p. 165, LO 2)* Refers to display screens that are much thinner, weigh less, and consume less power than CRTs. Flat-panel displays are made up of two plates of glass with a substance between them that is activated in different ways. One common type of technology used in flat-panel display screens is liquid-crystal display. Flat-panel screens are either active-matrix or passive-matrix displays. Images are represented on the screen by individual dots, or picture elements called *pixels*.

Flat-panel displays are used in portable computers.

font *(p. 168, LO 4)* Set of type characters in a particular type style and size.

Desktop publishing programs, along with laser printers, have enabled users to dress up their printed projects with many different fonts.

hardcopy *(p. 162, LO 1)* Refers to printed output (as opposed to softcopy). The principal examples are printouts, whether text or graphics, from printers. Film, including microfilm and microfiche, is also considered hardcopy output.

Hardcopy is convenient for people to use and distribute; it can be easily handled or stored.

impact printer *(p. 167, LO 4)* Type of printer that forms characters or images by striking a mechanism such as a print hammer or wheel against an inked ribbon, leaving an image on paper.

For microcomputer users, the most common impact printers are daisywheel printers and dot-matrix printers.

inkjet printer *(p. 168, LO 4)* Nonimpact printer that forms images with little dots. Inkjet printers spray small, electrically charged droplets of ink from four nozzles through holes in a matrix at high speed onto paper.

Because they produce high-quality images, they are often used by people in graphic design and desktop publishing. However, inkjet printers are slower than laser printers.

laser printer *(p. 167, LO 4)* Nonimpact printer similar to a photocopying machine; images are created on a drum, treated with a magnetically charged ink-like toner (powder), and then transferred from drum to paper.

Laser printers produce much better image quality than dot-matrix printers do and can print in many more colors; they are also quieter. Laser printers, along with page description languages, enabled the development of desktop publishing.

liquid-crystal display (LCD) *(p. 165, LO 2)* Flat-panel display that consists of a substance called *liquid crystal*, the molecules of which line up in a way that alters their optical properties. As a result, light—usually backlighting behind the screen—is blocked or allowed through to create an image.

LCD is useful not only for portable computers but also as a display for various electronic devices, such as watches and radios.

monochrome display screen *(p. 166, LO 2)* Refers to "single color"; a monochrome computer screen displays a single-color image on a contrasting background—usually black on white, amber on black, or green on black.

Monochrome display is suitable for nongraphics applications such as word processing or spreadsheets.

nonimpact printer *(p. 167, LO 4)* Printer that forms characters and images without making direct physical contact between printing mechanism and paper. Two types of nonimpact printers often used with microcomputers are laser printers and inkjet printers. A third kind, the thermal printer, is seen less frequently.

Nonimpact printers are faster and quieter than impact printers because they have fewer moving parts. They can print text, graphics, and color, but they cannot be used to print on multipage forms.

page description language *(p. 168, LO 4)* Software used in desktop publishing that describes the shape and position of characters and graphics to the printer.

Page description languages, such as PostScript and PCL, used along with laser printers, gave birth to desktop publishing. They allow users to combine different types of graphics with text in different fonts, all on the same page.

passive-matrix display *(p. 165, LO 2)* Type of flat-panel display in which each transistor controls a whole row or column of pixels.

Although passive-matrix displays are less bright and less sharp than active-matrix displays, they are less expensive and use less power.

pixel *(p. 165, LO 2)* Short for *picture element;* smallest unit on the screen that can be turned on and off or made different shades. A stream of bits defining the image is sent from the computer (from the CPU) to the CRT's electron gun, where the bits are converted to electrons.

Pixels are the building blocks that allow graphical images to be presented on a display screen.

plotter *(p. 169, LO 4)* Specialized hardcopy output device designed to produce high-quality graphics in a variety of colors. The three principal kinds of plotters are pen, electrostatic, and thermal.	Plotters are especially useful for creating maps and architectural drawings, although they may also produce less complicated charts and graphs.
PostScript *(p. 168, LO 4)* Printer language, or page description language, that has become a standard for printing graphics on laser printers.	PostScript printers are essential for users who need to print a lot of graphics or heavily designed pages or who want to generate different fonts in various sizes.
printer *(p. 167, LO 4)* Output device that prints characters, symbols, and perhaps graphics on paper.	Printers provide one of the principal forms of computer output.
resolution *(p. 165, LO 2)* Clarity or sharpness of a display screen; the more pixels there are per square inch, the better the resolution. Resolution is expressed in terms of the formula horizontal pixels × vertical pixels.	Users need to know what screen resolution is appropriate for their purposes.
robot *(p. 173, LO 5)* Automatic device that performs functions ordinarily ascribed to human beings or that operate with what appears to be almost human intelligence.	Robots are of several kinds—industrial robots, perception systems, and mobile robots. They are performing more and more functions in business and the professions.
softcopy *(p. 162, LO 1)* Refers to data that is shown on a display screen or is in audio or voice form. This kind of output is not tangible; it cannot be touched. Virtual reality and robots might also be considered softcopy devices.	This term is used to distinguish nonprinted output from printed output.
sound-output device *(p. 171, LO 5)* Audio-output device that produces digitized, nonverbal sounds, ranging from beeps and chirps to music. It includes software and a sound card or digital audio circuit board.	PC owners can customize their machines to greet each new program with particular sounds. Sound output is also used in multimedia presentations.
Super Video Graphics Array (SVGA) *(p. 166, LO 3)* Graphics board standard that supports 256 colors at higher resolution than VGA. SVGA has two graphics modes: 800 × 600 pixels and 1024 × 768.	Super VGA is a higher-resolution version of Video Graphics Array (VGA), introduced in 1987.
thermal printer *(p. 168, LO 4)* Nonimpact printer that uses colored waxes and heat to produce images by burning dots onto special paper.	The colored wax sheets are not required for black-and-white output because the thermal print head will register the dots on the paper.
video display adapter *(p. 166, LO 3)* Also called a *graphics adapter card;* circuit board that determines the resolution, number of colors, and how fast images appear on the display screen.	Video display adapters determine how fast the card processes images and how many colors it can display.
Video Graphics Array (VGA) *(p. 166, LO 3)* Graphics board standard that supports 16 to 256 colors, depending on resolution. At 320 × 200 pixels it will support 256 colors; at the sharper resolution of 640 × 480 pixels it will support 16 colors.	VGA is the most common video standard used today.
videoconferencing *(p. 171, LO 5)* A method of communicating whereby people in different geographical locations can have a meeting—and see and hear one another—using computers and communications technologies.	Videoconferencing technology enables people to conduct business meetings without having to travel.
virtual reality *(p. 172, LO 5)* Computer-generated artificial reality that projects user into sensation of three-dimensional space.	Virtual reality is used most in entertainment, as in arcade-type games, but has applications in architectural design and training simulators.
voice-output device *(p. 171, LO 5)* Audio-output device that converts digital data into speech-like sounds.	Voice-output devices are found in telephone systems, soft-drink machines, and toys and games.

(Selected answers appear at the back of the book.)

Short-Answer Questions

1. What is the difference between hardcopy and soft-copy?

2. What are the two types of display screen?

3. What is the difference between character-mapped displays and bitmapped displays?

4. What do VGA, SVGA, and XGA refer to, and how are they different?

5. How are active-matrix and passive matrix displays different?

6. What advantages does a laser printer have over other types of printers?

7. How does a display screen's refresh rate relate to screen clarity?

8. What is a page description language?

Fill-in-the-Blank Questions

1. The more pixels that can be displayed on the screen, the better the _____ of the image.

2. _____ is a computer-generated, 3-D environment that users can enter through the use of special hardware and software.

3. Output is available in two principal forms: _____ and _____.

4. PostScript and PCL are _____; they describe the shape and position of letters and graphics to laser printers.

5. _____ can produce high-quality graphics and are used most often for outputting maps and architectural drawings.

6. Whether for CRT or flat-panel display, screen clarity depends on the following three qualities:
   a. _____
   b. _____
   c. _____

7. _____ is a method whereby people in different geographical locations can have a meeting.

8. Voice-output devices convert _____ data into speech-like sounds.

9. A _____ is a computerized device that can operate with almost human intelligence.

Multiple-Choice Questions

1. To display graphics, a display screen must have a(n):
   a. CRT
   b. vacuum tube
   c. video display adapter
   d. plotter
   e. none of the above

2. For microcomputer users, the most common type of impact printer is the:
   a. dot-matrix printer
   b. laser printer
   c. inkjet printer
   d. thermal printer
   e. all of the above

3. Which of the following printers uses a combination of color waxes and heat to form images?
   a. laser printer
   b. dot-matrix printer
   c. inkjet printer
   d. thermal printer
   e. none of the above

4. Which of the following is a softcopy output device?
   a. display screen
   b. plotter
   c. printer
   d. robot
   e. none of the above

5. Which of the following determines the number of colors that can appear on a display screen?
   a. liquid crystal
   b. bitmapped capability
   c. video display adapter
   d. gas plasma
   e. none of the above

6. Virtual-reality is being used in:
   a. video games
   b. simulation (training)
   c. treatment of phobias
   d. hands-on learning
   e. all of the above

7. Which of the following would you use if you need to generate multipart forms?
    a. laser printer
    b. dot-matrix printer
    c. inkjet printer
    d. thermal printer
    e. none of the above

True/False Questions

T F 1. A picture element on the screen is called a *pixel.*

T F 2. *Display screen, CRT,* and *monitor* are different names for the same thing.

T F 3. CRTs are used on portable computers.

T F 4. Screen resolution is measured by vertical and horizontal lines of pixels.

T F 5. Hardcopy sometimes refers to information displaying on a display screen.

T F 6. Laser printers, inkjet printers, and thermal printers are all nonimpact printers that can output color graphics.

T F 7. You need the following special equipment to experience virtual reality: headgear, gloves, shoes, pants, and software.

T F 8. Audio-ouput devices can output only music.

Projects/Critical-Thinking Questions

1. Visit a local computer store to compare the output quality of the different printers on display. Then obtain output samples and a brochure on each printer sold. After comparing output quality and price, what printer would you recommend to a friend who needs a printer that can output resumes, research reports, and professional-looking correspondence with a logo? Why?

2. What hardware and software would you need to add to your PC in order for it to support videoconferencing? What companies sell videoconferencing systems? Who sells the best videoconferencing system? How is videoconferencing technology limited? Why do you think that videoconferencing technology isn't built into today's PCs? How could you use videoconferencing in your job, future line of work, or personal area of interest?

3. Computer magazines often sponsor tests to compare laser printers. By reviewing current computer magazines, identify the most highly-rated laser printer. How much does the printer cost? What capabilities does this printer have? Why was this printer rated above other printers in the study? Would you buy this printer for personal use? Why/why not?

4. Explore the state of the art of computer-generated 3-D graphics. What challenges are involved in creating photo-realistic 3-D images? What hardware and software are needed to generate 3-D graphics? Who benefits from this technology?

5. Explore the state of the art of liquid-crystal display (LCD) technology. What are the current limitations of LCD technology? Why do you think that billions of dollars are currently being invested in LCD development? Who will benefit from improved LCD technology?

(net) Exploring the Internet

Objective: *In this exercise we describe how to use Netscape Navigator to access an online magazine and newspaper.*

Before you continue*: We assume you have access to the Internet through your university, business, or commercial service provider and to the Web browser tool named Netscape Navigator 2.0 or 2.01. Additionally, we assume you know how to connect to the Internet and then load Netscape Navigator. If necessary, ask your instructor or system administrator for assistance.*

1. Make sure you have started Netscape. The home page for Netscape Navigator should appear on your screen.

2. Several magazines and newspapers are available to you online including the Network Observer (*http://communication.ucsd.edu/pagre/tno.html*), which emphasizes issues relating to networks and democracy, and the *Mercury Center* (*http://www.sjmercury.com/*), which carries top stories from the *San Jose Mercury News*, a Silicon Valley-based newspaper.

    To display the *Network Observers'* home page:
    CLICK: Open button (🖬)
    TYPE: http://communication.ucsd.edu/pagre/ tno.html
    PRESS: **Enter** or CLICK: Open

Your screen may appear similar to the following:

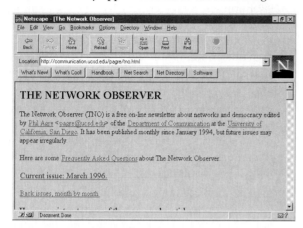

3. Drag the vertical scroll bar downward to see a list of articles. Notice that articles are grouped by section. Identify an article you're interested in reading and then click its link.

4. If you're connected to a printer, you can print the article and then read it at your leisure.

   CHOOSE: File, Print from the Menu bar (or click the Print button located on the Netscape toolbar)
   PRESS: **Enter** or CLICK: OK

5. To redisplay the list of topics:
   CLICK: Back button ( )

6. On your own, display additional articles.

7. To display the *Mercury Center's* home page:
   CLICK: Open button ( )
   TYPE: `http://www.sjmercury.com/`
   PRESS: **Enter** or CLICK: Open

   Your screen should appear similar to the following:

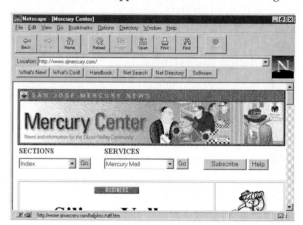

8. Drag the vertical scroll bar downward to see the contents of the home page. On your own, explore the contents of the newspaper by clicking links that are of interest to you. Remember that if you would rather read an article in printed form, you can print it by choosing File, Print from the Menu bar or by clicking the Print button.

9. Display Netscape's home page and then exit Netscape.

CHAPTER 7

Storage & Databases
Foundations for Interactivity, Multimedia, & Knowledge

Concepts You Should Know

After reading this chapter, you should be able to:

1. Define and explain units of storage capacity.
2. Distinguish between primary and secondary storage.
3. Explain why data compression is sometimes necessary.
4. Identify the basic criteria for rating secondary storage devices.
5. Explain how data is stored on diskette, hard disks, optical disks, and magnetic tape and describe the purposes of storage media.
6. Describe the parts of the data hierarchy and the role of key fields.
7. Distinguish batch from online processing, and online from offline storage.
8. Describe the difference between file management systems and database management systems.
9. Explain the best uses for sequential, direct, and indexed-sequential access storage, and the storage media associated with each.
10. Identify the advantages and disadvantages of the four database models.
11. Describe some ways database users can invade privacy and the methods or laws to fight such invasions.

See it. Hear it. Change it.

These are the promises frequently being heard, whether for today's videogames or tomorrow's smart TVs. Such claims have made "interactivity" and "multimedia" among the most overused marketing words of recent times. *Interactivity* refers to users' ability to have a back-and-forth dialogue with whatever is on their screen. *Multimedia* refers to the use of a variety of media—text, sound, video—to deliver information, whether on a computer disk or (as the phone and cable companies use it) via the anticipated union of computers, telecommunications, and broadcasting technologies.[1]

Interactive programs and multimedia programs require machines that can handle and store enormous amounts of data. For this reason, the capacities of storage hardware, called *secondary storage* or just *storage,* seem to be increasing almost daily. This chapter covers the most common types of secondary storage, as well as their uses. First, however, we need to go over a few concepts that relate to the whole topic of secondary storage: data capacity, data compression/decompression, and criteria for rating storage devices.

Storage Capacity

Preview & Review: Storage is categorized as primary or secondary. The capacity of storage devices is measured in bytes, kilobytes, megabytes, gigabytes, and terabytes.

You'll recall from Chapters 1 and 4 that *primary storage* is also known as *memory, main memory, internal memory,* or *RAM* (for random access memory). It is working storage that holds (1) data for processing, (2) instructions (the programs) for processing the data, and (3) processed data (that is, information) that is waiting to be sent to an output or secondary-storage device such as a printer.

Primary storage is in effect the computer's short-term capacity, determining the total size of the programs and data files it can work on at any given moment. Primary storage, which is contained on RAM chips (✓ p. 121), is also *temporary.* Once the power to the computer is turned off, all the data and programs within memory simply vanish. For this reason, primary storage is said to be *volatile. Volatile storage* is temporary storage; the contents are lost when the power is turned off. If you accidentally kick out the power cord underneath your desk, or a storm knocks down a power line to your house, whatever you are currently working on will immediately vanish.

In contrast, **secondary storage consists of devices that store data and programs permanently on disk or tape.** Secondary storage is **nonvolatile—that is, data and programs are permanent, or remain intact when the power is turned off.**

You'll also recall that computers are based on the principle that electricity may be "on" or "off," or "high-voltage" or "low-voltage," or "present" or "absent," or some similar two-state system. Thus, individual items of data are represented by 0 for off and 1 for on. A 0 or 1 is called a *bit.* A unit of 8 bits is called a *byte;* it may be used to represent a character, digit, or other value, such as A, ?, or 3.

Bits and bytes are the building blocks for representing data, whether it is being processed, stored, or telecommunicated.

Units of Measurement for Storage

We explained the meanings of kilobytes, megabytes, gigabytes, and terabytes in conjunction with the capabilities of processing hardware (✓ pp. 115–116). The same terms are also used to measure the data capacity of storage devices. To repeat:

- **Kilobyte:** A *kilobyte* (abbreviated *K or KB*) is equivalent to 1024 bytes. Kilobytes are a common unit of measure for storage capacity. One type-written page of text is the approximate equivalent of 1 KB.
- **Megabyte:** A *megabyte* (abbreviated *M or MB*) is about 1 million bytes.
- **Gigabyte:** A *gigabyte* (*G or GB*) is about 1 billion bytes.
- **Terabyte:** A *terabyte* (*T or TB*) is about 1 trillion bytes.

The amount of data being held in a file or database in your personal computer might be expressed in kilobytes, megabytes, or gigabytes. The amount of data being held by a far-flung database accessible to you over a communications line might be expressed in gigabytes or terabytes.

Because, as we said at the beginning of this chapter, new sophisticated software programs require huge storage capacities, the capacities of storage devices are steadily increasing. There is, however, a way to increase a system's storage capacity without buying new storage devices—that is, through compression programs.

Compression & Decompression

Preview & Review: Compression is a method of removing redundant elements from a computer file so that it requires less storage space.

"Like Gargantua, the computer industry's appetite grows as it feeds . . . ," says one writer. "The first symptoms of indigestion are emerging. So the smartest software engineers are now looking for ways to shrink the data-meals computers consume, without reducing their nutritional value."[2]

What this writer is referring to is the "digital obesity" brought on by the requirements of the new multimedia revolution for putting pictures, sound, and video onto disk or sending them over a communications line. For example, a 2-hour movie contains so much sound and visual information that, if stored on a standard CD-ROM, it would require 360 disk changes during a single showing. A broadcast of *Oprah Winfrey* that presently fits into one conventional, or analog, television channel would require 45 channels if sent in digital language.[3]

The solution for putting more data into less space comes from the mathematical process called compression. **Compression, or *digital-data compression*, is a method of removing redundant elements from a file or replacing repeated patterns with fewer characters so that the file requires less storage space and less time to transmit.** After the data is stored or transmitted and is to be used again, it is decompressed. Compression can be done with hardware expansion boards or with software programs like PKZIP, a shareware (✓ p. 46) program from PKWARE. Files compressed by PKZIP have names with the extension "ZIP" on them. They are decompressed using the PKUNZIP program.

Standards for Visual Compression

The major difficulty now is that several standards exist for compression, particularly of visual data. If you record and compress in one standard, you cannot play it back in another. The main reason for the lack of agreement is that different industries have different priorities. What will satisfy the users of still photographs, for instance, will not work for the users of moving images.

The principal compression schemes are as follows:

- **Still images—JPEG:** The leading standard for still images is *JPEG* (pronounced "jay-peg"), for the Joint Photographic Experts Group of the International Standards Organization. The JPEG compression scheme looks for a way to squeeze a single image, mainly by eliminating repetitive pixels, or picture-element dots, within the image. Unfortunately there are more than 30 kinds of JPEG programs. "Unless the decoder in your computer recognizes the version that was used to compress a particular image," noted one reporter, "the result on your computer screen will be multimedia applesauce."[4]

- **Moving images—MPEG:** People who work with videos are less concerned with the niceties of preserving details than are those who deal with still images. They are interested mainly in storing or transmitting an enormous amount of visual information in economical form. A group called *MPEG* ("em-peg"), for Motion Picture Experts Group, has been formed to set standards for weeding out redundancies between neighboring images in a stream of video. Three MPEG standards have been developed for compressing visual information—MPEG-1, MPEG-2, and MPEG-4.

The vast streams of bits and bytes of text, audio, and visual information threaten to overwhelm us. Compression/decompression has become a vital technology for rescuing us from the swamp of digital data.

Criteria for Rating Secondary-Storage Devices

Preview & Review: Storage capacity, access speed, transfer rate, size, and cost are all factors in rating secondary-storage devices.

Several kinds of secondary-storage devices are available, and they generally differ in *storage capacity, access speed, transfer rate, size,* and *cost.*

- **Storage capacity:** As we mentioned earlier, high-capacity storage devices are desirable or required for many sophisticated programs and large databases. However, as capacity increases, so does price. Some users find compression software to be an economical solution to storage-capacity problems. Hard disks can store more data than diskettes, and optical disks can store more than hard disks.

- **Access speed:** *Access speed* refers to the average time needed to locate data on a secondary storage device. Access speed is measured in milliseconds (thousandths of a second). Hard disks are faster than optical disks, which are faster than diskettes. Disks are faster than magnetic tape. However, the slower media are more economical.

- **Transfer rate:** *Transfer rate* refers to the speed at which data is transferred from secondary storage to main memory (✓ p. 113). It is measured in megabytes per second.

- **Size:** Some situations require compact storage devices, others don't. Users need to know what their options are.

- **Cost:** As we have intimated, the cost of a storage device is directly related to the previous four factors.

Now let's take an in-depth look at the following types of secondary storage devices.

- Diskettes
- Hard disks
- Optical disks
- Magnetic tape

Diskettes

Preview & Review: Diskettes are round pieces of flat plastic that store data and programs as magnetized spots. A disk drive copies, or reads, data from the disk and writes, or records, data to the disk.

Components of a diskette include tracks and sectors; diskettes come in various densities. All have write-protect features.

Care must be taken to avoid destroying data on disks, and users are advised to back up, or duplicate, the data on their disks.

A *diskette,* or *floppy disk,* **is a removable round, flat piece of mylar plastic that stores data and programs as magnetized spots.** More specifically, data is stored as electromagnetic charges on a metal oxide film that coats the mylar plastic. Data is represented by the presence or absence of these electromagnetic charges, following standard patterns of data representation (such as ASCII, ✓ p. 116). The disk is contained in a plastic case or square paper envelope to protect it from being touched by human hands. It is called "floppy" because the disk within the case or envelope is flexible, not rigid.

Two sizes of diskettes are commonly used for microcomputers. (■ *See Panel 7.1.)*

- **3½ inches:** The smaller size, now by far the most common, is 3½ inches across. This size comes inside a hard plastic jacket, so that no additional protective envelope is needed.
- **5¼ inches:** The older and larger size is 5¼ inches across. The disk is encased inside a flexible plastic jacket. The 5¼-inch disk is often inserted into a removable paper or cardboard envelope or sleeve for protection when it is not being used.

PANEL 7.1

Diskettes

(Top) 3½-inch diskettes.
(Bottom) 5¼-inch diskette.

The Diskette Drive

To use a diskette, you need a diskette drive, which is usually built into the computer's system unit (✓ p. 118). **A *diskette drive* is a device that holds, spins, and reads data from and writes data to a diskette.**

The words *read* and *write* have exact meanings:

- **Read:** *Read* means that the data represented in the magnetized spots on the disk (or tape) are converted to electronic signals and transmitted to the primary storage (memory) in the computer. That is, *read* means the disk drive *copies* data—stored as magnetic spots—from the diskette.
- **Write:** *Write* means that the electronic information processed by the computer is recorded magnetically onto disk (or tape). *Write* means that the disk drive *transfers* data—represented as electronic signals within the computer's memory—onto the storage medium.

How a Diskette Drive Works

A diskette is inserted into a slot, called the *drive gate* or *drive door*, in the front of the disk drive. (■ *See Panel 7.2.*) Sometimes a door or a latch must be closed after the disk is inserted. This clamps the diskette in place over the spindle of the drive mechanism so the drive can operate. Usually today, however, the diskette is simply pushed into the drive until it clicks into place. An access light goes on when the disk is in use. After using the disk, you can retrieve it either by pressing an eject button beside the drive or by opening the drive gate.

The device by which the data on a disk is transferred to the computer, and from the computer to the disk, is the disk drive's *read/write head*. The diskette spins inside its jacket, and the read/write head moves back and forth over the data access area. The *data access area* is an opening in the disk's jacket through which data is read or written.

Today most microcomputers have a single diskette drive for 3½-inch disks and a hard-disk drive. There may also, however, be an additional diskette drive that accepts the larger size, 5¼-inch disks.

Characteristics of Diskettes

Both 3½-inch and 5¼-inch disks work in similar ways, although there are some differences. The characteristics of diskettes are as follows:

● **Tracks and sectors:** On a diskette, **data is recorded in rings called *tracks*.** (■ *See Panel 7.3.*) Unlike on a phonograph record, these tracks are neither visible grooves nor a single spiral. Rather, they are closed concentric rings.

   Each track is divided into eight or nine *sectors*. **Sectors are invisible wedge-shaped sections used for storage reference purposes.** When you save data from your computer to a diskette, the data is distributed by tracks and sectors on the disk. That is, the systems software uses the point at which a sector intersects a track to reference the data location in order to spin the disk and position the read/write head.

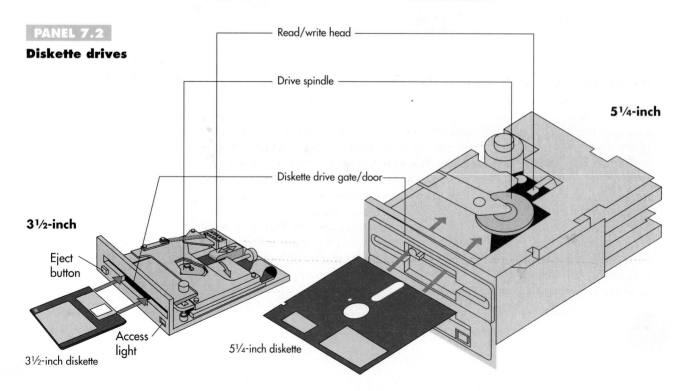

PANEL 7.2

Diskette drives

Read/write head

Drive spindle

Diskette drive gate/door

5¼-inch

3½-inch

Eject button

Access light

3½-inch diskette

5¼-inch diskette

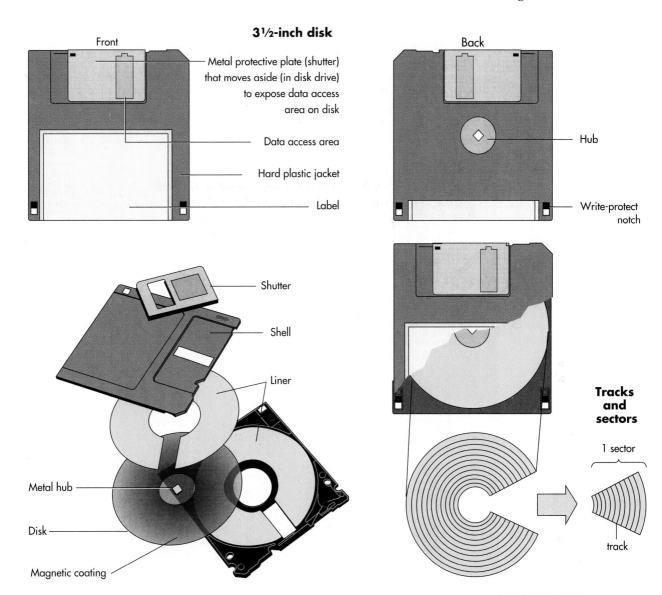

3½-inch disk

Front

Metal protective plate (shutter) that moves aside (in disk drive) to expose data access area on disk

Data access area

Hard plastic jacket

Label

Back

Hub

Write-protect notch

Shutter

Shell

Liner

Metal hub

Disk

Magnetic coating

Tracks and sectors

1 sector

track

PANEL 7.3

Diskette anatomy

- **Unformatted versus formatted disks:** When you buy a new box of diskettes to use for storing data, the box may state that it is "unformatted" (or say nothing at all). This means you have a task to perform before you can use the disks with your computer and disk drive. *Unformatted disks* are manufactured without tracks and sectors in place. **Formatting—or *initializing*, as it is called on the Macintosh—means that you must prepare the disk for use so that the operating system can write information on it. This includes defining the tracks and sectors on it.** Formatting is done quickly by using a few simple software commands.

  Alternatively, when you buy a new box of diskettes, the box may state that it is "formatted IBM." This means that you can simply insert a disk into the drive gate of your IBM or IBM-compatible microcomputer and use it without any effort. It's just like plunking an audiotape into a standard tape recorder.

• **Data capacity—sides and densities:** Not all disks hold the same amount of data, because the characteristics of microcomputer disk drives differ.

The first diskettes were *single-sided,* or diskettes that store data on one side only. Now all diskettes are *double-sided,* capable of storing data on both sides. They therefore hold twice as much data as single-sided disks. For double-sided diskettes to work, the disk drive must have read/write heads that will read both sides simultaneously. This is the case with current disk drives.

A disk's capacity also depends on its recording density. ***Recording density* refers to the number of bytes that can be written onto the surface of the disk.** Common densities are *double-density, high-density,* and *very-high-density.* A 3½-inch double-sided, double-density disk can store 720 kilobytes, or the equivalent of about 720 typewritten pages of text. A high-density 3½-inch disk can store 1.44 megabytes, or about 1440 typewritten pages. A very-high-density disk can store 2.88 megabytes of data, or 2880 pages.

• **Write-protect features:** Both 3½-inch and 5¼-inch disks have features to prevent someone from accidentally writing over—and thereby obliterating—data on the disk. (This is especially important if you're working on your only copy of a program or a document that you've transported from somewhere else.) This **write-protect feature allows you to protect a diskette from being written to.** (■ *See Panel 7.4.*)

Disks have additional features (such as the index hole, for positioning the disk over a photoelectric sensing mechanism within the disk drive). However, these are of no concern for our present purposes.

PANEL 7.4

Write-protect features

(Top)—3½-inch diskette: For data to be written to this disk, a small piece of plastic must be closed over the tiny window on one side of the disk. To protect the disk from being written to, you must open the window (using the tip of a pen helps). *(Bottom)* 5¼-inch diskette: For data to be written to this diskette, the write-protect notch must be uncovered, as shown at left. To protect the disk from being mistakenly written over, a small piece of tape must be folded around the notch. (The tape comes with the disks.)

Writable

Write-protect
window closed

Write-protected

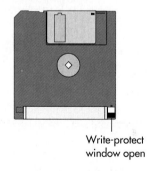

Write-protect
window open

Writable

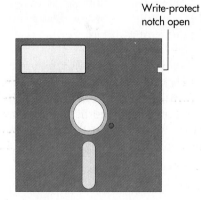

Write-protect
notch open

Write-protected

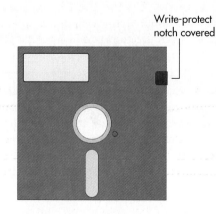

Write-protect
notch covered

Taking Care of Diskettes

Diskettes need at least the same amount of care that you would give to an audiotape or music CD. In fact, they need more care than that if you are dealing with difficult-to-replace data or programs. There are a number of rules for taking care of diskettes:

- **Don't touch disk surfaces:** Don't touch anything visible through the protective jacket, such as the data access area. Don't manipulate the metal shutter on 3½-inch disks.
- **Handle disks gently:** Don't bend them or put heavy weights on them.
- **Avoid risky physical environments:** Disks don't do well in sun or heat (such as in glove compartments or on top of steam radiators). They should not be placed near magnetic fields (including those created by nearby telephones or electric motors). They also should not be exposed to chemicals (such as cleaning solvents) or spilled coffee or alcohol.

The Importance of Backup

Even with the best of care, however, a disk can suddenly fail for reasons you can't understand. Many computer users have had the experience of being unable to retrieve data from a disk that worked perfectly the day before, because some defect has damaged a track or sector. Thus, you should always be thinking about backup. **Backup is the name given to a diskette (or tape) that is a duplicate or copy of another form of storage.** The best protection if you're writing, say, a make-or-break research paper is to make *two copies* of your data. One copy is on a diskette, certainly, but a duplicate should be on a second diskette or on your hard disk if you are using one.

Microcomputer Diskette Variations: Removable High-Capacity Diskettes

At present two kinds of removable high-capacity diskette drives seem to be emerging.

- **Zip and EZ diskette drives:** The first of this breed was the *Zip drive* from Iomega, an external drive about the size of a hardcover novel and weighing about 1 pound. It was soon followed by the *EZ135 drive* from SyQuest Technology. The EZ is about the size of two stacked VCR tapes, weighs about 2 pounds, and is twice as fast as the Zip.[5] Unfortunately, the two drives use different technology and incompatible disks.

   Both the Zip and the EZ use hard-shell diskettes about 4 inches square and a quarter-inch thick. The Zip diskette holds 100 megabytes, which is 70 times more than conventional diskettes. The EZ diskette holds 135 megabytes. Neither drive can read 3½- and 5¼-inch diskettes.

- **Backward-compatible high-capacity diskette drive:** In mid-1995, three computer-industry companies (Compaq, 3M, Matsushita) announced they had developed a 3½-inch diskette drive that could handle 83 times as much data as current diskette drives.[6] More important, the 120-megabyte drive is backward-compatible and so can read the 720-kilobyte and 1.44-megabyte diskettes that people currently use.

In spite of advances in diskette technology, diskettes still cannot store as much data as hard disks and optical disks, which we cover next.

Hard Disks

Preview & Review: Hard disks are rigid metal platters that, like diskettes, hold data as magnetized spots.

Usually a microcomputer hard-disk drive is built into the system unit, but external hard-disk drives are available, as are removable hard-disk cartridges.

Large computer systems use removable hard-disk packs, fixed-disk drives, or RAID storage systems.

Switching from a microcomputer that uses only diskettes to one containing a hard disk is like discovering the difference between moving your household in several trips in a small sportscar and doing it all at once with a moving van. Whereas a high-density 3½-inch diskette holds 1.44 megabytes, a hard disk in a personal computer may store as many as 9 *gigabytes*. Indeed, at first with a hard disk you may feel you have more storage capacity than you'll ever need. However, after a few months, you may worry that you don't have enough. This feeling may intensify if you're using graphics-oriented programs or multimedia programs, with pictures and other features requiring immense amounts of storage.

Diskettes are made out of flexible material, which makes them "floppy." By contrast, **hard disks are thin but rigid metal platters covered with a substance that allows data to be stored in the form of magnetized spots.** Hard disks are also tightly sealed within an enclosed unit to prevent any foreign matter from getting inside. Data may be recorded on both sides of the disk platters.

We'll now describe the following aspects of hard-disk technology:

* Internal microcomputer hard-disk drives
* Defragmentation to speed up hard disks
* Microcomputer hard-disk variations, including external drives
* Hard-disk technology for large computer systems

Internal Microcomputer Hard-Disk Drives

In microcomputers, **hard disks are one or more platters sealed inside a *hard-disk drive* that is built into the system unit and cannot be removed.** The drive is installed in a *drive bay*, a shelf or opening in the computer cabinet. From the outside of a microcomputer, a hard-disk drive is not visible; it looks simply like part of the front panel on the system cabinet. Inside, however, is a disk or disks on a drive spindle, read/write heads mounted on an actuator (access) arm that moves back and forth, and power connections and circuitry. (■ *See Panel 7.5.*) The disks may be 5¼ inches in diameter, although today they are more often 3½ inches, with some even smaller. The operation is much the same as for a diskette drive, with the read/write heads locating specific pieces of data according to track and sector.

Hard disks have a couple of real advantages over diskettes—and at least one significant disadvantage.

* **Advantages—capacity and speed:** We mentioned that hard disks have a data storage capacity that is significantly greater than that of diskettes. Microcomputer hard-disk drives typically hold 40–500 megabytes and newer ones hold several gigabytes.

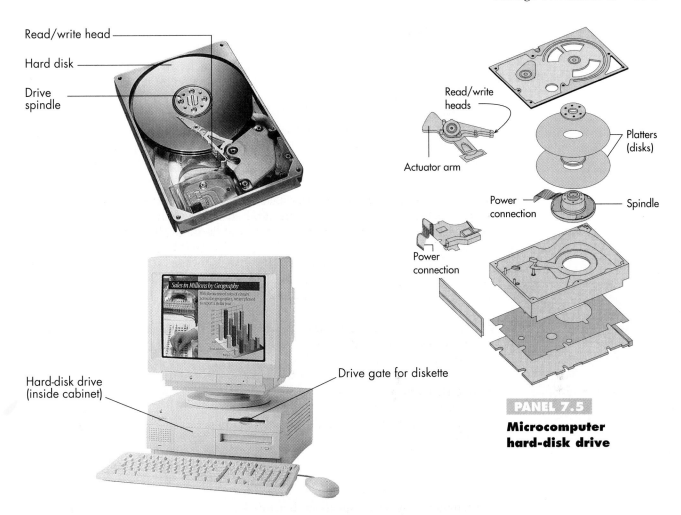

Read/write head

Hard disk

Drive spindle

Read/write heads

Actuator arm

Platters (disks)

Power connection

Spindle

Power connection

Sales in Millions by Geography

Hard-disk drive (inside cabinet)

Drive gate for diskette

PANEL 7.5

Microcomputer hard-disk drive

As for speed, hard disks allow faster access to data than do diskettes because a hard disk spins several times faster than a diskette. (A 2.1-gigabyte hard disk will spin at 7800 revolutions per minute [rpm], compared to 360 rpm for a diskette drive.)

• **Disadvantage—possible "head crash":** In principle a hard disk is quite a sensitive device. The read/write head does not actually touch the disk but rather rides on a cushion of air about 0.000001 inch thick. The disk is sealed from impurities within a container, and the whole apparatus is manufactured under sterile conditions. Otherwise, all it would take is a smoke particle, a human hair, or a fingerprint to cause what is called a head crash. **A *head crash* happens when the surface of the read/write head or particles on its surface come into contact with the disk surface, causing the loss of some or all of the data on the disk.** An incident of this sort could, of course, be a disaster if the data has not been backed up. There are firms that specialize in trying to retrieve (for a hefty price) data from crashed hard disks, though this cannot always be done.

Fragmentation & Defragmentation: Speeding Up Slow-Running Hard Disks

Like diskettes, for addressing purposes hard disks are divided into a number of invisible rings called *tracks* and typically nine invisible pie-shaped wedges called *sectors*. Data is stored within the tracks and sectors in groups of *clusters*. A *cluster* is the smallest storage unit the computer can access, and it always refers to a number of sectors. (The number varies among types of computers.)

With a brand-new hard disk, the computer will try to place the data in clusters that are *contiguous*—that is, that are adjacent (next to one another). Thus, data would be stored on track 1 in sectors 1, 2, 3, 4, and so on. However, as data files are updated and the disk fills up, the operating system stores data in whatever free space is available. Thus, files become fragmented. *Fragmentation* means that a data file becomes spread out across the hard disk in many noncontiguous clusters.

Fragmented files cause the read/write head to go through extra movements to find data, thus slowing access to the data. This means that the computer runs more slowly than it would if all the data elements in each file were stored in contiguous locations. To speed up the disk access, you must defragment the disk. *Defragmentation* means that data on the hard disk is reorganized so that data in each file is stored in contiguous clusters. Programs for defragmenting are available on some operating systems or as separate (external) software utilities (✓ p. 97).

PANEL 7.6

Removable hard-disk cartridge and portable hard-disk drive

Each cartridge has self-contained disks and read/write heads. The entire cartridge, which may store several gigabytes of data, may be removed for transporting or may be replaced by another cartridge.

Microcomputer Hard-Disk Variations: Power & Portability

If you have an older microcomputer or one with limited hard-disk capacity, some variations are available that can provide additional power or portability:

- **Miniaturization:** Newer hard-disk drives are less than half the height of older drives (1½ inches versus 3½ inches high) and so are called *half-height drives.* Thus, you could fit two disk drives into the bay in the system cabinet formerly occupied by one.

  In addition, the diameter of the disks has been getting smaller. Instead of 5¼ or 3½ inches, some platters are 2.5, 1.8, or even 1.3 inches in diameter. The half-dollar-size 1.3-inch Kittyhawk microdisk, which is actually designed for use in handheld computers, holds 21 megabytes of data.

- **External hard-disk drives:** An internal hard-disk drive may be the most convenient, but adding an external hard-disk drive is usually easy. Some detached external hard-disk drives, which have their own power supply and are not built into the system cabinet, can store gigabytes of data.

- **Hard-disk cartridges:** The disadvantages of hard disks are that they cannot be easily removed and that they have only a finite amount of storage. **Hard-disk cartridges consist of one or two platters enclosed along with read/write heads in a hard plastic case. The case is inserted into a detached external cartridge system connected to a microcomputer.** (■ *See Panel 7.6.*) A cartridge, which is removable and easily transported in a briefcase, may hold several gigabytes of data. An additional advantage of hard-disk cartridges is that they may be used for backing up data.

Hard-Disk Technology for Large Computer Systems

As a microcomputer user, you may regard secondary-storage technology for large computer systems with only casual interest. However, this technology forms the backbone of the revolution in making information available to you over communications lines. The large databases offered by such organizations as CompuServe, America Online, and Dialog, as well as the predicted movies-on-demand through cable and wireless networks, depend to a great degree on secondary-storage technology.

Secondary-storage devices for large computers consist of the following:

- **Removable packs:** A *removable-pack hard disk system* **contains 6–20 hard disks, of 10½- or 14-inch diameter, aligned one above the other in a sealed unit.** These removable hard-disk packs resemble a stack of phono-

graph records, except that there is space between disks to allow access arms to move in and out. Each access arm has two read/write heads—one reading the disk surface below, the other the disk surface above. However, only *one* of the read/write heads is activated at any given moment. The disk packs are inserted into receptacles in large, external drive units that can accommodate several disk packs at one time.

- **Fixed-disk drives:** ***Fixed-disk drives* are high-speed, high-capacity disk drives that are housed in their own cabinets.** Although not removable or portable, they generally have greater storage capacity and are more reliable than removable packs. A single mainframe computer might have 20 to 100 such fixed-disk drives attached to it.
- **RAID storage system:** A fixed-disk drive sends data to the computer along a single path. A *RAID storage system*, which consists of over 100 5¼-inch disk drives within a single cabinet, sends data to the computer along several parallel paths simultaneously. Response time is thereby significantly improved. *RAID* stands for Redundant Array of Inexpensive Disks.

  The advantage of a RAID system is that it not only holds more data than a fixed-disk drive within the same amount of space, but it also is more reliable, because if one drive fails, others can take over.

Optical Disks

Preview & Review: Optical disks are removable disks on which data is written and read using laser technology. Six types of optical disks are CD-ROM, CD Plus, CD-R, WORM, erasable, and DVD.

By now optical-disk technology is well known to most people. **An *optical disk* is a removable disk on which data is written and read through the use of laser beams.** The most familiar form of optical disk is the one used in the music industry. A *compact disk,* or *CD,* is an audio disk that uses digital code and that looks like a miniature phonograph record. A CD holds up to 74 minutes of high-fidelity stereo sound.

The optical-disk technology that has revolutionized the music business with music CDs is doing the same for secondary storage with computers. A single optical disk of the type called CD-ROM can hold up to about 700 megabytes of data. This works out to about 700,000 pages of text, or more than 7500 photos or graphics, or 20 hours of speech, or 77 minutes of video. Although some disks are used strictly for digital data storage, many combine text, visuals, and sound.

In the principal types of optical-disk technology, a high-power laser beam is used to represent data by burning tiny pits into the surface of a hard plastic disk. To read the data, a low-power laser light scans the disk surface: Pitted areas are not reflected and are interpreted as 0 bits; smooth areas are reflected and are interpreted as 1 bits. (■ *See Panel 7.7, next page.*) Because the pits are so tiny, a great deal more data can be represented than is possible in the same amount of space on a magnetic disk, whether flexible or hard.

The optical-disk technology used with computers consists of six basic types:

- CD-ROM disks
- CD Plus disks
- CD-R disks
- WORM disks
- Erasable optical disks
- DVD disks

Optical disks

(Top) In most cases, the disk producer uses a high-power laser beam in special recording equipment to burn tiny pits, in an encoded pattern, onto the disk's surface. *(Bottom)* The user's optical disk drive uses a low-power laser beam to read the code by reflecting the beam off smooth areas—interpreted as 1 bits. The beam does not reflect off pitted areas—interpreted as 0 bits.

Recording data

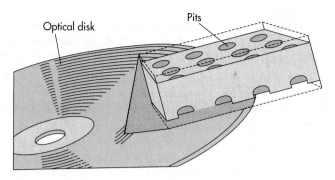

Reading data

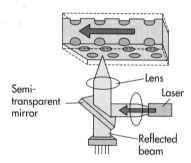
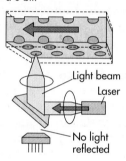

Reading "1": The laser beam reflects off the smooth surface, which is interpreted as a 1 bit.

Reading "0": The laser beam enters a pit and is not reflected, which is interpreted as a 0 bit.

CD-ROM drive in a microcomputer

CD-ROM

For microcomputer users, the best-known type of optical disk is the CD-ROM. **CD-ROM, which stands for *compact disk–read-only memory*, is an optical-disk format that is used to hold prerecorded text, graphics, and sound.**

Like music CDs, a CD-ROM is a read-only disk. *Read-only* means the disk cannot be written on or erased by the user. You as the user have access only to the data imprinted by the disk's manufacturer.

More and more microcomputers are being made with built-in CD-ROM drives. (■ *See Panel 7.8.*) However, many microcomputer users buy their CD-ROM drives separately and connect them to their computers.

At one time a CD-ROM drive was only a single-speed drive. Now double-speed and quadruple-speed drives—abbreviated 2X and 4X—are standard, and 6X and 8X drives are rapidly becoming affordable. A single-speed drive will access data at 150 kilobytes per second, a double-speed drive at 300 kilobytes per second. This means that a double-speed drive spins the compact disk twice as fast. "Quad-speed" (4X) CD-ROM drives access data at 600 kilobytes per second. The faster the drive spins, the more quickly it can deliver data to the processor.

Some CD-ROMs are designed to run only with Microsoft Windows on IBM-compatible computers or only on Apple Macintoshes. However, some are "hybrid" disks that include versions of the same program for both Macs and Windows.

There are many uses for CD-ROMs, including the following ones:

- **Entertainment and games:** Examples are 25 years of Garry Trudeau's comic strip, *Doonesbury* (on one disk), as well as games such as *Myst, Doom II, Dark Forces,* and *Sherlock Holmes, Consulting Detective.*

- **Music, culture, and films:** In *Xplora 1: Peter Gabriel's Secret World,* you can not only hear rock star Gabriel play his songs but also create "jam sessions" in which you can match up musicians from around the world and hear the result.[7] Other examples of such CD-ROMs are *Bob Dylan: Highway 61 Interactive; Multimedia Beethoven, Mozart, Schubert; Art Gallery; American Interactive; A Passion for Art;* and *Robert Mapplethorpe: An Overview.* Developers have also released several films on CD-ROM, such as the 1964 Beatles movie *A Hard Day's Night, This Is Spinal Tap,* and *The Day After Trinity.*

- **Encyclopedias, atlases, and reference works:** The principal CD-ROM encyclopedias are *The Grolier Multimedia Encyclopedia, Compton's Interactive Encyclopedia,* and *Microsoft Encarta Encyclopedia.* Each packs the entire text of a traditional multivolume encyclopedia onto a single disk, accompanied by pictures, maps, animation, and snippets of audio and video. All have pull-down menus and buttons to trigger various search functions.

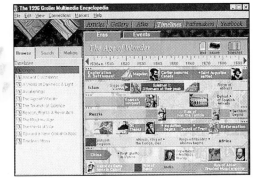

There are also street atlases (*Street Atlas U.S.A., Street-Finder*), which give detailed maps that can pinpoint addresses and show every block in a city or town, and trip planners (*TripMaker, Map 'n' Go*), which suggest routes, attractions, and places to eat and sleep.

Examples of other types of CD-ROM reference works are *Eyewitness History of the World, Eyewitness Encyclopedia of Nature, Skier's Encyclopedia,* and *The Way Things Work.* You can get the full text of 1750 great works of literature and other books and documents on *Library of the Future.*

- **Catalogs:** Publishers have also discovered that CD-ROMs can be used as electronic catalogs, or even "megalogs."

- **Education and training:** Want to learn photography? You could buy a pair of CD-ROMs by Bryan Peterson called *Learning to See Creatively* (about composition) and *Understanding Exposure* (discussing the science of exposure). When you pop these disks in your computer, you can practice on screen with lenses, camera settings, film speeds, and the like.[8] CD-ROMs (*Score Builder for the SAT, Inside the SAT*) are also available to help students raise their scores on the Scholastic Aptitude Test.

- **Edutainment:** *"Edutainment" software* consists of programs that look like games but actually teach, in a way that feels like fun.[9] An example for children ages 3–6 is *Yearn 2 Learn Peanuts,* which teaches math, geography, and reading.

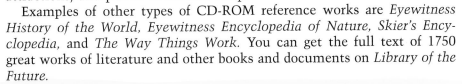

- **Books and magazines:** Book publishers have hundreds of CD-ROM titles, ranging from *Discovering Shakespeare* and *The Official Super Bowl Commemorative Edition* to business directories such as *ProPhone Select* and *11 Million Businesses Phone Book.*

Clearly CD-ROMs are not just a mildly interesting technological improvement. They have evolved into a full-fledged mass medium of their own, on the way to becoming as important as books or films.

CD Plus/Enhanced CD: The Hybrid CD for Computers & Stereos

In the fall of 1995, disks began to appear with a new technical standard representing the convergence of software and music—*CD Plus*, also called *enhanced CD*. Many people believe the disks "represent a transformation in the way we use music—from a passive activity to an engaging interactive experience," said one writer. "Skeptics contend it's just the latest scheme to separate you and your money."[10]

A *CD Plus*, or *enhanced CD*, is a digital disk that is a hybrid of audio-only compact disk and multimedia CD-ROM. It can be played like any music CD in a stereo system, then put in a computer's CD-ROM drive to provide additional audio, video, text, and graphics.

A problem at present is the number of technical snags.[11,12] With some CD Plus disks, for instance, you may have to skip track 1 manually when playing the disk on an audio CD player because "noise" from the computer data on that track can damage the speakers on your stereo (or your ears). Some CD Plus disks won't perform on users' present CD-ROM drives at all, such as the single-speed models. Indeed, it is variously estimated that CD Plus won't work in one-third to one-half of today's installed CD-ROM drives. The bottom line, then, is that the standard is really designed for *future* computer systems, which will all be CD Plus compatible.

CD-R: Recording Your Own CDs

***CD-R*, which stands for *compact disk–recordable*, is a CD format that allows you to write data onto a specially manufactured disk that can then be read by a standard CD-ROM drive.** Like CD-ROM disks, CD-R disks cannot be erased.

Until quite recently, CD-R technology was far too expensive and complicated for ordinary computer users. Now systems for recording your own CD-ROMs have dropped to under $1000, and by early 1996 blank disks had become available in bulk for about $7 each. Says one article, "With an in-house production system consisting of the CD-R hardware and software,

a fast PC, and a dedicated hard drive, an experienced user can record a CD-ROM in an hour or less."[13]

One of the most interesting examples of CD-R technology is the Photo CD system. Developed by Eastman Kodak, *Photo CD* is a technology that allows photographs taken with an ordinary 35-millimeter camera to be stored digitally on an optical disk. You can view the disk on one of Kodak's own Photo CD players, which attaches directly to a television set. Or the disk will play in most CD-ROM-equipped personal computers running Windows or the Macintosh Operating System.

Photo CD is used mostly by professional photographers, multimedia developers, and designers.[14] Photo CD is particularly significant for the impact it will have on the manipulation of photographs. With the right software, you can flip, crop, and rotate photos and incorporate them into desktop-publishing materials and print them out on laser printers.[15] Commercial photographers and graphics professionals can manipulate images at the level of pixels (✓ p. 165). For example, photographers can easily do *morphing*—merge images from different sources, such as superimpose the heads of show-business figures in the places of the U.S. presidents on Mount Rushmore. "Because the image is digital," says one writer, "it can be taken apart pixel by pixel and put back together in many ways."[16] This helps photo professionals further their range, although at the same time it presents a danger that photographs will be compromised in their credibility.

WORM Disks

WORM stands for *write once, read many*. A *WORM disk* can be written, or recorded, onto just once and then cannot be erased; it can be read many times. In principle, WORM technology is similar to CD-R technology; however, it is a bit older and more expensive than CD-R technology. WORM technology is useful for storing data for backup and archival purposes because it can store greater volumes of data than other types of CD-ROMs (and magnetic tape). A mainframe-based WORM disk may hold 200 gigabytes of data; a microcomputer-based WORM disk can hold about 600 megabytes. (WORM disks require special WORM drives.)

R E A D M E

Case Study: The Magic of Morphing

Ever wish you could turn your local congressperson into the rodent of your choice? Give it a whirl with HSC Software's *Digital MORPH.* In fact, you might want to try turning your spouse into a squash or your landlord into an attractive ceiling fixture while you're at it. . . .

Morphing, based on the Greek word *morphe,* meaning "shape," is a type of technology that stretches and distorts one image to another. Digital MORPH is a special-effects program that produces shape-shifting effects such as those seen in the movie *Terminator 2* (where the antagonist shape-shifts into different objects) and Michael Jackson's *Black or White* video in which faces morph from one race to another.

The program was originally created in a higher-end version to help plastic and oral surgeons and orthodontists determine the after-surgery appearances of their patients. Today, it is available from HSC Software, a developer of graphic imaging, special effects, and multimedia software, for those with Microsoft Windows-capable computers and a desire for some creative fun.

Digital MORPH works its magic upon scanned images—images that have been digitally recorded by a scanner (a device that reads images using light beams much like a photocopier). To morph an image, the user need only select source and destination images (a starting and finishing image), select field lines (these determine where on the image changes will take place), and select the number of frames in which the change will take place. The program takes care of the rest.

"It's a lot of fun playing with images on a computer," says HSC CEO John Wilcazk. "Imagine morphing your boss into a gorilla, your sibling into the family dog, or your car into a leopard. The creativity is really endless, not only for individual users but also for multimedia producers, graphic artists, animators, and business-presentation creators."

—Tosca Moon Lee, "Shape-Shifting with Digital MORPH," *PC Novice*

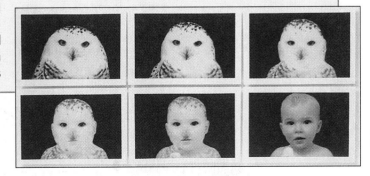

Erasable Optical Disks

An *erasable optical disk* allows users to erase data so that the disk can be used over and over again. The most common type of erasable and rewritable optical disk is probably the *magneto-optical (MO) disk,* which uses aspects of both magnetic-disk and optical-disk technologies. Most personal computer users haven't used MO disk drives, but that may change. One recently released MO drive writes on removable disks that hold 4.6 gigabytes of data in all. However, such drives cost about twice as much as a 2-gigabyte hard-disk drive.

DVD & DVD-ROM Disks: The "Digital Convergence" Disk

According to the various industries sponsoring it, *DVD* isn't an abbreviation for anything. The letters used to stand for "digital video disk" and later, when its diverse possibilities became obvious, for "digital versatile disk."[17] But DVD is the designation that Sony/Philips and Toshiba/Time Warner agreed to in late 1995, when they avoided a "format war" by joining forces to meld their two advanced disk designs into one. Suffice it to say that *DVD* is a 5-inch optically readable digital disk that looks like an audio compact disk but can store 4.8 gigabytes of data on a side, allowing great data storage, studio-quality video images, and theater-like surround sound. Actually, the home-entertainment version is called simply the *DVD.* **The computer version of the DVD is called the *DVD-ROM disk.* It represents a new generation of high-density CD-ROM disks, with either write-once or rewritable capabilities.** To play a DVD-ROM, you need a DVD-ROM drive, which can also play regular CD-ROM disks.

DVD, says *San Jose Mercury News* computer writer Mike Langberg,

> is clearly destined for greatness as the first information storage technology designed from the ground up to work with all types of electronics, televisions, stereos, computers, and videogame players. That means DVD will finally launch the much-heralded concept of "convergence," where the lines blur between all these devices.[18]

Interactive & Multimedia CD-ROM Formats

As use of CD-ROMs has burgeoned, so has the vocabulary, creating difficulty for consumers. Much of this confusion arises in conjunction with the words *interactive* and *multimedia.*

As we mentioned earlier, **interactive means that the user controls the direction of a program or presentation on the storage medium.** That is, there is back-and-forth interaction, as between a player and a videogame. You could create an interactive production of your own on a regular 3½-inch diskette. However, because of its limited capacity, you would not be able to present full-motion video and high-quality audio. The best interactive storage media are CD-ROMs and DVD-ROMs.

Multimedia (an adjective or noun) refers to technology that presents information in more than one medium, including text, graphics, animation, video, sound, and voice. As used by telephone, cable, broadcasting, and entertainment companies, *multimedia* also refers to the so-called Information Superhighway. On this avenue various kinds of information and entertainment will be delivered to your home through wired and wireless communication lines.

There are perhaps 20 different CD-ROM formats. The majority of nongame CD-ROM disks are available for Macintosh or for Windows- or DOS-based microcomputers. However, there are a host of other formats—CD-I, CDTV, Video CD, Sega CD, 3DO, CD+G, CD+MIDI, CD-V—that are not mutually compatible. Most will probably disappear as the DVD format takes over.

Now we leave the topic of optical disk storage and turn to a much older type of storage, presently used principally for backup—magnetic tape.

Magnetic Tape

Preview & Review: Magnetic tape is thin plastic tape on which data can be represented with magnetized spots. On large computers, tapes are used on magnetic-tape units. On microcomputers, tapes are used in cartridge tape units.

Magnetic tape **is thin plastic tape that has been coated with a substance that can be magnetized; data is represented by the magnetized or nonmagnetized spots.** Today "mag tape" is used mainly to provide backup, or duplicate storage. It's a much slower form of secondary storage, but it's also less expensive than disk storage.

The two principal forms of tape storage of interest to us are *magnetic-tape units*, traditionally used with mainframes and minicomputers, and *cartridge tape units*, which are often used for backup on microcomputers.

Magnetic-Tape Units for Large Computers

The kind of cassette tapes you use for an audiotape recorder are 200 feet long and record 200 bytes per inch. By contrast, a reel of magnetic tape used in mainframe and minicomputer storage systems is ½ inch wide, 3600 feet long, and can hold 1600–6250 bpi. A traditional 10½-inch tape reel can hold up to 250 megabytes of data.

Tapes are used on *magnetic-tape units* or *magnetic-tape drives*, which consist of a read/write head and spindles for mounting two tape reels: a supply reel and a take-up reel. The tape is reeled off the supply reel, fed through pulleys that regulate its speed and hold it still long enough for data to be read from it or written to it by the read/write heads, and then wound up on the take-up reel. During the writing process, any existing data on the tape is automatically written over, or erased.

Large organizations, such as public utilities, often use reels of mag tape for storing backup records of essential data, such as customer names and account numbers. Usually these reels are housed in *tape libraries*, or special rooms, and there are strict security procedures governing their use.

Cartridge Tape Units

"Sometimes I think computing without backup should be against the law," says a computer journalist. "Once you could copy to a few floppy disks. In an era of gigabyte drives and enormous multimedia files, that just isn't practical because that could mean shuffling hundreds of floppies each time you back up."[19]

You could use removable disks, such as Iomega's and Syquest's, but even they would require 10 disks each (at $20 apiece) to back up a 1-gigabyte hard-disk drive. Here is the problem for which cartridge tape units are the solution.

Cartridge tape units, **also called** *tape streamers,* **are used to back up data from a microcomputer hard disk onto a tape cartridge.** These tapes may hold

500–1600 megabytes; some hold 2 gigabytes or more. Is backup really that important? "Tape drives are similar to life insurance," says a computer columnist, "in that you hope you never need it and resent the premiums— until you need the coverage."[20] (■ *See Panel 7.9.*)

Now that we have covered the basic forms of secondary storage, we'll proceed to look at *what* is stored on them—the topic of databases.

Organizing Data in Secondary Storage: Databases

Preview & Review: Databases are integrated collections of files. Organizations usually appoint a database administrator to manage the database.

Many types of files exist—for example, the letter you just wrote and saved is stored as a document file; the bank's daily transactions are stored in a transaction file; and software instructions are stored in program files. Here we are concerned with *database files*, the type used for keeping records. However, databases are not just an interesting way to computerize filing systems. Databases are *integrated* collections of files, which, as we shall see, makes them usable in more ways than traditional filing systems (computerized or not).

A database may be small, contained entirely within your own personal computer. Or it may be massive, available online through computer and telephone connections. Such online databases are of special interest to us in this book because they offer us phenomenal resources that until recently were unavailable to most ordinary computer users.

Microcomputer users can set up their own databases using popular database management software like that we discussed earlier (✓ p. 58). Examples are Paradox, Access, dBASE 5, and FoxPro. Such programs are used, for example, by graduate students to conduct research, by salespeople to keep track of clients, by purchasing agents to monitor orders, and by coaches to keep watch on other teams and players.

Some databases are so large that they cannot possibly be stored in a microcomputer. Some of these can be accessed by going online through a microcomputer, or other computer. Such databases, sometimes called *information utilities*, represent enormous compilations of data, any part of which is available, for a fee, to the public.

PANEL 7.9

How much is backup worth?

"'When you lose a disk, you're not only losing the hardware and software,' said John L. Copen, president of Integ, an information protection company in Manhattan. 'The information has to be reproduced, and if you have to reproduce it without a backup. . . .'

Mr. Copen demonstrates the point by holding up a digital audiotape (DAT) cassette, one of the newer technologies used for backing up data on larger hard disk drives, the kind that act as hubs for networks of personal computers in an office.

'I ask people in the audience what it's worth,' he said. 'It's a little cassette about the size of a credit card. The cassette costs about $16. I ask them to guess how much it can store. Forty megs? Eighty megs? It stores four gigabytes.' A gigabyte is roughly a thousand megabytes, or a billion characters of information.

'How much information can you put in four gigs?' Mr. Copen continued. 'About 20,000 big spreadsheets, which translates to about 100,000 days of work, or 800,000 hours. At $20 an hour, that's $16 million. Never before have people been able to reach down, pick up a cassette and walk out the door with $16 million of data in their pocket.'"

—Peter H. Lewis, "Finding an Electronic Safe-Deposit Data Box," *New York Times*

Examples of well-known information utilities—more commonly known as *online services*—are America Online, CompuServe, and Prodigy. As we describe in Chapter 8, these services offer access to news, weather, travel information, home shopping services, reference works, and a great deal more. Some public-access databases are specialized, such as Lexis, which gives lawyers access to local, state, and federal laws.

Other types of large databases are private—collections of records shared or distributed throughout a company or other organization. Generally, the records are available only to employees or selected individuals and not to outsiders.

For example, many university libraries have been transforming drawers of catalog cards into electronic databases for use by their students and faculty. Libraries at Yale, Johns Hopkins, and other universities have contracted with a Virginia company called The Electronic Scriptorium, which employs monks and nuns at six monasteries to convert card catalogs to an electronic system.[21,22]

One thing that large databases have in common is that they must be managed. This is done by the database administrator.

The Database Administrator

The information in a large database—such as a corporation's patents, formulas, advertising strategies, and sales information—is the organization's lifeblood. Someone, then, needs to manage all activities related to the database. This person is the *database administrator (DBA)*, a person who coordinates all related activities and needs for a corporation's database. The responsibilities include the following:

- **Database design, implementation, and operation:** At the beginning, the DBA helps determine the design of the database. Later he or she determines how space will be used on secondary-storage devices, how files and records may be added and deleted, and how losses may be detected and remedied.

- **Coordination with users:** The DBA determines user access privileges, assists in establishing priorities for requests, and adjudicates conflicting user needs.

- **System security:** The DBA sets up and monitors a system for preventing unauthorized access to the database.

- **Backup and recovery:** Because loss of data or a crash in the database could vitally affect the organization, the DBA needs to make sure the system is regularly backed up. He or she also needs to develop plans for recovering data or operations should a failure or disaster occur.

- **Performance monitoring:** The DBA monitors the system to make sure it is serving users appropriately. A standard complaint is that the system is too slow, usually because too many users are trying to access it.

The Data Storage Hierarchy & the Concept of the Key Field

Preview & Review: Data in storage is organized as a hierarchy: bits, bytes, fields, records, and files, which are the elements of a database. In data organization, the role of the key field, which uniquely identifies a record, is important.

How does a database actually work? To understand this, first we need to consider how stored data is structured—the *data storage hierarchy* and the concept of *key field*.

The Data Storage Hierarchy

Data can be grouped into a hierarchy of categories, each increasingly more complex. **The *data storage hierarchy* consists of the levels of data stored in a computer database: bits, bytes (characters), fields, records, and files.** (■ *See Panel 7.10.*)

As mentioned, **0 or 1 is called a *bit*. A unit of 8 bits is called a *byte;* it may be used to represent a character, digit, or other value,** such as A, ?, or 3. Bits and bytes are the building blocks for representing data, whether it is being processed, stored, or telecommunicated. Bits and bytes are what the computer hardware deals with, but you need not be concerned with them when you are working with databases. You will, however, in this case be dealing with characters, fields, records, files, and databases.

- Byte (character): A *byte* is a group of 8 bits. **A *character* may be—but is not necessarily—the same as a byte. A character is a single letter, number, or special character** such as ;, $, or %.
- Field: **A *field* is a unit of data consisting of one or more characters.** An example of a field is your name, your address, or your Social Security number.

   Note: One reason the Social Security number is often used in computing—for good or for ill—is that, perhaps unlike your name, it is a *distinctive* (unique) field. Thus, it can be used to easily locate information about you. Such a field is called a *key field*. More on this below.
- Record: **A *record* is a collection of related fields.** An example of a record would be your name *and* address *and* Social Security number.
- File: **A *file* is a collection of related records.** An example of a file is collected data on everyone employed in the same department of a company, including all names, addresses, and Social Security numbers.
- Database: **A *database* is an integrated collection of files.** A company database might include files on all past and current employees in all departments. There would be various files for each employee: payroll, retirement benefits, sales quotas and achievements (if in sales), and so on.

The Key Field

An important concept in data organization is that of the *key field*. **A *key field* contains unique data used to identify a record so that it can be easily retrieved and processed.** The key field is often an identification number, Social Security number, customer account number, or the like. As mentioned, the primary characteristic of the key field is that it is *unique*. Thus, numbers are clearly preferable to names as key fields because there are many people with common names like James Johnson, Kim Lee, Susan Williams, Ann Wong, or Roberto Sanchez, whose records might be confused.

Before databases existed there were files. To understand database management, therefore, we first need to understand file management and file management systems.

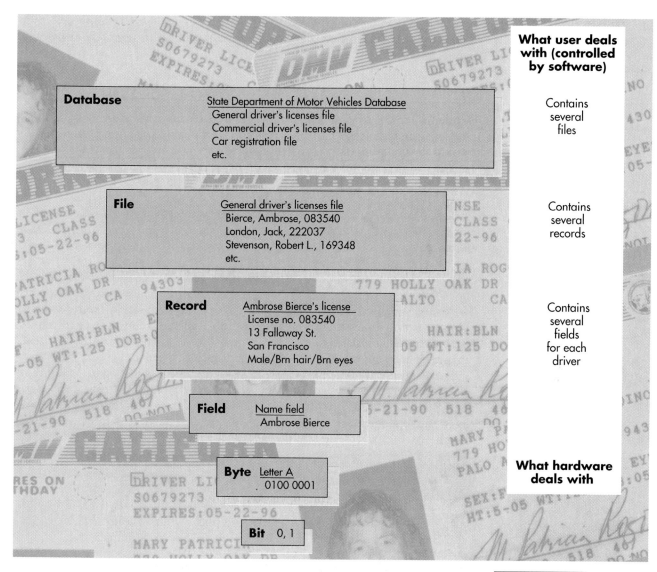

What user deals with (controlled by software)

Database
State Department of Motor Vehicles Database
General driver's licenses file
Commercial driver's licenses file
Car registration file
etc.

Contains several files

File
General driver's licenses file
Bierce, Ambrose, 083540
London, Jack, 222037
Stevenson, Robert L., 169348
etc.

Contains several records

Record
Ambrose Bierce's license
License no. 083540
13 Fallaway St.
San Francisco
Male/Brn hair/Brn eyes

Contains several fields for each driver

Field
Name field
Ambrose Bierce

Byte
Letter A
0100 0001

Bit 0, 1

What hardware deals with

Data storage hierarchy: how data is organized

Bits are organized into bytes, bytes into fields, fields into records, records into files. Related files may be organized into a database.

File Management: Basic Concepts

Preview & Review: Master files and transaction files are commonly used to update data.

Files may be processed by batch processing or by real-time processing. Batch processing tends to favor offline storage; online, or real-time, processing requires online storage.

Three methods of file organization are sequential access, direct access, and indexed-sequential.

Master File & Transaction File

Among the several types of files two are traditionally used to update data: a master file and a transaction file.

- **Master file:** The *master file* is a data file containing relatively permanent records that are generally updated periodically. An example of a master file would be the address-label file for all students currently enrolled at your college.

- **Transaction file:** The *transaction file* is a temporary holding file that holds all changes to be made to the master file: additions, deletions, and revisions.

    For example, in the case of the address labels for your college, a transaction file would hold new names and addresses to be added (because over time new students enroll) and names and addresses to be deleted (because students leave). It would also hold revised names and addresses (because students change their names or move). Each month or so, the master file would be *updated* with the changes called for in the transaction file.

Batch Versus Online Processing

Data may be taken from secondary storage and processed in one of two ways: (1) "later," via *batch processing*, or (2) "right now," via *online (real-time) processing.*

- **Batch processing:** **In *batch processing*, data is collected over several days or weeks in a transaction file and then processed all at one time, as a "batch," against a master file.** Thus, if users need to make some request of the system, they must wait until the batch has been processed. Batch processing is less expensive than online processing and is suitable for work in which immediate answers to queries are not needed.

    An example of batch processing is that done by banks for balancing checking accounts. When you deposit a check in the morning, the bank will make a record of it. However, it will not compute your account balance until the end of the day, after all checks have been processed in a batch.

- **Online processing:** *Online processing*, **also called *real-time processing*, means entering transactions into a computer system as they take place and updating the master files as the transactions occur.** For example, when you use your ATM card to withdraw cash from an automated teller machine, the system automatically computes your account balance then and there. Airline reservation systems also use online processing.

Offline Versus Online Storage

Whether it's on magnetic tape or on some form of disk, data may be stored either offline or online.

- **Offline:** *Offline storage* **means that data is not directly accessible for processing until the tape or disk it's on has been loaded onto an input device.** That is, the storage is not under the direct, immediate control of the central processing unit.

- **Online:** *Online storage* **means that stored data is directly accessible for processing.** That is, storage is under the direct, immediate control of the central processing unit. You need not wait for a tape or disk to be loaded onto an input device.

For processing to be online, the storage must be online and *fast.* Generally, this means storage on disk rather than magnetic tape. With magnetic tape, it is not possible to go directly to the required record; instead, the read/write head has to search through all the records that precede it, which takes time. With disk, however, the system can go directly and quickly to the record—just as a CD player can go directly to a particular spot on a music CD.

File Organization: Three Methods

In general, tape storage falls in the category of sequential access storage. *Sequential access storage* **means that information is stored in sequence,** such as alphabetically. Thus, you would have to search a tape past all the information from A to J, say, before you got to K. This process may require running several inches or feet off a reel of tape, which, as we said, takes time.

Disk storage, by contrast, generally falls into the category of direct access storage (although data *can* be stored sequentially). *Direct access storage* **means that the system can go directly to the required information.** Because you can directly access information, retrieving data is much faster with magnetic or optical disk than it is with magnetic tape.

From these two fundamental forms, computer scientists devised three methods of organizing files for secondary storage: *sequential*, *direct*, and *indexed-sequential*. (■ *See Panel 7.11, next two pages.*)

- **Sequential file organization:** *Sequential file organization* stores records in sequence, one after the other. This is the only method that can be used with magnetic tape. Records can be retrieved only in the sequence in which they were stored. The method can also be used with disk.

  Sequential file organization is useful, for example, when a large portion of the records needs to be accessed, as when a mail-order house is sending out catalogs to all names on a mailing list. The method also is less expensive than other methods because it uses magnetic tape, which is cheaper than magnetic or optical disk.

  The disadvantage of sequential file organization is that records must be ordered in a particular way and so searching for data is slow.

- **Direct file organization:** Instead of storing records in sequence, *direct file organization*, or *random file organization*, stores records in no particular sequence. A record is retrieved according to its key field, or unique element of data. This method of file organization is used with hard or optical disks. It is ideal for applications such as airline reservations systems or computerized directory-assistance operations. In these cases, records need to be retrieved only one at a time, and there is no fixed pattern to the requests for records.

  Direct file organization is much faster than sequential file organization for finding a specific record. However, because the method requires hard-disk or optical-disk storage, it is more expensive than magnetic tape. Moreover, it is not as efficient as sequential file organization for listing large numbers of records.

- **Indexed-sequential file organization:** A compromise has been developed between the preceding two methods. *Indexed-sequential file organization*, or simply *indexed file organization*, stores records in sequential order. However, the file in which the records are stored contains an index that lists each record by its key field and identifies its physical location on the disk. The method requires magnetic or optical disk.

  This method is slower than direct file organization because it requires an index search. The indexed-sequential method is best when large batches of transactions occasionally must be updated, yet users want frequent, rapid access to records. For example, bank customers and tellers want to have up-to-the-minute information about checking accounts, but every month the bank must update bank statements to send to customers.

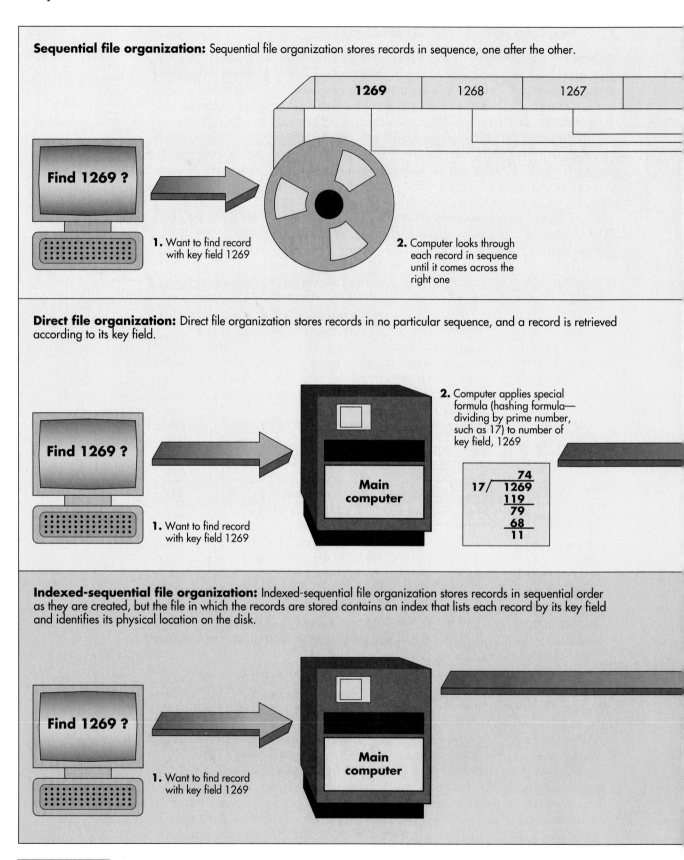

Sequential file organization: Sequential file organization stores records in sequence, one after the other.

1269	1268	1267

Find 1269 ?

1. Want to find record with key field 1269

2. Computer looks through each record in sequence until it comes across the right one

Direct file organization: Direct file organization stores records in no particular sequence, and a record is retrieved according to its key field.

Find 1269 ?

1. Want to find record with key field 1269

Main computer

2. Computer applies special formula (hashing formula—dividing by prime number, such as 17) to number of key field, 1269

```
         74
17/   1269
      119
       79
       68
       11
```

Indexed-sequential file organization: Indexed-sequential file organization stores records in sequential order as they are created, but the file in which the records are stored contains an index that lists each record by its key field and identifies its physical location on the disk.

Find 1269 ?

1. Want to find record with key field 1269

Main computer

PANEL 7.11

Three methods of file organization: sequential, direct, and indexed-sequential

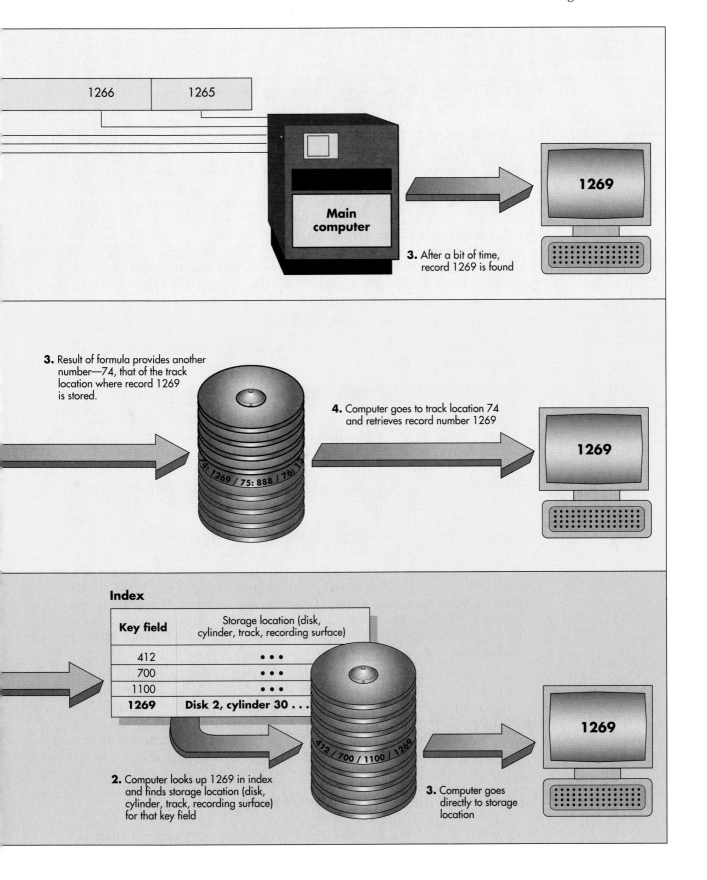

1266   1265

Main computer

1269

3. After a bit of time, record 1269 is found

3. Result of formula provides another number—74, that of the track location where record 1269 is stored.

74: 1269 / 75: 888 / 76: 1...

4. Computer goes to track location 74 and retrieves record number 1269

1269

Index

Key field	Storage location (disk, cylinder, track, recording surface)
412	• • •
700	• • •
1100	• • •
1269	**Disk 2, cylinder 30 . . .**

412 / 700 / 1100 / 1269

2. Computer looks up 1269 in index and finds storage location (disk, cylinder, track, recording surface) for that key field

3. Computer goes directly to storage location

1269

File Management Systems

Preview & Review: Files may be retrieved through a file management system, one file at a time. Disadvantages of a file management system are data redundancy, lack of data integrity, and lack of program independence.

In the 1950s, when commercial use of computers was just beginning, magnetic tape was the storage medium and records and files were stored sequentially. To work with these files, a user needed a file management system.

A *file management system,* or *file manager,* **is software for creating, retrieving, and manipulating files, one file at a time.** Traditionally, a large organization such as a university would have different files for different purposes. For you as a student, for example, there might be one file on you for course grades, another for student records, and a third for tuition billing. Each file would be used independently to produce its own separate reports. If you changed your address, someone had to make the change separately in each file.

Disadvantages of File Management Systems

File management systems worked well enough for their time, but they had several disadvantages:

- **Data redundancy:** *Data redundancy* means that the same data fields appear in many different files and often in different formats. Thus, separate files tend to repeat some of the same data over and over. A student's course grades file and tuition billing file would contain some of the same data (name, address, telephone number). When data fields are repeated in different files, they waste storage space.

- **Lack of data integrity:** *Data integrity* means that data is accurate, consistent, and up to date. However, when the same data fields (a student's address and phone number, for example) must be changed in different files, some files may be missed and mistakes will be made. The result is that some reports will be produced with erroneous information.

- **Lack of program independence:** With file management systems, different files were often written by different programmers using different file formats. Thus, the files were not *program-independent.* The arrangement meant more time was required to maintain files. It also prevented a programmer from writing a single program that would access all the data in multiple files.

As computers became more and more important in daily life, the frustrations of working with separate, redundant files lacking data integrity and program independence began to be overwhelming. Fortunately, magnetic disk began to supplant magnetic tape as the most popular medium of secondary storage, leading to new possibilities for managing data, which we discuss next.

Database Management Systems

Preview & Review: Database management systems are an improvement over file management systems. They use database management system software, which controls the structure of the database and access to the data. The advantages of databases are reduced data redundancy, improved data integrity, more program independence, increased user productivity, and increased security.

When magnetic tape began to be replaced by magnetic disk, sequential access storage began to be replaced by direct access storage. The result was a new technology and new software: the database management system.

As mentioned, a *database* is a collection of integrated files, meaning that the file records are logically related, or cross-referenced, to one another. Thus, even though all the pieces of data on a topic are kept in records in different files, they can easily be organized and retrieved with simple requests.

The software for manipulating databases is **database management system (DBMS) software, or a database manager, a program that controls the structure of a database and access to the data.** With a DBMS, then, a large organization such as a university might still have different files for different purposes. As a student, you might have the same files as you would have had in a file management system (one for course grades, another for student records, and a third for tuition billing). However, in the database management system, data elements are integrated (cross-referenced) and shared among different files. Thus, your address data would need to be in only one file, because it can be automatically accessed by the other files.

Advantages & Disadvantages of a DBMS

The advantages of databases and DBMS software are as follows:

- **Reduced data redundancy:** Instead of the same data fields being repeated in different files, in a database the information appears just once. The single biggest advantage of a database is that the *same* information is available to *different* users. Moreover, reduced redundancy lowers the expense of storage media and hardware, because more data can be stored on the media.

- **Improved data integrity:** Reduced redundancy increases the chances of data integrity—that the data is accurate, consistent, and up to date—because each updating change is made in only one place.

- **More program independence:** With a database management system, the program and the file formats are the same, so that one programmer or even several programmers can spend less time maintaining files.

- **Increased user productivity:** Database management systems are fairly easy to use, so that users can get their requests for information answered without having to resort to technical manipulations. In addition, users don't have to wait for a computer professional to provide what they need.

- **Increased security:** Although various departments may share data in common, access to specific information can be limited to selected users. Thus, through the use of passwords, a student's financial, medical, and grade information in a university database is made available only to those who have a legitimate need to know.

Although there are clear advantages to having databases, there are still some disadvantages:

- **Cost issues:** Installing and maintaining a database is expensive, particularly in a large organization. In addition, there are costs associated with training people to use it correctly.

- **Security issues:** Although databases can be structured to restrict access, it's always possible unauthorized users will get past the safeguards. And when they do, they may have access to *all* the files, not just a few. In addition, if a database is destroyed by fire, earthquake, theft, or hardware or software problems, it could be fatal to an organization's business activities—unless steps have been taken to regularly make backup copies of the files and store them elsewhere.

• **Privacy issues:** Databases may hold information they should not and be used for unintended purposes, perhaps intruding on people's privacy. Medical data, for instance, may be used inappropriately in evaluating an employee for a job promotion.

Types of Database Organization

Preview & Review: Types of database organization are hierarchical, network, relational, and object-oriented.

Just as files can be organized in different ways (sequentially or directly, for example), so can databases. The four most common arrangements for database management systems are *hierarchical*, *network*, *relational*, and *object-oriented*.

Hierarchical Database

In a *hierarchical database*, fields or records are arranged in related groups resembling a family tree, with lower-level records subordinate to higher-level records. (■ *See Panel 7.12.*) A lower-level record is called a *child*, and a higher-level record is called a *parent*. The parent record at the top of the database is called the *root record*.

Hierarchical DBMSs are the oldest of the four forms of database organization, and are still used in some reservation systems. Also, accessing or updating data is very fast, because the relationships have been predefined. However, because the structure must be defined in advance, it is quite rigid. There may be only one parent per child and no relationships among the child records. Moreover, adding new fields to database records requires that the entire database be redefined.

PANEL 7.12

Hierarchical database: example of a cruise ship reservation system

Records are arranged in related groups resembling a family tree, with "child" records subordinate to "parent" records. Cabin numbers (A-1, A-2, A-3) are children of the parent July 15. Sailing dates (April 15, May 30, July 15) are children of the parent The Love Boat. The parent at the top, Miami, is called the "root record."

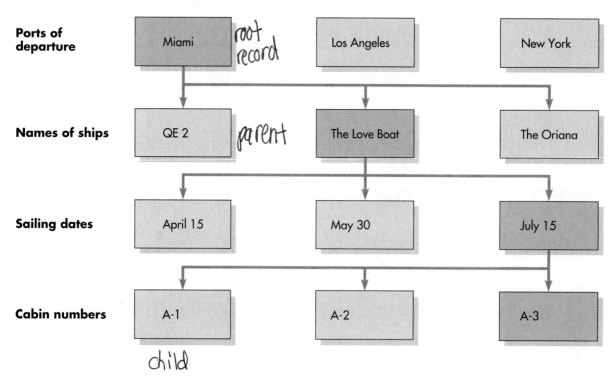

Network Database

A *network database* is similar to a hierarchical DBMS, but each child record can have more than one parent record. (■ *See Panel 7.13.*) Thus, a child record, which in network database terminology is called a *member*, may be reached through more than one parent, which is called an *owner*.

This arrangement is more flexible than the hierarchical one, because different relationships may be established between different branches of data. However, it still requires that the structure be defined in advance. Moreover, there are limits to the number of links that can be made.

Relational Database

More flexible than hierarchical and network database models, **the *relational database* relates, or connects, data in different files through the use of a key field, or common data element.** (■ *See Panel 7.14, next page.*) In this arrangement there are no access paths down through a hierarchy. Instead, data elements are stored in different tables made up of rows and columns. In database terminology, the tables are called *relations*, the rows are called *tuples*, and the columns are called *attributes*. All related tables must have a key field that uniquely identifies each row.

The advantage of relational databases is that the user does not have to be aware of any "structure." Thus, they can be used with little training. Moreover, entries can easily be added, deleted, or modified. A disadvantage is that some searches can be time consuming. Nevertheless, the relational model has become popular for microcomputer DBMSs, such as Paradox and Access.

Object-Oriented Database

Relational databases are good for storing and manipulating business data. Object-oriented databases can handle not only numerical and text data but any type of data, including graphics, audio, and video. Object-oriented databases, then, are important in the new world of video servers and technological convergence.

PANEL 7.13

Network database: example of a college class scheduling system

This is similar to a hierarchical database, but each child, or "member," record can have more than one parent, or "owner." For example, Student B's owners are instructors D. Barry and R. DeNiro. The owner Broadcasting 210 has three members—D. Barry, R. DeNiro, and D. Rather.

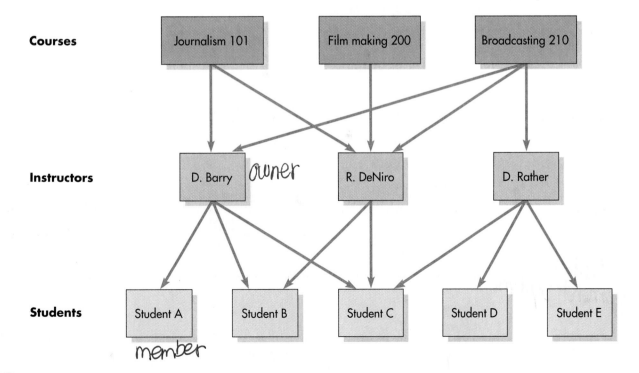

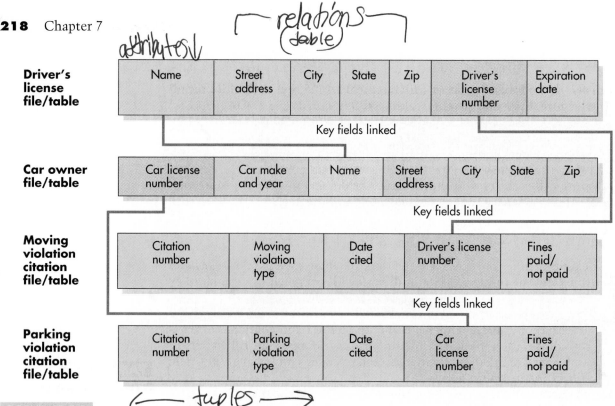

relations (table)

attributes

← tuples →

PANEL 7.14

Relational database: example of a state department of motor vehicles database

This kind of database relates, or connects, data in different files through the use of a key field, or common data element. The relational database does not require predefined relationships.

An *object-oriented database* uses "objects" as elements within database files. An object consists of (1) text, sound, video, and pictures and (2) instructions on the action to be taken on the data. A hierarchical, network, or relational database would contain only numeric and text data about a student —identification number, name, address, and so on. By contrast, an object-oriented database might also contain the student's photograph, a "sound bite" of his or her voice, and even a short piece of video. Moreover, the object would store operations, called *methods*, that perform actions on the data— for example, how to calculate the student's grade-point average and how to display or print the student's record.

Features of a DBMS

Preview & Review: Features of a database management system include (1) data dictionary, (2) utilities, (3) query language, (4) report generator, (5) access security, and (6) system recovery.

A database management system may have a number of components, including the following. (■ *See Panel 7.15.*)

Data Dictionary

Some databases have a *data dictionary*, **a file that stores the data definitions or a description of the structure of data used in the database.** The data dictionary may monitor the data being entered to make sure it conforms to the rules defined during data definition, such as field name, field size, type of data (text, numeric, date, and so on). The data dictionary may also help protect the security of the database by indicating who can access it.

DBMS - database management system

Utilities

The **DBMS utilities are programs that allow the maintenance of the database** by creating, editing, and deleting data, records, and files. The utilities allow

PANEL 7.15 **Some important features of a database management system**

Component	Description
Data dictionary	Describes files and fields of data
Utilities	Help maintain the database by creating, editing, and monitoring data input
Query language	Enables users to make queries to a database and retrieve selected records
Report generator	Enables nonexperts to create readable, attractive on-screen or hardcopy reports of records
Access security	Specifies user access privileges
System (data) recovery	Enables contents of database to be recovered after system failure

people to establish what is acceptable input data, to monitor the types of data being input, and to adjust display screens for data input.

Query Language

Also known as a *data manipulation language*, a **query language is a computer language for making queries to a database and for retrieving selected records,** based on the particular criteria and format indicated. Typically, the query is in the form of a sentence or near-English typed command, using such basic words as SELECT, DELETE, or MODIFY. There are several different query languages, each with its own vocabulary and procedures. One of the most popular is *Structured Query Language*, or *SQL*. An example of an SQL query is as follows:

```
SELECT    PRODUCT-NUMBER, PRODUCT-NAME
FROM      PRODUCT
WHERE     PRICE < 100.00
```

This query selects all records in the product file for products that cost less than $100.00 and displays the selected records according to product number and name—for example, like this:

```
A-34    Mirror
C-50    Chair
D-168   Table
```

Report Generator

A *report generator* is a program users may employ to produce an on-screen or printed-out document from all or part of a database. You can specify the format of the report in advance—row headings, column headings, page headers, and so on. With a report generator, even nonexperts can create attractive, readable reports on short notice.

Access Security

At one point in the Michael Douglas/Demi Moore movie *Disclosure*, the Douglas character, the beleaguered division head suddenly at odds with his company, types SHOW PRIVILEGES into his desktop computer, which is tied to the corporate network. To his consternation, the system responds by showing him downgraded from PRIOR USER LEVEL: 5 to CURRENT USER LEVEL: 0, shutting him out of files to which he formerly had access.

This is an example of the use of *access security,* a feature allowing database administrators to specify different access privileges for different users of a DBMS. The purpose of this security feature, of course, is to protect the database from unauthorized access and sabotage.

System Recovery

Some advanced database management systems have a *system recovery* feature, which enables the DBA to recover contents of the database in the event of a hardware or software failure.

E   The Ethics of Using Databases: Concerns About Accuracy & Privacy

Preview & Review: Databases may contain inaccuracies or be incomplete. They also may endanger privacy—in the areas of finances, health, employment, and commerce.

"The corrections move by bicycle while the stories move at the speed of light," says Richard Lamm, a former governor of Colorado.

Lamm was lamenting that he was quoted out of context by a Denver newspaper in a speech he made in 1984. Yet even 10 years afterward—long after the paper had run a correction—he still saw the error repeated in later newspaper articles.[23]

How do such mistakes get perpetuated? The answer, suggests journalist Christopher Feola, is the Misinformation Explosion. "Fueled by the growing popularity of both commercial and in-house computerized news databases," he says, "journalists have found it that much easier to repeat errors or rely on the same tired anecdotes and experts."[24]

If news reporters—who are supposed to be trained in careful handling of the facts—can continue to repeat inaccuracies found in databases, what about those without training who have access to computerized facts? How can you be sure that databases with essential information about you— medical, credit, school, employment, and so on—are accurate and, equally important, are secure in guarding your privacy? We examine the topics of *information accuracy and completeness* and of *privacy* in this section.

Matters of Accuracy & Completeness

Databases—including public databases such as Nexis, Lexis, Dialog, and Dow Jones News/Retrieval—can provide you with *more* facts and *faster* facts but not always *better* facts. Penny Williams, professor of broadcast journalism at Buffalo State College in New York and formerly a television anchor and reporter, suggests there are five limitations to bear in mind when using databases for research:[25]

- **You can't get the whole story:** For some purposes, databases are only a foot in the door. There may be many facts or aspects to the topic you are looking into that are not in a database.
- **It's not the gospel:** Just because you see something on a computer screen doesn't mean it's accurate. Numbers, names, and facts may need to be verified in other ways.

- **Know the boundaries:** One database service doesn't have it all. For example, you can find full text articles from the *New York Times* on Lexis/Nexis, from the *Wall Street Journal* on Dow Jones News/Retrieval, and from the *San Jose Mercury News* on America Online, but no service carries all three.

- **Find the right words:** You have to know which key words (search words) to use when searching a database for a topic. As Lynn Davis, a professional researcher with ABC News, points out, in searching for stories on guns, the key word "can be guns, it can be firearms, it can be handguns, it can be pistols, it can be assault weapons. If you don't cover your bases, you might miss something."[26]

- **History is limited:** Most public databases, Davis says, have information going back to 1980, and a few into the 1970s, but this poses problems if you're trying to research something that was written about earlier.

Matters of Privacy

Privacy is the right of people to not reveal information about themselves. However, the ease with which databases and communications lines may pull together and disseminate information threatens privacy.

As you've no doubt discovered, it's no trick at all to get your name on all kinds of mailing lists. Theo Theoklitas, for instance, has received applications for credit cards, invitations to join video clubs, and notification of his finalist status in Ed McMahon's $10 million sweepstakes. Theo is a 6-year-old black cat who's been getting mail ever since his owner sent in an application for a rebate on cat food.[27] A whole industry has grown up of professional information gatherers and sellers, who collect personal data and sell it to fund-raisers, direct marketers, and others.

In the 1970s, the Department of Health, Education, and Welfare developed a set of five Fair Information Practices. These rules have since been adopted by a number of public and private organizations. The practices also led to the enactment of a number of later laws to protect individuals from invasion of privacy. (■ *See Panel 7.16.*)

Privacy concerns don't stop with the use or misuse of information in databases. As you will see (Chapter 8), they also extend to privacy in communications. Although the government is constrained by several laws on acquiring and disseminating information and listening in on private conversations, privacy advocates still worry. In recent times, the government has tried to impose new technologies that would enable law-enforcement agents to gather a wealth of personal information. Proponents have urged that Americans must be willing to give up some personal privacy in exchange for safety and security. We discuss this matter in Chapter 10.

◥ Onward

When Cynthia Schoenbrun was laid off as a research administrator from a computer software company, she found something even better. She used her own personal computer to link up with people she knew in Russia and become part of a new field known as *information brokering.*

What is an information broker? "Part librarian, part private eye, and part computer nerd," one writer explains, "an information broker searches for everything written and published on a given subject, be it an obscure corner of the biomedical market or the whereabouts of a German engineering expert."[28] Schoenbrun, for instance, searches computer databases and her

Fair Information Practices

1. There must be no personal data record-keeping systems whose existence is a secret from the general public.
2. People have the right to access, inspect, review, and amend data about them that is kept in an information system.
3. There must be no use of personal information for purposes other than those for which it was gathered without prior consent.
4. Managers of systems are responsible and should be held accountable and liable for the reliability and security of the systems under their control, as well as for any damage done by those systems.
5. Governments have the right to intervene in the information relationships among private parties to protect the privacy of individuals.

Important Federal Privacy Laws

Freedom of Information Act (1970): Gives you the right to look at data concerning you that is stored by the federal government.

Fair Credit Reporting Act (1970): Bars credit agencies from sharing credit information with anyone but authorized customers. Gives you the right to review and correct your records and to be notified of credit investigations for insurance or employment. A drawback is that credit agencies may share information with anyone they reasonably believe has a "legitimate business need." *Legitimate* is not defined.

Privacy Act (1974): Prohibits federal information collected about you for one purpose from being used for a different purpose. Allows you the right to inspect and correct records.

Family Educational Rights and Privacy Act (1974): Gives students and their parents the right to review, and to challenge and correct, students' school and college records; limits sharing of information in these records.

Right to Financial Privacy Act (1978): Sets strict procedures that federal agencies must follow when seeking to examine customer records in banks; regulates financial industry's use of personal financial records.

Privacy Protection Act (1980): Prohibits agents of federal government from making unannounced searches of press offices if no one there is suspected of a crime.

Cable Communications Policy Act (1984): Restricts cable companies in the collection and sharing of information about their customers.

Computer Fraud and Abuse Act (1986): Allows prosecution for unauthorized access to computers and databases. A drawback is that people with legitimate access can still get into computer systems and create mischief without penalty.

Electronic Communications Privacy Act (1986): Makes eavesdropping on private conversations illegal without a court order.

Computer Security Act (1987): Makes actions that affect the security of computer files and telecommunications illegal.

Computer Matching and Privacy Protection Act (1988): Regulates computer matching of federal data; allows individuals a chance to respond before government takes adverse actions against them.

Video Privacy Protection Act (1988): Prevents retailers from disclosing video-rental records without the customer's consent or a court order.

PANEL 7.16

The five Fair Information Practices and important federal privacy laws

The Fair Information Practices were developed by the U.S. Department of Health, Education, and Welfare in the early 1970s. They have been adopted by many public and private organizations since.

network of contacts to find business and investment information about Russia and other countries formerly in the Soviet Union. She then sells this information to clients.

The majority of information brokers, who are mainly in one- or two-person firms, are people who have seen a chance to own a business without making a heavy investment. Among other advantages, the profession gives people a lot of flexibility in setting their own hours.

One need not become an information broker, however, to benefit from being able to search a database. Doctors, lawyers, and other professionals are turning to databases in order to keep up with the information explosion within their fields. In the new world of computers and communications, everyone should at least know the rudiments of this skill.

SUMMARY

What It Is / What It Does	Why It's Important
backup *(p. 195, LO 5)* Name given to a diskette (or tape or hard-disk cartridge) that contains duplicates, or copies, of files on another form of storage.	Because secondary storage media can fail or be destroyed, users (and companies) should always make backup copies of their files so that they don't lose them.
batch processing *(p. 210, LO 7)* Method of processing whereby data is collected over several days or weeks and then processed all at one time, as a "batch."	With batch processing, if users need to make a request of the system, they must wait until the batch has been processed. Batch processing is less expensive than online processing and is suitable for work in which immediate answers to queries are not needed.
bit *(p. 208, LO 6)* Binary digit—0 or 1.	The bit is the fundamental element of all data and information stored and manipulated in a computer system. It represents the lowest level in the data hierarchy.
byte *(p. 208, LO 6)* Unit of 8 bits.	Bytes represent the second-lowest level in the data hierarchy.
cartridge tape units *(p. 205, LO 5)* Also called *tape streamers;* secondary storage used to back up data from a hard disk onto a tape cartridge.	Cartridge tape units are often used with microcomputers.
CD Plus/enhanced CD *(p. 202, LO 5)* Digital disk that is a hybrid of audio-only compact disk and multimedia CD-ROM.	A CD Plus can be played like any music CD in a stereo system, but it also can be used for multimedia presentations on a computer.
character *(p. 208, LO 6)* May be—but is not necessarily—the same as a byte (8 bits); a single letter, number, or special character such as ;, $, or %.	*See byte.*
compact disk-read only memory (CD-ROM) *(p. 200, LO 5)* Optical-disk form of secondary storage that holds more data, including photographs, art, sound, and video, than diskettes and many hard disks. Like music CDs, a CD-ROM is a read-only disk. CD-ROM disks will not play in a music CD player.	CD-ROM disks are used in computer systems to create multimedia presentations and do research, among other things.
compact disk-recordable (CD-R) *(p. 202, LO 5)* CD format that allows users to write data onto a specially manufactured disk that can then be read by a standard CD-ROM drive.	Home users can do their own recordings in CD format.
data dictionary *(p. 218, LO 10)* File that stores data definitions and descriptions of database structure. It also monitors new entries to the database as well as user access to the database.	The data dictionary monitors the data being entered to make sure it conforms to the rules defined during data definition. The data dictionary may also help protect the security of the database by indicating who has the right to gain access to it.
data storage hierarchy *(p. 208, LO 6)* Defines the levels of data stored in a computer database: bits, bytes, fields, records, and files.	Bits and bytes are what the computer hardware deals with, so users need not be concerned with them. They will, however, deal with characters, fields, records, files, and databases.

database *(p. 208, LO 6)* Integrated collection of files in a computer system.

Businesses and organizations build databases to help them keep track of and manage their affairs. In addition, users with online connections to database services have enormous research resources at their disposal.

database management system (DBMS) (database manager) *(p. 215, LO 8)* Software that controls the structure of a database and access to the data; allows users to manipulate more than one file at a time (as opposed to file managers).

This software enables: sharing of data (same information is available to different users); economy of files (several departments can use one file instead of each individually maintaining its own files, thus reducing data redundancy, which in turn reduces the expense of storage media and hardware); data integrity (changes made in the files in one department are automatically made in the files in other departments); security (access to specific information can be limited to selected users).

DBMS utilities *(p. 219, LO 10)* Programs that allow the maintenance of databases by creating, editing, and deleting data, records, and files.

DBMS utilities allow people to establish what is acceptable input data, to monitor the types of data being input, and to adjust display screens for data input.

direct access storage *(p. 211, LO 9)* Storage media that allows the computer direct access to a storage location without having to go through what's in front of it.

Direct access storage (disk) is much faster than sequential storage (tape).

diskette (disk) drive *(p. 191, LO 5)* Computer hardware device that holds, spins, reads from, and writes to magnetic or optical disks.

Users need disk drives in order to use their disks. Disk drives can be internal (built into the computer system cabinet) or external (connected to the computer by a cable).

diskette *(p. 191, LO 5)* Also called *floppy disk;* secondary storage medium; removable round, flexible mylar disk, usually 3½ inches in diameter, that stores data as electromagnetic charges on a metal oxide film that coats the mylar plastic. Data is represented by the presence or absence of these electromagnetic charges, following standard patterns of data representation (such as ASCII).

Diskettes are used on all microcomputers.

DVD-ROM disk *(p. 204, LO 5)* Five-inch optical disk that looks like a regular audio CD but can store 4.7 gigabytes of data on a side.

DVD-ROMs provide great storage capacity, studio-quality images, and theater-like surround sound.

erasable optical disk *(p. 204, LO 5)* Optical disk that allows users to erase data so that the disk can be used over and over again (as opposed to CD-ROMs, which can be read only).

The most common type of erasable and rewritable optical disk is probably the magneto-optical disk, which uses aspects of both magnetic-disk and optical-disk technologies. Such disks are useful to people who need to save successive versions of large documents, handle enormous databases, or work in multimedia production.

field *(p. 208, LO 6)* Unit of data consisting of one or more characters (bytes). An example of a field is your name, your address, *or* your Social Security number.

A collection of fields make up a record. *Also see key field.*

file *(p. 208, LO 6)* Collection of related records. An example of a file is collected data on everyone employed in the same department of a company, including all names, addresses, and Social Security numbers.

Integrated files make up a database.

file-management system (file manager) *(p. 214, LO 8)* Software for creating, retrieving, and manipulating files, one file at a time.

In the 1950s, magnetic tape was the storage medium and records and files were stored sequentially. File managers were created to work with these files. Today, however, database managers are more common.

fixed-disk drive *(p. 199, LO 5)* High-speed, high-capacity disk drive housed in its own cabinet.

Although fixed disks are not removable or portable, these units generally have greater storage capacity and are more reliable than removable disk packs. A single mainframe computer might have 20–100 such fixed disk drives attached to it.

formatting (initializing) *(p. 193, LO 5)* Process by which users prepare diskettes so that the operating system can write information on them. This includes defining the tracks and sectors (the storage layout). Formatting is carried out by one or two simple computer commands.

Diskettes cannot be used until they have been formatted.

gigabyte (G or GB) *(p. 189, LO 1)* Approximately 1 billion bytes (1,073,741,824 bytes); a measure of storage capacity.

Gigabytes are used to express the storage capacity of some microcomputers and many large computers, such as mainframes.

hard disk *(p. 196, LO 5)* Secondary storage medium; generally nonremovable disk made out of metal and covered with a magnetic recording surface. It holds data in the form of magnetized spots. Hard disks are tightly sealed within an enclosed unit to prevent any foreign matter from getting inside.

Hard disks hold much more data than diskettes do. Nearly all microcomputers now use hard disks as their principal secondary storage medium.

hard-disk cartridge *(p. 198, LO 5)* One or two hard-disk platters enclosed along with read/write heads in a hard plastic case. The case is inserted into an external cartridge system connected to a microcomputer.

A hard-disk cartridge, which is removable and easily transported in a briefcase, may hold gigabytes of data. Hard-disk cartridges are often used for transporting large graphics files and for backing up data.

hard-disk drive *(p. 196, LO 5)* One or more hard-disk platters sealed along with read/write heads inside the computer's system unit; it cannot be removed.

Nearly all microcomputers now use hard disks as their principal secondary storage medium.

head crash *(p. 197, LO 5)* Disk disturbance that occurs when the surface of a read/write head or particles on its surface come into contact with the disk surface, causing the loss of some or all of the data on the disk.

Head crashes can spell disaster if the data on the disk has not been backed up.

hierarchical database *(p. 216, LO 10)* One of the three arrangements for database management systems; fields or records are arranged in related groups resembling a family tree, with "child" records subordinate to "parent" records. A parent may have more than one child, but a child always has only one parent. To find a particular record, one starts at the top with a parent and traces down the chart to the child.

Hierarchical DBMSs work well when the data elements have an intrinsic one-to-many relationship, as might happen with a reservations system. The difficulty, however, is that the structure must be defined in advance and is quite rigid. There may be only one parent per child and no relationships among the child records.

interactive *(p. 204, LO 5)* Refers to a situation in which the user is able to make an immediate response to what is going on and modify processes; that is, there is back-and-forth interaction between the user and the computer or communications device. The best interactive storage media are those with high capacity, such as CD-ROMs.

Interactive devices allow the user to be an active participant in what is going on instead of just observing it.

key field *(p. 208, LO 6)* Field that contains unique data used to identify a record so that it can be easily retrieved and processed. The key field is often an identification number, Social Security number, customer account number, or the like. The primary characteristic of the key field is that it is *unique*.

Key fields are needed to identify and retrieve specific records in a database.

kilobyte (K or KB) *(p. 189, LO 1)* 1024 bytes (often rounded off to 1000 bytes).

Kilobytes are a common unit of measure for storage capacity. The amount of data stored in a file or database might be expressed in kilobytes, megabytes, or gigabytes.

magnetic tape *(p. 205, LO 7)* Thin plastic tape coated with a substance that can be magnetized; data is represented by the magnetized or nonmagnetized spots. Tape can store files only sequentially.

Tapes are used in reels, cartridges, and cassettes. Today "mag tape" is used mainly to provide backup, or duplicate storage.

megabyte (M or MB) *(p. 189, LO 1)* Unit for measuring storage capacity; equals approximately 1 million bytes.

The storage capacities of many microcomputer hard disks are measured in megabytes. Users need to know how much data their hard disks can hold and how much space new software programs will take so that they do not run out of disk space.

multimedia *(p. 204, LO 5)* Refers to technology that presents information in more than one medium, including text, graphics, animation, video, sound effects, music, and voice.

Use of multimedia is becoming more common in business, the professions, and education as a means of improving the way information is communicated. Multimedia systems have also added greater depth and variety to presentations, such as those in entertainment and education. As used by telephone, cable, broadcasting, and entertainment companies, *multimedia* also refers to the so-called Information Superhighway. On this avenue various kinds of information and entertainment will be delivered to users' homes through wired and wireless communication lines.

network database *(p. 217, LO 10)* One of the three common arrangements for database management systems; it is similar to a hierarchical DBMS, but each child record can have more than one parent record. Thus, a child record may be reached through more than one parent.

This arrangement is more flexible than the hierarchical one. However, it still requires that the structure be defined in advance. Moreover, there are limits to the number of links that can be made among records.

nonvolatile storage *(p. 188, LO 2)* Permanent storage, as in secondary storage.

Data and programs are permanent; they remain intact when the power to the computer is turned off.

object-oriented database *(p. 218, LO 10)* Database structure that uses objects as elements within database files. An object consists of (1) text, sound, video, and pictures and (2) instructions on the action to be taken on the data.

In addition to textual data, an object-oriented database can store, for example, a person's photo, "sound bites" of her voice, and a video clip.

offline storage *(p. 210, LO 7)* Refers to data that is not directly accessible for processing until a tape or disk has been loaded onto an input device.

The storage medium and data are not under the immediate, direct control of the central processing unit.

online processing *(p. 210, LO 7)* Also called *real-time processing*; means entering transactions into a computer system as they take place and updating the master files as the transactions occur; requires direct access storage.

Online processing gives users accurate information from an ATM machine or an airline reservations system, for example.

online storage *(p. 210, LO 7)* Refers to stored data that is directly accessible for processing.

Storage is under the immediate, direct control of the central processing unit; users need not wait for a tape or disk to be loaded onto an input device before they can access stored data.

optical disk *(p. 199, LO 5)* Removable disk on which data is written and read through the use of laser beams. The most familiar form of optical disk is the one used in the music industry.

Optical disks hold much more data than magnetic disks. Optical disk storage is expected to dramatically affect the storage capacity of microcomputers.

query language *(p. 219, LO 10)* Easy-to-use computer language for making queries to a database and retrieving selected recods.

Query languages make it easier for users to deal with databases. To retrieve information from a database, users make queries—that is, they use a query language. These languages have commands such as SELECT, DELETE, and MODIFY.

read/write head *(p. 192, LO 5)* The device that transfers data on a disk to the computer and from the computer to the disk. The diskette spins inside its jacket, and the read/write head moves back and forth over the data access area.

The read/write head locates the specific area on the disk on which the user is seeking to find a file.

record *(p. 208, LO 6)* Collection of related fields. An example of a record would be your name *and* address *and* Social Security number.

Related records make up a file.

recording density *(p. 194, LO 5)* Refers to the number of bytes per inch of data that can be written onto the surface of the disk. There are three diskette densities: single-density, double-density, and high-density.

Users need to know what types of disks their system can use.

relational database *(p. 217, LO 10)* One of the three common arrangements for database management systems; relates, or connects, data in different files through the use of a key field, or common data element. In this arrangement there are no access paths down through a hierarchy. Instead, data elements are stored in different tables made up of rows and columns. The tables are called *relations*, the rows are called *tuples*, and the columns are called *attributes*.

The relational database is the most flexible arrangement. The advantage of relational databases is that the user does not have to be aware of any "structure." Thus, they can be used with little training. Moreover, entries can easily be added, deleted, or modified. A disadvantage is that some searches can be time consuming. Nevertheless, the relational model has become popular for microcomputer DBMSs.

removable-pack hard disk system *(p. 198, LO 5)* Secondary storage with 6–20 hard disks, of 10½- or 14-inch diameter, aligned one above the other in a sealed unit. These removable hard-disk packs resemble a stack of phonograph records, except that there is space between disks to allow access arms to move in and out. Each access arm has two read/write heads—one reading the disk surface below, the other the disk surface above. However, only one of the read/write heads is activated at any given moment.

Such secondary storage systems enable a large computer system to store massive amounts of data.

secondary storage *(p. 188, LO 2)* Consists of devices that store data and programs permanently on disk or tape.

Secondary storage is nonvolatile—that is, data and programs are permanent or remain intact when the power is turned off. Secondary storage is also needed because computer users require far greater storage capacity than is available through primary storage.

sectors *(p. 192, LO 5)* On a diskette, eight or nine invisible wedge-shaped sections used by the computer for storage reference purposes.

When users save data from computer to diskette, it is distributed by tracks and sectors on the disk. That is, the systems software uses the point at which a sector intersects a track to reference the data location in order to spin the disk and position the read/write head.

sequential access storage *(p. 211, LO 9)* Storage method, like magnetic tape, whereby data is stored in sequence, such as alphabetically.

With sequential access, the system must search through all the preceding data on a tape before reaching the desired item. This process may require running several inches or feet off a reel of tape, which takes time. Disk storage, by contrast, generally falls into the category of direct access storage.

terabyte (T or TB) *(p. 189, LO 1)* Unit for measuring storage capacity; equals approximately 1 trillion bytes.

The storage capacities of supercomputers are measured in terabytes, as is also the amount of data being held in remote databases accessible to users over a communications line.

See sectors.

tracks *(p. 192, LO 5)* The rings on a diskette along which data is recorded. Unlike on a phonograph record, these tracks are neither visible grooves nor a single spiral. Rather, they are closed concentric rings. Each track is divided into eight or nine sectors.

write once, read many (WORM) *(p. 203, LO 5)* Refers to an optical disk that can be written, or recorded, onto just once and then cannot be erased; it can be read many times. WORM disks hold more data than other types of optical disks.

WORM technology is useful for storing data that needs to remain unchanged, such as that used for archival purposes.

write-protect feature *(p. 194, LO 5)* Feature of 3½-inch and 5¼-inch disks that prevents someone from accidentally writing over—and thereby obliterating—data on a disk.

This feature allows users to protect data on diskettes from accidental change or erasure.

(Selected answers appear at the back of the book.)

Short-Answer Questions

1. What does "read-only" mean?

2. Why is the process of formatting (initializing) necessary?

3. What is the difference between primary storage and secondary storage?

4. What is a hard-disk cartridge, and how can it be useful to a microcomputer user?

5. What is the difference between volatile and non-volatile storage?

6. As they apply to disk storage, what is meant by the terms *fragmentation* and *defragmentation*?

7. What is a collection of related fields called?

8. What is the main advantage of direct access storage (as compared to sequential storage)?

9. What is the main difference between a file-management system and a database management system?

10. What is the difference between batch processing and online processing?

Fill-in-the-Blank Questions

1. The two standard diskette sizes are
   _____ and
   _____.

2. A _____ feature allows you to protect a diskette from being written on.

3. _____ refers to the average time needed for the computer to locate data on a disk.

4. A(n) _____ is a removable disk on which data is written and read through the use of laser beams.

5. _____ is a method of removing redundant elements from a computer file so that the file takes up less space.

6. The _____ is the person who coordinates all activities and needs relating to the company's database.

7. In ascending order, the levels of data stored in a database are _____,
   _____,
   _____,
   _____, and
   _____.

8. A(n) _____ is a collection of integrated files.

9. A(n) _____ is a particular field chosen to uniquely identify a record so that it can be easily retrieved and processed.

10. _____ is the right of individuals to not reveal information about themselves.

Multiple-Choice Questions

1. Which of the following provides you with the most storage?
   a. 3½-inch diskette
   b. hard disk
   c. disk cartridge
   d. CD-ROM
   e. 5¼-inch diskette

2. Which of the following disk devices is used with microcomputers?
   a. removable pack
   b. fixed-disk drive
   c. external hard-disk drive
   d. RAID storage system
   e. none of the above

3. Which of the following can be written on only once?
   a. diskette
   b. hard disk
   c. disk cartridge
   d. CD-ROM
   e. none of the above

4. A(n) _____ holds, spins, and reads data from and writes data to a diskette.
   a. drive gate
   b. sector
   c. disk drive
   d. disk cartridge
   e. none of the above

5. Which of the following is used mainly for backup purposes?
   a. diskette
   b. magnetic tape
   c. CD-ROM
   d. removable-pack hard disk
   e. all of the above

6. Which of the following *isn't* a disadvantage of a file-management system?
   a. expensive to develop
   b. data redundancy
   c. lack of data integrity
   d. lack of program independence
   e. all of the above

7. Which of the following are parts of the data hierarchy?
   a. bit
   b. byte
   c. field
   d. file
   e. all of the above

8. Which of the following *doesn't* describe a method whereby files are organized?
   a. sequential
   b. direct
   c. indexed-sequential
   d. batch
   e. all of the above

True/False Questions

T F 1. WORM disks are used principally for storing databases, documents, and other archival information.

T F 2. Secondary storage is nonvolatile.

T F 3. Most of today's CD-ROM drives are single-speed.

T F 4. Diskettes are divided into tracks and sectors.

T F 5. Hard-disk cartridges are used mainly with microcomputer systems.

T F 6. Hard disks are read-only.

T F 7. Fields are larger than files.

T F 8. With online storage, data is directly accessible for processing.

T F 9. One of the disadvantages of database management systems is the cost of installing and maintaining the database.

T F 10. In a direct file organization, records are stored in no particular sequence.

T F 11. A database administrator must be concerned with backup and recovery issues.

Projects/Critical-Thinking Questions

1. What types of storage hardware are currently being used in the computer you use at school or at work? What is the storage capacity of this hardware? Would you recommend alternate storage hardware be used? Why? Why not?

2. You want to purchase a hard disk for use with your microcomputer. Because you don't want to have to upgrade your secondary-storage capacity in the near future, you are going to buy one with the highest storage capacity you can find. Use computer magazines or visit computer stores and find a hard disk you would like to buy. What is its capacity? How much does it cost? Who is the manufacturer? What are the system compatibility requirements? Is it an internal or an external drive? Why have you chosen this unit?

3. What would you do if your computer had a "head crash"—a failure of your hard-disk drive—that seemed to wipe out a major project? Assume you had not been foresighted enough to back up your data on diskettes or tape. Go to the library (or use the Internet) and investigate articles on data retrieval or hard-disk salvage methods. Or look in the telephone book Yellow Pages to see if you can find an organization that specializes in retrieving data from damaged hard disks; then call them up and ask what they do and what their rates are.

4. Do you think books on CD-ROMs will ever replace printed books? Why or why not? Look up some recent articles on this topic and prepare a short report.

5. What types of databases are used at your school or business? Who has access to them? Who is responsible for keeping them current? How are the databases organized? Who is responsible for administering the database?

6. Describe a personal database that you would use to organize some aspect of your personal or business life. What fields would you include in the file structure? What types of queries would you like to perform on the database? Would you need to use relational operators? How often would data need to be updated?

net Using the Internet

Objective: *In this exercise you use Netscape Navigator to shop on the World Wide Web for a multimedia computer that costs under $2000.*

Before you continue: *We assume you have access to the Internet through your university, business, or commercial service provider and to the Web browser tool Netscape Navigator 2.0 or 2.01 Additionally, we assume you know how to connect to the Internet and then load Netscape Navigator. If necessary, ask your instructor or system administrator for assistance.*

1. Make sure you have started Netscape. The home page for Netscape Navigator should appear on your screen.

2. Although the commercial side of the Web is still in its formative stages, the Web does provide an excellent means for accessing some resources. Today, most Web stores provide the same functionality as a mail-order catalog with an associated toll-free 800 number.

Not surprisingly, many mail-order companies use their store on the Web as an additional way to get customers. The following is a small sampling of popular Web shops:

a. *Noteworthy Music* (*http://www.netmarket.com/noteworthy/bin/ main*)—Noteworthy Music, a discount CD dealer, uses its Web site to supplement its mail-order business. As you order CDs, a running total of your charges appears on the page.

b. *Internet Shopping Network* (*http://shop.internet.net/*)—For hardware and software at discounted prices, the Internet Shopping Network (ISN) may be the shop for you. Because you can use key words to search for items, they are easier to find than in a paper catalog.

c. *Computer Literacy* (*http://www.clbooks.com/*)— Computer Literacy, a discount computer book dealer, uses its Web site to supplement its mail-order business. Computer Literacy's search feature makes it easy to find the book you want by typing in a topic or an author name.

d. *The Internet Mall* (*http://www.mecklerweb.com/imall/*)—The Internet Mall, like the shopping malls we're accustomed to, provides access to more than one company with something to sell. The Internet Mall is organized into floors, each of which is sponsored by a different company. Companies (or individuals) who have a Web site and/or an electronic e-mail address are listed (for free) in the The Internet Mall.

3. In the following steps you'll use the Internet Shopping Network to find a multimedia computer that costs under $2000.

To display the Internet Shopping Network's home page:
CLICK: Open button (⬚)
TYPE: http://shop.internet.net/
PRESS: **Enter** or CLICK: Open
Your screen may appear similar to the following:

4. Drag the vertical scroll bar downward until you see the ISN Directory list. The list should appear similar to the following:

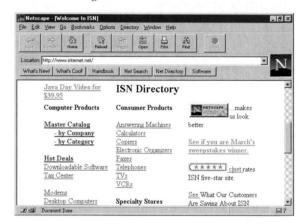

5. On your own, shop for a computer by clicking links in the ISN Directory. Remember that the computer should have multimedia capabilities and cost under $2000.

6. Print out a description of the computer. (*Note:* If you were actually purchasing the computer, you would click the Buy link, located after the product's description.)

7. Display Netscape's home page and then exit Netscape.

CHAPTER 8

Communications

Starting Along the Information Highway

Concepts You Should Know

After reading this chapter, you should be able to:

1. Discuss examples of usage of communications technology: telephone-related services, online information services, the Internet, videoconferencing, shared resources, portable work, and the information appliance.

2. Explain the communications components of a microcomputer, describing analog and digital signals, modems, and communications software.

3. Identify and explain the various communications channels, both wired and wireless.

4. Discuss the three types of networks.

5. List and discuss the factors affecting data transmission.

6. Discuss ethical matters of netiquette, free speech, censorship, and privacy.

"Computers and communications: These are the parents of the Information Age," says one writer. "When they meet, the fireworks begin."[1]

What kind of fireworks are we talking about? Maybe it is that portable information and communications technologies *are changing conventional meanings of time and space.* As one expert pointed out (during a round-table discussion on an online network), "the physical locations we traditionally associate with work, leisure and similar pursuits are rapidly becoming meaningless."[2]

Alaska salmon fisherman Blanton Fortson exemplifies this blurring of traditional boundaries between work and leisure, isolation and availability. Fortson has so much portable technology on his boat and at home that he wears baggy pants to carry it all around. "I find I very rarely need to be tied to any specific location in order to take care of business," he says.[3]

Not everyone is as enthusiastic about being so accessible. Said one observer back in 1991: "The movable office—a godsend for workaholics, a nightmare for those who live with them—has only just begun."[4] Through communications technologies, computers, telephones, and wireless devices are being linked to create invisible networks everywhere.

The Practical Uses of Communications & Connectivity

Preview & Review: The area of communications is expanding rapidly through connectivity, the ability to connect devices via communications lines to other devices and sources of information. Communications devices and services range from telephone-related communication to video communication to computer-related communications.

Clearly, communications is extending into every nook and cranny of civilization—the "plumbing of cyberspace," as it has been called. The term *cyberspace* was coined by William Gibson in his novel *Neuromancer* to refer to a futuristic computer network that people use by plugging their brains into it. Today *cyberspace* has come to mean the computer online world and the Internet in particular, but it is also used to refer to the whole wired and wireless world of communications in general.

Communications & Connectivity

Communications, also called *telecommunications*, is the electronic transfer of information from one location to another. It also refers to the devices and systems for communicating. The data being communicated may consist of voice, sound, text, video, graphics, or all these together. The instruments sending and receiving the data may be telegraph, telephone, cable, microwave, radio, television, or computer. The distance may be as close as the next room or as far away as the outer edge of the solar system. **The ability to connect devices by communications technology to other devices and sources of information is known as *connectivity*.** the ability to connect computers to one another by modem or network

Tools of Communications & Connectivity

What kinds of options do communications and connectivity give you? Let us consider the possibilities. We will take them in order, more or less, from *low-skill activities* to *high-skill activities*. That is, we will begin with those that demand relatively little training and proceed to those that require more training. (■ *See Panel 8.1, pages 234–235.*) They include:

* Telephone-related communications services: fax messages, voice mail, and e-mail
* Video/voice communication: videoconferencing and picture phones
* Online information services
* The Internet
* Shared resources: workgroup computing, Electronic Data Interchange (EDI), and intranets
* Portable work: telecommuting, mobile workplaces, and virtual offices
* Information (Internet) appliances

Telephone-Related Communications Services

Preview & Review: Telephone-related communications include fax messages, transmitted by dedicated fax machines and fax modems; voice mail, storing voice messages in digitized form; and e-mail (electronic mail), transmitting written messages.

Phone systems and computer systems have begun to fuse together. Services available through telephone connections, whether the conventional wired kind or the wireless cellular-phone type, include *fax messages, voice mail, and e-mail.*

Fax Messages

Asking "What is your fax number?" is about as common a question in the work world today as asking for someone's telephone number. Indeed, the majority of business cards include both a telephone number and a fax number. *Fax* stands for "facsimile transmission," or reproduction.

A fax may be sent by dedicated fax machine or by fax modem. (■ *See Panel 8.2, next page.*)

* **Dedicated fax machines: *Dedicated fax machines* are specialized devices that do nothing except send and receive copies of documents over transmission lines to and from other fax machines.** These are the stand-alone machines nowadays found everywhere, from offices to airports to instant-printing shops.
* **Fax modems: A *fax modem*, which is installed as a circuit board inside a computer's system cabinet, is a modem with fax capability. It enables you to send and receive signals directly between your computer and someone else's fax machine or fax modem.**

Voice Mail

A service offered by telephone companies, ***voice mail* digitizes incoming voice messages and stores them in the recipient's "voice mailbox" in digitized form. It then converts the digitized versions back to voice messages when they are retrieved.** Messages can be copied and forwarded to other people, saved, or erased.

PANEL 8.1

The world of connectivity

Wired or wireless communications links offer several options for information and communications.

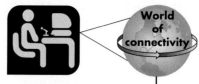

World of connectivity

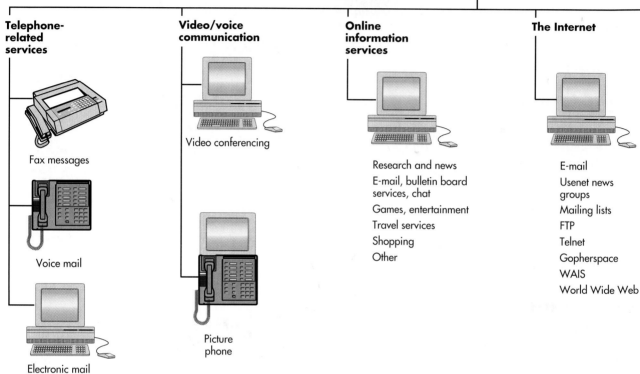

Telephone-related services

Fax messages

Voice mail

Electronic mail

Video/voice communication

Video conferencing

Picture phone

Online information services

Research and news

E-mail, bulletin board services, chat

Games, entertainment

Travel services

Shopping

Other

The Internet

E-mail

Usenet news groups

Mailing lists

FTP

Telnet

Gopherspace

WAIS

World Wide Web

PANEL 8.2

Two types of fax hardware

(Left) A fax modem. This circuit board is installed in an expansion slot on the motherboard inside the system cabinet of a microcomputer. *(Right)* A dedicated fax machine.

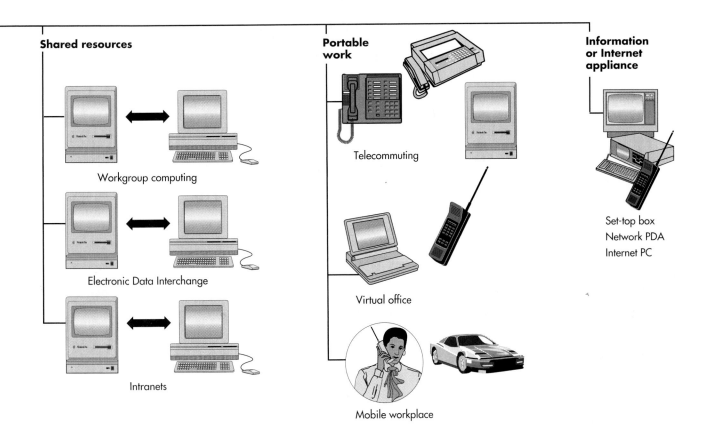

Shared resources

Workgroup computing

Electronic Data Interchange

Intranets

Portable work

Telecommuting

Virtual office

Mobile workplace

Information or Internet appliance

Set-top box
Network PDA
Internet PC

E-Mail

"E-mail is so clearly superior to paper mail for so many purposes," writes *New York Times* computer writer Peter Lewis, "that most people who try it cannot imagine going back to working without it."[5] Says another writer, e-mail "occupies a psychological space all its own: It's almost as immediate as a phone call, but if you need to, you can think about what you're going to say for days and reply when it's convenient."[6]

E-mail, or *electronic mail*, links computers by wired or wireless connections and allows users, through their keyboards, to post messages and to read responses on their display screens. With e-mail, you dial the e-mail system's telephone number, type in the recipient's "mailbox" address, and then type in the message and click on the "Send" button. The message will be sent to the recipient's mailbox. When you access your own mailbox, you are presented with a list of senders and topics; for example, you can read messages, print them out, delete them, or download (transfer) to your hard disk.

If you're part of a company, university, or other large organization, you may get e-mail for free. Otherwise you can sign up with a commercial online service (America Online, CompuServe, Prodigy), e-mail service (such as MCI Mail), or Internet access provider (such as Netcom's NetCruiser or PSI's Pipeline USA).

What, however, if you want to meet *face-to-face* with someone who is far away? Then you can use videoconferencing or picture phones.

Video/Voice Communication: Videoconferencing & Picture Phones

Preview & Review: Videoconferencing is the use of television, sound, and computer technology to enable people in different locations to see, hear, and talk with one another.

Want to have a meeting with people on the other side of the country or the world but don't want the hassle of travel? You may have heard of or participated in a *conference call*, also known as *audio teleconferencing*, a meeting in which more than two people in different geographical locations talk on the telephone. A variation on this meeting format is e-mail-type *computer conferencing*, sometimes called "chat sessions." Computer conferencing is a keyboard conference among several users at microcomputers or terminals linked through a computer network. Now we have video/voice communication, specifically *videoconferencing* and *picture phones*.

Videoconferencing

Videoconferencing, **also called** *teleconferencing*, **is the use of television video and sound technology as well as computers to enable people in different locations to see, hear, and talk with one another.** Videoconferencing equipment can be set up on people's desks, with a camera and microphone to capture the people talking, and a monitor and speakers for the listeners. The *picture phone* also allows communication with video. This device is a telephone with a TV-like screen and a built-in camera that allow you to see the person you're calling, and vice versa. With improvement in the quality of telephone-company transmission lines will come improved picture quality; thus the picture phone should become more popular in the near future.

Online Information Services

Preview & Review: Online information services provide computer users access to, among other things, research and news resources; games, entertainment, and clubs; and travel and shopping services.

For subscribers equipped with telephone-linked microcomputers, *online information services* **provide access to all kinds of databases and electronic meeting places.** Says one writer:

> Online services are those interactive news and information retrieval sources that can make your computer behave more like a telephone; or a TV set; or a newspaper; or a video arcade, a stock brokerage firm, a bank, a travel agency, a weather bureau, a department store, a grocery store, a florist, a set of encyclopedias, a library, a bulletin board, and more.[7]

Several online services exist, but those known as the Big Three have the most subscribers and are considered the most mainstream. They are *America Online (AOL)*, with approximately 5 million subscribers; *CompuServe*, with about 4.2 million; and *Prodigy*, with an estimated 1.4 million.[8] (These numbers are increasing daily.)

To gain access to online services, you need a *microcomputer* (with hard disk and printer for storing and printing downloaded messages and information). You also need a *modem*, the hardware that, when connected to your phone line, enables data to be transmitted to your computer. Finally, you need *communications software*, so your computer can communicate via modem and interact with distant computers that have modems. (We discuss modems and communications software later in the chapter.) America Online, CompuServe, and Prodigy provide subscribers with their own software for going online, but you can also buy communications programs, such as ProComm Plus, separately.

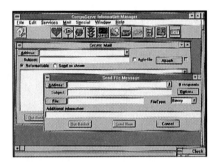

Opening an account with an online service requires a credit card, and billing policies resemble those used by cable-TV and telephone companies. As with cable-TV, you may be charged a fee for basic service, with additional fees for specialized services. In addition, the online service may charge you for the time spent on the line. Finally, you will also be charged by your telephone company for your time on the line, just as when making a regular phone call. However, most information services offer local access numbers. Thus, unless you live in a rural area, you will not be paying long-distance phone charges.

The Offerings of Online Services

What kinds of things could you use an online service for? Here are a few:

- **People connections—e-mail, bulletin boards, chat rooms:** Online services can provide a community through which you can connect with people with similar interests (without identifying yourself, if you prefer). The primary means for making people connections are via e-mail, bulletin boards, and "chat rooms."

  E-mail is basically the same as we described earlier.

  Bulletin boards, or *message boards,* allow you to post messages on any of thousands of special topics and read and reply to messages posted by other people.

  Chat rooms are discussion areas in which you join others in a real-time "conversation," typed in via your keyboard. The topic may be general or specific, and the collective chat-room conversation scrolls on the screen.

- **Research and news:** The only restriction on the amount of research you can do online is the limit on whatever credit card you are charging your time to. Depending on the online service, you can avail yourself of several encyclopedias. Many online services provide users with access to huge databases of unabridged text from newspapers, magazines, and journals. The information resources available online are mind-boggling, impossible to describe in this short space.

- **Games, entertainment, and clubs:** Online computer games are extremely popular. In single-player games, you play against the computer. In multiplayer games, you play against others, whether someone in your household or someone overseas. Other entertainments include cartoons, sound clips, pictures of show-business celebrities, and reviews of movies and CDs. You can also join online clubs with others who share your interests, whether science fiction, punk rock, or cooking.

- **Travel services:** Online services offer Eaasy Sabre or Travelshopper, streamlined versions of the reservations systems travel agents use. You can search for flights and book reservations through the computer and have tickets sent to you by FedEx. You can also refer to weather maps, and you can review hotel directories and restaurant guides.

- **Downloading:** *Downloading* means transferring or copying. Many users obtain download freeware, shareware (✓ p. 46), and commercial demonstration programs from online sources. (This can also be done via the Internet.)

- **Shopping:** CompuServe, for instance, offers 24-hour shopping with its Electronic Mall. This feature lists products from over 100 retail stores, discount wholesalers, specialty shops, and catalog companies. You can scan through listings of merchandise, order something on a credit card with a few keystrokes, and have the goods delivered by UPS or U.S. mail.

Will Online Services Survive the Internet?

Commercial services represent only one way to go online. You can also access the Internet directly. In fact, the Internet and particularly the World Wide Web, which we cover next, threaten to swamp the online services.[9] As the Internet and the Web have become easier to navigate, online services have begun to lose customers and content providers—even as they have added their own arrangements for accessing the Internet.

Still, the online services have a lot to offer. One survey, for instance, found that *half* of the people on the Internet got there through commercial services, which suggests they may be among the easiest ways to get to the Web.[10] In addition, the online services package information so that you can more quickly and easily find what you're looking for. It's also easier to conduct a live "chat" session on an online service than it is on the Web and it is easier for parents to exert control over the kinds of materials their children may view.[11]

The Internet

Preview & Review: The Internet, the world's biggest network, uses a protocol called TCP/IP to allow computers to communicate. Users can connect to the Internet via direct connections, online information services, and Internet service providers. There are many tools available to navigate the Internet.

Called "the mother of all networks," **the Internet, or simply "the Net," is an international network connecting approximately 36,000 smaller networks.** (■ *See Panel 8.3.*) To connect with it, you need pretty much the same things you need to connect with online services: a computer, modem, telephone line, and appropriate communications software.

Created by the U.S. Department of Defense in 1969 (under the name ARPAnet—ARPA was the department's Advanced Research Project Agency), the Internet was built to serve two purposes. The first was to share research among military, industry, and university sources. The second was to provide a system for sustaining communication among military units in the event of nuclear attack. Thus, the system was designed to allow many routes among many computers, so that a message could arrive at its destination by many possible ways, not just a single path. In 1973 the first international connections were made with England and Norway. By 1977 many more international connections had been made—thus the name *Internet*.

With the many different kinds of computers being connected, engineers had to find a way for the computers to speak the same language. The solution developed was *TCP/IP*, the standard since 1983 and the standard language of the Internet. *TCP/IP,* for *Transmission Control Protocol/Internet Protocol,* is the standardized set of guidelines (protocols) that allow different computers on different networks to communicate with each other efficiently, no matter how they gained access to the Net, the topic of the next section.

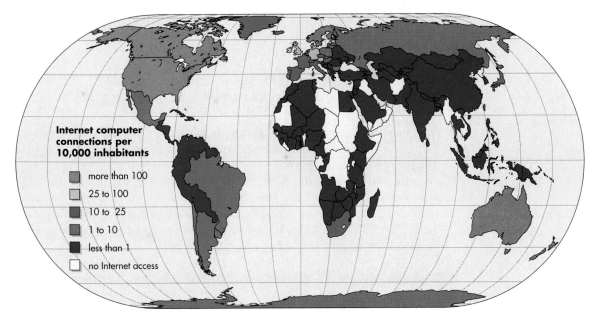

Internet computer
connections per
10,000 inhabitants

- more than 100
- 25 to 100
- 10 to 25
- 1 to 10
- less than 1
- no Internet access

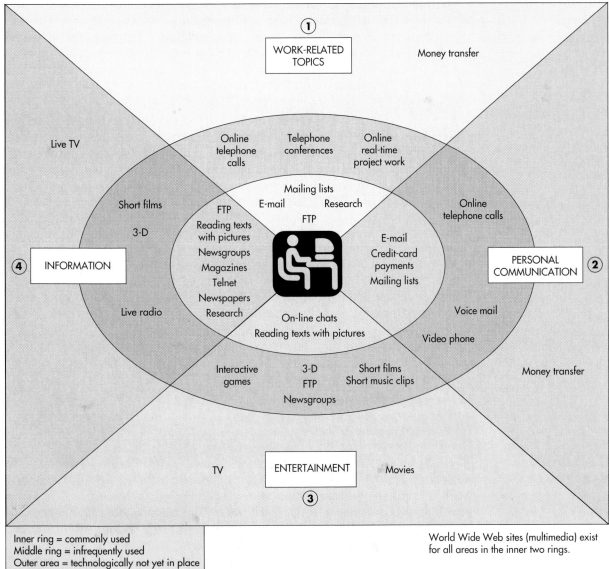

① WORK-RELATED TOPICS

Money transfer

Live TV

Online telephone calls | Telephone conferences | Online real-time project work

Short films

Mailing lists
E-mail | Research
FTP

Online telephone calls

3-D

FTP
Reading texts with pictures
Newsgroups
Magazines
Telnet
Newspapers
Research

④ INFORMATION

E-mail
Credit-card payments
Mailing lists

② PERSONAL COMMUNICATION

Live radio

Voice mail

On-line chats
Reading texts with pictures

Video phone

Money transfer

Interactive games | 3-D FTP | Short films Short music clips

Newsgroups

TV | ENTERTAINMENT | Movies

③

Inner ring = commonly used
Middle ring = infrequently used
Outer area = technologically not yet in place

World Wide Web sites (multimedia) exist
for all areas in the inner two rings.

PANEL 8.3

The Internet

(Top) Approximate numbers of users connected to the Internet worldwide computer network
(numbers increase daily). *(Bottom)* What's available through the Internet.

Connecting to the Internet

There are three ways to connect your PC with the Internet:

- **Through school or work:** Many universities, colleges, and large businesses have dedicated, high-speed phone lines that provide a direct connection to the Internet. If you're a student, this may be the best deal because the connection is free or low cost. However, if you live off-campus and want to get this Internet connection from home, you probably won't be able to do so.[12] To use a direct connection, your microcomputer must have TCP/IP software and be connected to the local area network that has the direct-line connection to the Net.

The next two types of connection are called "dial-up" connections.

- **Through online information services:** As mentioned, subscribing to a commercial online information service, which provides you with its own communications software, may not be the cheapest way to connect to the Internet, but it may well be the most trouble-free. America Online, CompuServe, and Prodigy all offer such access—that is, they provide an electronic "gateway" to the Internet. However, these types of connections do not always provide *complete* Internet services.

- **Through Internet service providers (ISPs):** To obtain complete Internet services through a dial-up connection, you use an ISP. **Internet service providers (ISPs) are local or national companies that provide unlimited public access to the Internet and World Wide Web for a flat rate.** Using your computer and the ISP's communications software, you dial up your ISP's phone number. The ISP's main computer uses SLIP (serial line Internet protocol) or PPP (point-to-point protocol) software to connect you to the Internet. Forrester Research predicts that ISPs could claim as many as 32 million online subscribers in the U.S. by 2000 versus 12.7 million for commercial online services.[13]

  You can ask someone who is already on the Web to access the world-wide list of ISPs at *http://www.thelist.com.* Besides giving information about each provider in your area, "the list" provides a rating (on a scale of 1 to 10) by users of different ISPs.

Once you're on the Net, how do you get where you want to go? That topic is next.

Internet Addresses

To send and receive e-mail on the Internet and interact with other networks, you need an Internet address. Internet addresses use the domain system. In the *domain name system,* the Internet's addressing scheme, an Internet address usually has two sections. For example, consider the address

president@whitehouse.gov.us

The first section, the user ID, tells "who" is at the address—in this case, *president.* The second section, after the @ (called "at") symbol, tells "where" the address is—subdomain (if required), domain, domain type, and country (if required)—in this case, *whitehouse.gov.us,* The White House, government, United States. Components of the second part of the address are separated by periods (called "dots"). (■ *See Panel 8.4.)*

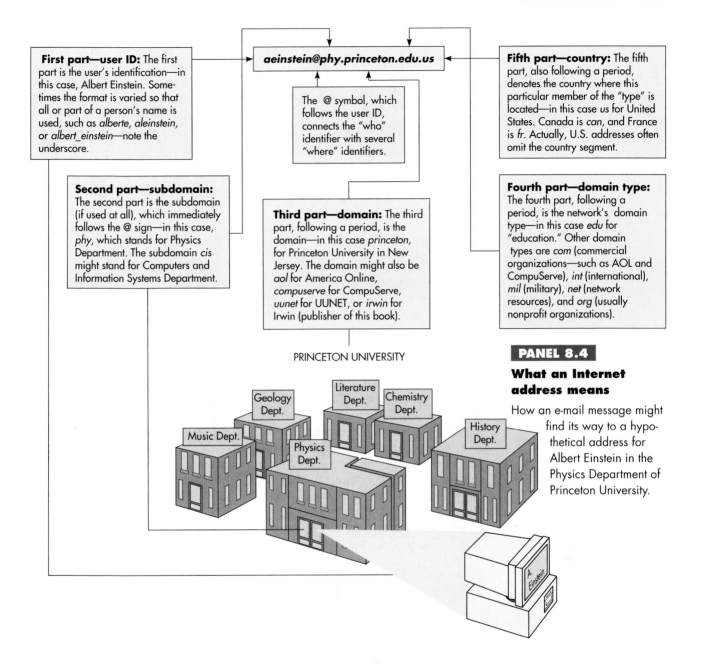

First part—user ID: The first part is the user's identification—in this case, Albert Einstein. Sometimes the format is varied so that all or part of a person's name is used, such as *alberte, aleinstein,* or *albert_einstein*—note the underscore.

aeinstein@phy.princeton.edu.us

The @ symbol, which follows the user ID, connects the "who" identifier with several "where" identifiers.

Fifth part—country: The fifth part, also following a period, denotes the country where this particular member of the "type" is located—in this case *us* for United States. Canada is *can,* and France is *fr.* Actually, U.S. addresses often omit the country segment.

Second part—subdomain: The second part is the subdomain (if used at all), which immediately follows the @ sign—in this case, *phy,* which stands for Physics Department. The subdomain *cis* might stand for Computers and Information Systems Department.

Third part—domain: The third part, following a period, is the domain—in this case *princeton,* for Princeton University in New Jersey. The domain might also be *aol* for America Online, *compuserve* for CompuServe, *uunet* for UUNET, or *irwin* for Irwin (publisher of this book).

Fourth part—domain type: The fourth part, following a period, is the network's domain type—in this case *edu* for "education." Other domain types are *com* (commercial organizations—such as AOL and CompuServe), *int* (international), *mil* (military), *net* (network resources), and *org* (usually nonprofit organizations).

PRINCETON UNIVERSITY

PANEL 8.4

What an Internet address means

How an e-mail message might find its way to a hypothetical address for Albert Einstein in the Physics Department of Princeton University.

Features of & Tools for Navigating the Internet

The principal features of the Internet are e-mail, discussion groups, file transfer, remote access, and information searches.

• **E-mail:** "The World Wide Web is getting all the headlines, but for many people the main attraction of the Internet is electronic mail," says technology writer David Einstein.[14] There are 90 million users of e-mail in the world, and although half of them are on private corporate networks, a great many of the rest are on the Internet.

Foremost among the Internet e-mail programs is Qualcomm's Eudora software, which has about 10 million users worldwide and is used by 60% of the educational institutions on the Internet.

- **Usenet newsgroups—electronic discussion groups:** One of the Internet's most interesting features goes under the misleading name *newsgroups*, although they don't have much to do with news. Actually, *newsgroups* are electronic discussions held by groups of people focusing on a specific topic, the equivalents of CompuServe's or America Online's forums. They are one of the most lively and heavily trafficked areas of the Net.

  Usenet is a public access network of dispersed newsgroups that exchange e-mail and messages ("news"). "Users post questions, answers, general information, and FAQ files on Usenet," says one online specialist. "The flow of messages, or 'articles,' is phenomenal, and you can get easily hooked."[15] An FAQ, for *frequently asked questions*, is a file that contains basic information about a newsgroup. It's always best to read a newsgroup's FAQ before joining the discussion or posting (asking) questions.

- **Mailing lists—e-mail-based discussion groups:** Combining e-mail and newsgroups, mailing lists allow you to subscribe (generally free) to an e-mail mailing list on a particular subject or subjects. The mailing-list sponsor then sends the identical message to everyone on that list. There are probably 3000-plus electronic mailing-list discussion groups.

- **FTP—for copying all the free files you want:** Many Net users enjoy "FTPing"—cruising the system and checking into some of the tens of thousands of so-called FTP sites offering interesting free files to copy (download). *FTP,* for *file transfer protocol,* is a method whereby you can connect to a remote computer and transfer publicly available files to your own PC.

- **Telnet—to connect to remote computers:** *Telnet* is a cooperative system that allows you to connect (log on) to remote computers. This feature enables you to tap into Internet computers and public-access files as though you were connected directly. It is especially useful for perusing large databases or library card catalogs. There are perhaps 1000 library catalogs accessible through the Internet.

- **Gopherspace—the easy menu system:** Several other software tools exist to help sift through the staggering amount of information on the Internet, but one of the most important is Gopher. *Gopher* is a uniform system of menus, or series of lists, that allows users to easily browse through and retrieve files stored on different computers.

- **WAIS—ways of searching by content:** Pronounced "wayz," *WAIS,* for *Wide Area Information Server,* is a system for searching Internet databases by subject, using specific words or phrases rather than sorting through a hierarchy of menus. Unfortunately, WAIS is offered only by certain information sites (servers) and so can be applied to only a limited number of files.

- **World Wide Web—for multimedia and hypertext:** The fastest-growing part of the Internet—and many times larger than any online service—the World Wide Web is the most graphically inviting and easily navigable section of the Internet. **The *World Wide Web*, or simply "the Web," consists of an interconnected system of computer sites, or places, all over the world that can store information in multimedia form—sounds, photos, video, as well as text. The sites share a form consisting of a hypertext series of links that connect similar words and phrases.** Web software was developed in 1990 by Tim Berners in Cern, Switzerland.

  Note two distinctive features:

  (1) Whereas Gopher and WAIS deal with text, the Web provides information in *multimedia* form—it contains graphics, video, and audio as well as text.

(2) Whereas Gopher is a menu-based approach to accessing Net resources, the Web uses a hypertext (✓ p. 69) format. *Hypertext is a system in which documents scattered across many Internet sites are directly linked, so that a word or phrase in one document becomes a connection to an entirely different document.* In particular, the format used on the Web is called *hypertext markup language (HTML)* and swaps information using *hypertext transfer protocol (HTTP).* When you use your mouse to point-and-click on a hypertext link (a highlighted word or phrase), it may become a doorway to another place within the same document or another document on another computer thousands of miles away.

The places you visit on the Web are called *Web sites,* and the estimated number of such sites throughout the world ranges between 90,000 and 265,000.[16] More specifically, **a *Web site* is a file stored on a computer (server or host computer).** For example, the Parents Place Web site (*http://www.parentsplace.com*) is a resource run by mothers and fathers that includes links to other related sites, such as the Computer Museum Guide to the Best Software for Kids and the National Parenting Center.[17]

Information on a Web site is stored on "pages" (✓ p. 129). **The *home page* is the main page or first screen you see when you access a Web site,** but there are often other pages or screens. *Web site* and *home page* tend to be used interchangeably, although a site may have many pages. (There might be a total of 16 million individual pages.[18] Some of them are simply abandoned because their creators have not updated or deleted them, the online equivalent of space-age debris orbiting the earth.[19])

To find a particular Web site (home page), you need its *URL.* **The *URL,* for Uniform Resource Locator, is an address that points to a specific resource on the Web.** Often it looks something like this: *http://www.blah.blah.html.* (Here *http* stands for "hypertext transfer protocol," *www* for "World Wide Web," and *html* for "hypertext markup language.")

To get to this address, you need a *Web browser*—**software that helps you get information you want by clicking your mouse pointer on words or pictures on the screen.** Four popular Web browsers are Netscape Navigator, NCSA Mosaic, and Microsoft Internet Explorer. With the browser you can browse (search through) the Web. When you connect with a particular Web site, the screenful of information (the home page) is sent to you. You can easily skip from one page to another by using your mouse to click on the hypertext links. ISPs and online services provide their subscribers with Web browsers.

We offered some suggestions for exploring the Web in the Experience Box at the end of Chapter 4.

Shared Resources: Workgroup Computing, Electronic Data Interchange, & Intranets

Preview & Review: Workgroup computing enables teams of co-workers to use networked microcomputers to share information and cooperate on projects.

Electronic data interchange (EDI) is the direct electronic exchange of standard business documents between organizations' computer systems.

Intranets are internal corporate networks that use the infrastructure and standards of the Internet and the Web.

When they were first brought into the workplace, microcomputers were used simply as another personal-productivity tool, like typewriters or calculators. Gradually, however, companies began to link a handful of microcomputers together on a network, usually to share an expensive piece of hardware, such as a laser printer. Then employees found that networks allowed them to share files and databases as well. Networking using common software also allowed users to buy equipment from different manufacturers—a mix of workstations from both Sun Microsystems and Hewlett-Packard, for example. The possibilities for sharing resources have led to workgroup computing.

Workgroup Computing & Groupware

Workgroup computing, **also called** *collaborative computing,* **enables teams of co-workers to use networks of microcomputers to share information and cooperate on projects.** Workgroup computing is made possible not only by networks and microcomputers but also by *groupware.* **Groupware** **is software that allows two or more people on a network to work on the same information at the same time.**

In general, groupware, such as Lotus Notes, permits office workers to collaborate with colleagues and tap into company information through computer networks. It also enables them to link up with crucial contacts outside their organization.

Electronic Data Interchange

Electronic data interchange (EDI) **is the direct electronic exchange between organizations' computer systems of standard business documents,** such as purchase orders, invoices, and shipping documents. For example, Wal-Mart has electronic ties to major suppliers like Procter & Gamble, allowing both companies to track the progress of an order or other document through the supplier company's computer system.

To use EDI, organizations wishing to exchange transaction documents must have compatible computer systems, or else go through an intermediary. For example, more than 500 colleges are now testing or using EDI to send transcripts and other educational records to do away with standard paper handling and its costs.

Intranets

It had to happen: First, businesses found that they could use the World Wide Web to get information to customers, suppliers, or investors. FedEx, for example, saved millions by putting up a server in 1994 that enabled customers to click through Web pages to trace their parcels, instead of having FedEx customer-service agents do it. It was a short step from that to companies starting to use the same technology inside—in internal Internet-like networks called *intranets.*[20,21] **Intranets** **are internal corporate networks that use the infrastructure and standards of the Internet and the World Wide Web.**

One of the greatest considerations of an intranet is security—making sure that sensitive company data accessible on intranets is protected from the outside world. The means for doing this is security software called *firewalls.* A *firewall* is a security program that connects the intranet to external networks, such as the Internet. It blocks unauthorized traffic from entering the intranet and can also prevent unauthorized employees from accessing the intranet.

We consider security matters at length in Chapter 10. Now we move on to look at some of the changes that connectivity has brought about in people's work lives.

Portable Work: Telecommuting & Virtual Offices

Preview & Review: Working at home with computer and communications connections between office and home is called *telecommuting*.

The virtual office is a nonpermanent and mobile office run with computer and communications technology.

"In a country that has been moaning about low productivity and searching for new ways to increase it," observed futurist Alvin Toffler, "the single most anti-productive thing we do is ship millions of workers back and forth across the landscape every morning and evening."[22]

Toffler was referring, of course, to the great American phenomenon of physically commuting to and from work. More than 108 million Americans commute to work by car and another 6 million by public transportation. Information technology has responded to the cry of "Move the work instead of the workers!" Computers and communications tools have led to telecommuting and telework centers, the mobile workplace, and the virtual office and "hoteling."

Telecommuting & Telework Centers

Working at home with telecommunications between office and home is called *telecommuting*. In 1994, the number of part-time and full-time telecommuters reached 9.1 million, up 20% from the year before.[23] The reasons for telecommuting are quite varied. One may be to eliminate the daily drive, reducing traffic congestion, energy consumption, and air pollution. Another may be to take advantage of the skills of homebound workers with physical disabilities (especially since the passage of the Americans with Disabilities Act). Parents with young children, as well as "lone eagles" who prefer to live in resort areas or other desirable locations, are other typical telecommuter profiles.

Another term for telecommuting is *telework*. However, *telework* includes not only those who work at least part time from home but also those who work at remote or satellite offices, removed from organizations' main offices. Such satellite offices are sometimes called *telework centers*. An example of a telework center is the Riverside Telecommuting Center, in Riverside, California, supported by several companies and local governments. The center provides office space that helps employees who live in the area avoid lengthy commutes to downtown Los Angeles. However, these days an office can be virtually anywhere.

The Virtual Office

The term *virtual office* borrows from "virtual reality" (artificial reality that projects the user into a computer-generated three-dimensional space). **The *virtual office* is an often nonpermanent and mobile office run with computer and communications technology.** Employees work not in a central office but from their homes, cars, and other new work sites. They use pocket pagers, portable computers, fax machines, and various phone and network services to conduct business.

Could you stand not having a permanent office at all? Here's how one variant, called *hoteling*, would work: You call ahead to book a room and speak to the concierge. However, your "hotel" isn't a Hilton, and the "concierge" isn't a hotel employee who handles reservations, luggage, and local tours. Rather, the organization is your employer, and the concierge is an administrator who handles scheduling of available office cubicles—of which there is only one for every three workers.

Hoteling works for Ernst & Young, an accounting and management consulting firm. Its auditors and management consultants spend 50–90% of their time in the field, in the offices of clients. When they need to return to E&Y headquarters, they call a few hours in advance. The concierge consults a computerized scheduling program and determines which cubicles are available on the days requested. He or she chooses one and puts the proper nameplate on the office wall. The concierge then punches a few codes into the phone to program its number and voice mail. When employees come in, they pick up personal effects and files from lockers and then take them to the cubicles they will use for a few days or weeks.[24,25]

What makes hoteling possible, of course, is computer and communications technology. Computers handle the cubicle scheduling and reprogramming of phones. They also allow employees to carry their work around with them stored on the hard drives of their laptops. Cellular phones, fax machines, and e-mail permit employees to stay in touch with supervisors and co-workers.

As we stated at the outset of this chapter, information technology is blurring time and space, eroding the barriers between work and private life. Some people thrive on it, but others hate it.

◢ The Coming Information, or Internet, Appliance

Preview & Review: The coming convergence of computers, communications, consumer electronics, and mass media will likely produce an "Internet appliance," which will combine PC capabilities with TV, telephone, and cable.

At the beginning of this book, we discussed the phenomenon of *technological convergence,* or *digital convergence,* the merger of several industries—computers, communications, consumer electronics, and mass media—through various devices that exchange information in digital form. We suggested that the embodiment of this convergence is the *information appliance* (✓ p. 29). This gadget would presumably receive a digitized stream of sound, video, text, and data from some sort of electronic delivery system, and we would be able to "talk back" or interact with it.

How close are we to realizing this device? The answer is: Maybe we're practically there now.

The Information/Internet Appliance

At present there seem to be three possible variations on what is called either an *information appliance* or *Internet appliance:* the set-top box, the network PDA, and the network PC.[26]

* **Set-top box:** A *set-top box,* or "Net-top box," is a keypad that allows cable-TV viewers to change channels or, in the case of interactive systems, to exercise other commands. A videogame console could double as a set-top box that would let consumers surf the Web on their TV set over phone lines or cable. Companies such as Philips, Sega, Sony, and Thomson are working on new developments in this area.

- **Network PDAs:** Different kinds of *network personal digital assistants (PDAs)* (✓ p. 27)—variously known as personal communicators, pocket PCs, and smart phones—could provide low-cost, wireless access to the Internet. With these handheld devices, connections to the Net could be as easy and portable as cell-phone telephone calls are today. Companies such as Apple, Sun, and Toshiba are trying to bring such products to market.

- **Internet PCs:** This is the under-$500 "hollow PC" or "network computer" (also called a "Web PC"). A hollow PC does not include a monitor (you use one you already own or a TV set) or any disk drives (diskette, hard, or CD-ROM). Instead, you connect the box to your phone or cable line and get all your software from computers located elsewhere.[27-29] Low-cost Internet PCs have been announced by Oracle, Sun Microsystems, and Apple Computer.

We are still, of course, not at the point where your microcomputer has become the true information appliance just described. Why has it taken so long for computers and communications systems—telephone, radio, television, and so on—to come together? Why can't voice, data, and images be easily transmitted via a telephone/television/computer "information appliance" between your home and that of a friend? To understand this is to understand why you need the equipment you do to communicate using your present PC.

Using a Microcomputer to Communicate: Analog & Digital Signals, Modems & Datacomm Software, ISDN Lines, & Cable Modems

Preview & Review: To communicate online through a microcomputer, users need communications software and a modem to send and receive computer-generated messages over telephone lines. Modems translate the computers' digital signals of discrete bursts into analog signals of continuous waves, and vice versa. ISDN lines and cable modems allow faster data transmission than do traditional modems.

The principal reason the hollow PC is still not quite here lies in the fact that information is transmitted by two types of signals, each requiring different kinds of communications technology. The two types of signals are *analog* and *digital*. (■ *See Panel 8.5, next page.*) In a way they resemble analog and digital watches. As we illustrated in Chapter 1 (✓ p. 7), an analog watch shows time as a continuum. A digital watch shows time as discrete numeric values.

Analog Signals: Continuous Waves

Telephones, radios, and televisions—the older forms of communications technology—were designed to work with an analog signal. **An *analog signal* is a continuous electrical signal in the form of a wave.** The wave is called a *carrier wave*.

Two characteristics of analog carrier waves that can be altered are frequency and amplitude.

- **Frequency:** *Frequency* is the number of times a wave repeats during a specific time interval—that is, how many times it completes a *cycle* in a second.

PANEL 8.5

Review of analog and digital signals

An analog signal represents a continuous electrical signal in the form of a wave. A digital signal is discontinuous, expressed as discrete bursts in on/off electrical pulses.

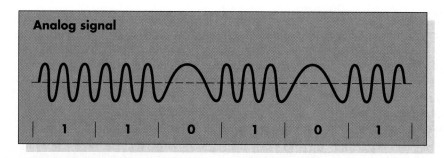

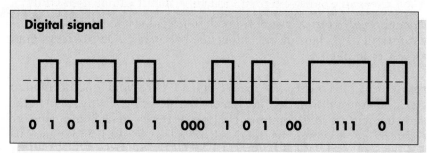

• **Amplitude:** *Amplitude* is the height of a wave within a given period of time. Amplitude is actually the strength or volume—the loudness—of a signal.

Both frequency and amplitude can be modified by making adjustments to the wave. Indeed, it is by such adjustments that an analog signal can be altered to represent a digital signal, as we shall explain.

Digital Signals: Discrete Bursts

A *digital signal* **uses on/off or present/absent electrical or light pulses in discontinuous, or discrete, bursts, rather than a continuous wave.** This two-state kind of signal is used to represent the two-state binary language of 0s and 1s that computers use. That is, the presence of an electrical/light pulse can represent a 1 bit, its absence a 0 bit.

Modems

Digital signals are better—that is, faster and more accurate—at transmitting computer data. However, many of our present communications connections and devices, such as telephone and microwave, are still analog. To get around this problem, we need a device called a *modem*. **A *modem*—short for *modulater/demodulater*—converts digital signals into analog form (a process known as *modulation*) to send over phone lines. A receiving modem at the other end of the phone line then converts the analog signal back to a digital signal (a process known as *demodulation*).** (■ *See Panel 8.6.*)

PANEL 8.6

How modems work

A sending modem translates digital signals into analog waves for transmission over phone lines. A receiving modem translates the analog signals back into digital signals.

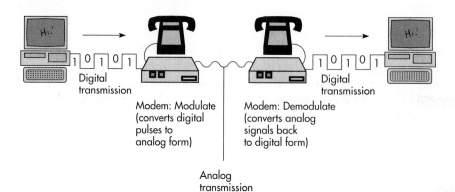

Two criteria for choosing a modem are whether you want an internal or external one, and what transmission speed you wish:

- **External versus internal:** Modems are either internal or external.

  An *external modem* is a box that is separate from the computer. The box may be large or it may be portable, pocket size. A line connects the modem to a port in the back of the computer. A second line connects the modem to a standard telephone jack. There is also a power cord that plugs into a standard AC wall socket.

  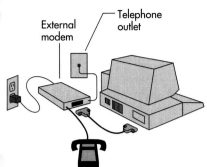

  The advantage of the external modem is that it can be used with different computers. Thus, if you buy a new microcomputer, you will probably be able to use your old external modem. Also, external modems help isolate the computer's internal circuitry from phone-line conducted lightning surges.

  An *internal modem* is a circuit board that plugs into a slot inside the system cabinet. Nowadays many new microcomputers come with an internal modem already installed. Advantages of the internal modem are that it doesn't take up extra space on your desk, it is less expensive than an external modem, and it doesn't have a separate power cord.

- **Transmission speed:** Because most modems use standard telephone lines, users are charged the usual rates by phone companies, whether local or long-distance. Users are also often charged by online services for time spent online. Accordingly, *transmission speed*—the speed at which modems transmit data—becomes an important consideration. The faster the modem, the less time you need to spend on the telephone line.

  Today users refer to **bits per second (bps) or, more likely, *kilobits per second (kbps)* to express data transmission speeds.** A 14,400-bps modem, for example, is a 14.4-kbps modem.

  Today's modems transmit at 1200, 2400, and 4800 bps (considered slow, and not really worth using anymore); 9600 and 14,400 bps (moderately fast); and 19,200 and 28,800 bps (high speed). A 10-page single-spaced letter can be transmitted by a 2400-bps modem in 2½ minutes. It can be transmitted by a 9600-bps modem in 38 seconds and by a 19,200-bps modem in 19 seconds. Video data takes much longer to transmit than text.

Communications Software

To communicate via a modem, your microcomputer requires communications software. **Communications software, or "datacomm software," manages the transmission of data between computers or video display terminals.** Macintosh users have Smartcom; Windows users have Smartcom, Crosstalk, ProComm Plus, Wincom, CommWorks, Telix, and HyperAccess; OS/2 Warp users have HyperAccess. Often the software comes on diskettes bundled with (sold along with) the modem.

ISDN Lines & Cable Modems

Users who found themselves banging the table in frustration as their 14.4-kbps modem took 45 minutes to transmit a 1-minute low-quality video from a Web site are about to get some relief. Probably the two most immediate contenders to standard phone modems are *ISDN lines* and *cable modems*. (■ *See Panel 8.6.*)

How long it takes to send a 1-megabyte file:	
Standard phone modem, 28.8 kbps	4.6 minutes
ISDN line, 64–128 kbps	2.1 minutes or less
Cable modem, 3000 kbps or more	2.6 seconds or less

- **ISDN lines:** ISDN stands for Integrated Services Digital Network. *ISDN consists of hardware and software that allow voice, video, and data to be communicated as digital signals over traditional copper-wire telephone lines.* Capable of transmitting up to 128 kbps, ISDN lines are up to five times faster than conventional phone modems.[30]

  ISDN is not cheap, costing perhaps two or three times as much per month as regular phone service. Installation could also cost $200 or more if you need a phone technician to wire your house and install the software in your PC.[31] Nevertheless, with the number of people now working at home and/or surfing the Internet, demand has pushed ISDN orders off the charts. Forecasts are for 7 million U.S. installations by 2000, from the current 450,000 lines today.[32]

  Even so, ISDN's time may have come and gone. The reason: Cable modem and other technologies threaten to render it obsolete.

- **Cable modems:** Cable companies say that cable and cable modems can carry digital data 1000 times faster than plain old telephone system (POTS) lines, and they've found that usage shoots up when the service is connected.

  A cable modem is a modem that connects a personal computer to a cable-TV system that offers online services. The gadgets are still fairly exotic, and it will probably be 1997 before internationally standardized cable modems go on sale.[33] The reason? So far probably 90% of U.S. cable subscribers are served by networks that don't permit much in the way of two-way data communications. "The vast majority of today's . . . cable systems can deliver a river of data downstream," says one writer, "but only a cocktail straw's worth back the other way."[34] Nevertheless, Forrester Research predicts about 6.8 million American homes will have cable modems by 2000.[35]

In addition, competition may also be expected from fiber-optic lines and digital satellites, discussed next.

Communications Channels: The Conduits of Communications

Preview & Review: A channel is the path, either wired or wireless, over which information travels. Various channels occupy various radio-wave bands on the electromagnetic spectrum. Types of wired channels include twisted-pair wire, coaxial cable, and fiber-optic cable. Two principal types of wireless channels are microwave and satellite systems.

Today there are many kinds of communications channels, wired and wireless. **A *channel* is the path over which information travels in a telecommunications system from its source to its destination.** (Channels are also called *links*, *lines*, or *media*.) The basis for all telecommunications channels, both wired and wireless, is the electromagnetic spectrum.

The Electromagnetic Spectrum

Telephone signals, radar waves, and the invisible commands from a garage-door opener all represent different waves on what is called the electromagnetic spectrum. **The *electromagnetic spectrum* consists of fields of electrical energy and magnetic energy, which travel in waves.** (■ *See Panel 8.7.*)

PANEL 8.7

The electromagnetic spectrum

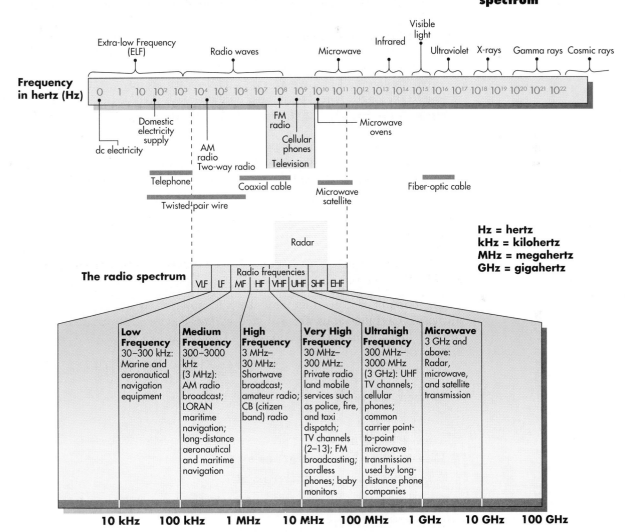

Cellular	Private land mobile	Narrowband PCS	Industrial	Common carrier paging	Point-to-multipoint Point-to-point	PCS	Industrial
824–849 MHz 869–894 MHz	896–901 MHz 930–931 MHz Includes RF packet radio services	901–902 MHz 930–931 MHz	902–928 MHz Unlicensed commercial use such as cordless phones and LANs	931–932 MHz Includes national paging services	932–935 MHz 941–944 MHz	1850–1970 MHz 2130–2150 MHz 2180–2200 MHz	2400–2483.5 MHz Unlicensed commercial use such as LANs

Frequencies for wireless data communications

All radio signals, light rays, and X-rays radiate an energy that behaves like rippling waves. The waves can be characterized according to frequency and wavelength:

- **Frequency:** As we've seen, *frequency* is the number of times a wave repeats (makes a cycle) in a second. Frequency is measured in *hertz (Hz)*, with 1 Hz equal to 1 cycle per second. One thousand hertz is called a *kilohertz (kHz)*, 1 million hertz is called a *megahertz (MHz)*, and 1 billion hertz is called a *gigahertz (GHz)*.

  Ranges of frequencies are called *bands* or *bandwidths.* The bandwidth is the difference between the lowest and highest frequencies transmitted. Bandwidths are usually referred to by the range they cover. For example, cellular phones are on the 800–900 megahertz band—that is, their bandwidth is 100 megahertz.

- **Wavelength:** Waves also vary according to their length—their *wavelength*. We hear references to wavelength in "shortwave radio" and "microwave oven."

  At the low end of the spectrum, the waves are of low frequency and of long wavelength (such as domestic electricity). At the high end, the waves are of high frequency and short wavelength (such as cosmic rays).

The electromagnetic spectrum can be represented by the appliances and machines that emit or detect particular wavelengths. We could start on the left, at the low-frequency end, with video display terminals and hair dryers. We would then range up through AM and FM radios, shortwave radios, VHF and UHF television, and cellular phones. Next we would proceed through radar, microwave ovens, infrared "nightscope" binoculars, and ultraviolet-light tanning machines. Finally, we would go through X-ray machines and end up with gamma-ray machines for food irradiation at the high-frequency end. The part of the spectrum of interest to us is that area in the middle—between 3 million and 300 billion hertz (3 megahertz to 300 gigahertz). This is the portion that is regulated by the government for communications purposes. Certain bands are assigned by the Federal Communications Commission (FCC) for certain purposes—that is, to be controlled by different classes or groups of users.

Let us now look more closely at the various types of channels:

- Twisted-pair wire
- Coaxial cable
- Fiber-optic cable
- Microwave and satellite systems
- Other wireless communications

Twisted-Pair Wire

The telephone line that runs from your house to the pole outside is probably twisted-pair wire. ***Twisted-pair wire* consists of two or more strands of insulated copper wire, twisted around each other in pairs.** They are then covered in another layer of plastic insulation. (■ *See Panel 8.8.*)

Because so much of the world is already served by twisted-pair wire, it will no doubt continue to be used for years, both for voice messages and for modem-transmitted computer data. However, it is relatively slow and does not protect well against electrical interference. As a result, it will certainly be superseded by better communications channels, wired or wireless.

twisted wire

coaxial cable

Coaxial Cable

Coaxial cable, **commonly called "co-ax," consists of insulated copper wire wrapped in a solid or braided metal shield, then in an external cover.** Co-ax is widely used for cable television. Coaxial cable is much better at resisting noise than twisted-pair wiring. Moreover, it can carry voice and data at a faster rate (perhaps 200 megabits per second, compared to 10 megabits per second for twisted-pair wire).

Fiber-Optic Cable

A *fiber-optic cable* **consists of hundreds or thousands of thin strands of glass that transmit not electricity but rather pulsating beams of light.** These strands, each as thin as a human hair, can transmit billions of pulses per second, each "on" pulse representing one bit. When bundled together, fiber-optic strands in a cable 0.12 inch thick can support a quarter- to a half-million voice conversations at the same time. Moreover, unlike electrical signals, light pulses are not affected by random electromagnetic interference in the environment. Thus, they have much lower error rates than normal telephone wire and cable. In addition, fiber-optic cable is lighter and more durable than twisted-pair and coaxial cable.

Microwave & Satellite Systems

Wired forms of communications, which require physical connection between sender and receiver, will not disappear any time soon, if ever. For one thing, fiber-optic cables can transmit data communications 10,000 times faster than microwave and satellite systems can. Moreover, they are resistant to illegal data theft.

Still, some of the most exciting developments are in wireless communications. After all, there are many situations in which it is physically difficult to run wires. Here let us consider microwave and satellite systems.

PANEL 8.8

Three types of wired communications channels

(Top) Twisted-pair wire. This type does not protect well against electrical interference. *(Middle left)* Coaxial cable. This type is shielded against electrical interference. It also can carry more data than twisted-pair wire. *(Right)* When coaxial cable is bundled together, as here, it can carry more than 40,000 conversations at once. *(Bottom left)* Fiber-optic cable. Thin glass strands transmit pulsating light instead of electricity. These strands can carry computer and voice data over long distances.

- **Microwave systems:** *Microwave systems* **transmit voice and data through the atmosphere as super-high-frequency waves.** Microwave systems transmit microwaves, of course. *Microwaves* are the electromagnetic waves that vibrate at 1 gigahertz (1 billion hertz) per second or higher. These frequencies are used not only to operate microwave ovens but also to transmit messages between ground-based earth stations and satellite communications systems.

  Nowadays you see dish- or horn-shaped microwave antennas nearly everywhere—on towers, buildings, and hilltops. Why, you might wonder, do people have to interfere with nature by putting a microwave dish on top of a mountain? The reason is that microwaves cannot bend around corners or around the earth's curvature; they are *line-of-sight*. *Line-of-sight* means that there must be an unobstructed view between transmitter and receiver. Thus, microwave stations need to be placed within 25–30 miles of each other, with no obstructions in between. The size of the dish varies with the distance (perhaps 2–4 feet in diameter for short distances, 10 feet or more for long distances). A string of microwave relay stations will each receive incoming messages, boost the signal strength, and relay the signal to the next station.

Line-of-sight signal   Microwave relay station

  More than half of today's telephone system uses dish microwave transmission. However, the airwaves are becoming so saturated with microwave signals that future needs will have to be satisfied by other channels, such as satellite systems.

- **Satellite systems:** To avoid some of the limitations of microwave earth stations, communications companies have added microwave "sky stations"—communications satellites. **Communications satellites are microwave relay stations in orbit around the earth.** Traditionally, the orbit has been 22,300 miles above the earth (although newer systems will be much lower). Because they travel at the same speed as the earth, they appear to an observer on the ground to be stationary in space—that is, they are *geostationary.* Consequently, microwave earth stations are always able to beam signals to a fixed location above. The orbiting satellite has solar-powered receivers and transmitters (transponders) that receive the signals, amplify them, and retransmit them to another earth station. The satellite contains many communications channels and receives both analog and digital signals from earth stations. Note that it can take more than one satellite to get a message delivered, which can slow the delivery process down.

Cellular Phones

Of course, mobile wireless communications have been around for some time. The Detroit Police Department started using two-way car radios in 1921. Mobile telephones were introduced in 1946. Today, however, we are witnessing an explosion in mobile, wireless cellular-phone use.

Analog cellular phones **are designed primarily for communicating by voice through a system of cells.** Each *cell* is hexagonal in shape, usually 8 miles or less in diameter, and is served by a transmitter-receiving tower. Calls are directed between cells by a mobile telephone switching office (MTSO). Movement between cells requires that calls be "handed off" by the MTSO. (■ *See Panel 8.9.)*

- *Calling from a cellular phone:* When you dial a call on a cellular phone, whether on the street or in a car, the call moves as radio waves to the transmitting-receiving tower that serves that particular cell. The call then moves by wire or microwaves to the mobile telephone switching office (MTSO), which directs the call from there on—generally to a regular local phone exchange, after which it becomes a conventional phone call.
- *Receiving a call on a cellular phone:* The MTSO transmits the number dialed to all the cells it services. Once it finds the phone, it directs the call to it through the nearest transmitting-receiving tower.
- *On the move:* When you make calls to or from phones while on the move, as in a moving car, the MTSO's computers sense when a phone's signal is becoming weaker. The computers then figure out which adjacent cell to "hand off" the call to and find an open frequency in that new cell to switch to.

1. A call originates from a mobile cellular phone.
2. The call wirelessly finds the nearest cellular tower using its FM tuner to make a connection.
3. The tower sends the signal to a Mobile Telephone Switching Office (MTSO) using traditional telephone network land lines.
4. The MTSO routes the call over the telephone network to a land-based phone or initiates a search for the recipient on the cellular network.
5. The MTSO sends the recipient's phone number to all its towers, which broadcast the number via radio frequency.
6. The recipient's phone "hears" the broadcast and establishes a connection with the nearest tower. A voice line is established via the tower by the MTSO.

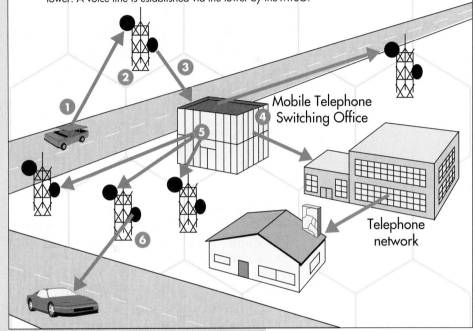

PANEL 8.9

Cellular connections

Communications Networks

Preview & Review: Communications channels and hardware may have different layouts or networks, varying in size from large to small: wide area networks (WANs), metropolitan area networks (MANs), and local networks.

Networks allow users to share peripheral devices, programs, and data; to have better communications; to have more secure information; and to have access to databases.

Whether wired, wireless, or both, all the channels we've described can be used singly or in mix-and-match fashion to form networks. **A *network,* or *communications network,* is a system of interconnected computers, telephones, or other communications devices that can communicate with one another and share applications and data.** It is the tying together of so many communications devices in so many ways that is changing the world we live in.

Here let us consider the following:

- Types of networks—wide area, metropolitan area, and local
- Some network features
- Advantages of networks

Types of Networks: Wide Area, Metropolitan Area, & Local

Networks are categorized principally in the following three sizes:

- **Wide area network:** A *wide area network (WAN)* **is a communications network that covers a wide geographical area, such as a state or a country.** Some examples of computer WANs are Tymnet, Telenet, Uninet, and Accunet. The international pathway Internet links together hundreds of computer WANs. Most telephone systems—long-distance and local—are WANs.
- **Metropolitan area network:** A *metropolitan area network (MAN)* **is a communications network covering a geographic area the size of a city or suburb.** The purpose of a MAN is often to bypass local telephone companies when accessing long-distance services. Cellular phone systems are often MANs.
- **Local network:** A *local network* **is a privately owned communications network that serves users within a confined geographical area.** The range is usually within a mile—perhaps one office, one building, or a group of buildings close together, as a college campus. Local networks are of two types: private branch exchanges (PBXs) and local area networks (LANs), as we discuss shortly.

All these networks may consist of various combinations of computers, storage devices, and communications devices.

Some Features: Hosts & Nodes, Downloading & Uploading

Many computer networks, particularly large ones, are served by a host computer. **A *host computer*, or simply a *host*, is the main computer—the central computer that controls the network.** On a local area network, some or all of the functions of the host may be performed by a computer called a *server*. **A *server* is a computer shared by several users in a network.**

A *node* is simply a device that is attached to a network. A node may be a microcomputer, terminal, or some peripheral device. (As you'll recall, a *peripheral device* is any piece of hardware that is connected to a computer.)

As a network user you can download and upload files. *Download* means that you retrieve files from another computer and store them in your computer. *Upload* means that you send files from your computer to another computer.

Advantages of Networks

The following advantages are particularly true for local networks, although they apply to MANs and WANs as well.

- **Sharing of peripheral devices:** Laser printers, disk drives, and scanners are examples of peripheral devices. Any newly introduced piece of hardware is often quite expensive, as was the case with laser or color printers.

To justify their purchase, companies want them to be shared by many users. Usually the best way to do this is to connect the peripheral device to a network serving several computer users.

* **Sharing of programs and data:** In most organizations, people use the same software and need access to the same information. It could be expensive for a company to buy one copy of, say, a word processing program for each employee. Rather, the company will usually buy a network version of that program that will serve many employees.

  Organizations also save a great deal of money by letting all employees have access to the same data on a shared storage device. This way the organization avoids such problems as some employees updating customer addresses on their own separate machines while other employees remain ignorant of such changes. It is much easier to update (maintain) software on the server than it is to update it on each user's individual system.

  Finally, network-linked employees can, using workgroup software, work together online on shared projects.

* **Better communications:** One of the greatest features of networks is electronic mail, as we have seen. With e-mail everyone on a network can easily keep others posted about important information. Thus, the company eliminates the delays encountered with standard interoffice mail delivery or telephone tag.

* **Security of information:** Before networks became commonplace, an individual employee might be the only one with particular information, stored in his or her desktop computer. If the employee was dismissed—or if a fire or flood demolished the office—no one else in the company might have any knowledge of that information. Today such data would be backed up or duplicated on a networked storage device shared by others.

* **Access to databases:** Networks also enable users to tap into numerous databases, whether the private databases of a company or the public databases of online services.

Local Networks

Preview & Review: Local networks may be private branch exchanges (PBXs) or local area networks (LANs).

LANs may be client/server or peer-to-peer, and may take one of five forms: star, ring, bus, hybrid, or FDDI.

Although large networks are useful, many organizations need to have a local network—an in-house network—to tie together their own equipment. Here let's consider the following aspects of local networks:

* Types of local networks—PBXs and LANs
* Types of LANs—client/server and peer-to-peer
* Components of a LAN
* Topology of LANs—star, ring, bus, hybrid, and FDDI
* Impact of LANs

Types of Local Networks: PBXs & LANs

The most common types of local networks are PBXs and LANs.

- **Private branch exchange (PBX):** **A *private branch exchange (PBX)* is a private or leased telephone switching system that connects telephone extensions in-house.** It also connects them to the outside phone system.

  A public telephone system consists of "public branch exchanges"— thousands of switching stations that direct calls to different "branches" of the network. A private branch exchange is essentially the old-fashioned company switchboard. You call in from the outside, the switchboard operator says "How may I direct your call?" and you are connected to the extension of the person you wish to talk to.

  Newer PBXs can handle not only analog telephones but also digital equipment, including computers. However, because older PBXs use existing telephone lines, they may not be able to handle the volume of electronic messages found in some of today's organizations. These companies may be better served by LANs.

- **Local area network (LAN):** PBXs may share existing phone lines with the telephone system. Local area networks usually require installation of their own communication channels, whether wired or wireless. ***Local area networks (LANs)* are local networks consisting of a communications link, network operating system, microcomputers or workstations, servers, and other shared hardware.** Such shared hardware might include printers, scanners, and storage devices. Unlike larger networks, LANs do not use a host computer.

Types of LANs: Client/Server & Peer-to-Peer

Local area networks are of two principal types: client/server and peer-to-peer. (■ *See Panel 8.10.*)

- **Client/server LANs:** **A *client/server LAN* consists of requesting microcomputers, called *clients*, and supplying devices that provide a service, called *servers*.** The server is a computer that manages shared devices, such as laser printers. The server microcomputer is usually a powerful one, running on a powerful chip such as a Pentium Pro. Client/server networks, such as those run under Novell's NetWare operating system, are the most common type of LAN.

  There may be different servers for managing different tasks—files and programs, databases, printers. The one you may hear about most often is the file server. **A *file server* is a computer that stores the programs and data files shared by users on a LAN.** It acts like a disk drive but is in a remote location.

- **Peer-to-peer:** The word *peer* denotes one who is equal in standing with another (as in the phrases "peer pressure" or "jury of one's peers"). **A *peer-to-peer LAN* is one in which all microcomputers on the network communicate directly with one another without relying on a server.** Peer-to-peer networks are less expensive than client/server networks and work effectively for up to 25 computers. Beyond that they slow down under heavy use. They are thus appropriate for networking in small groups, as for workgroup computing.

Many LANs mix elements from client/server and peer-to-peer models.

PANEL 8.10

Two types of LANs: client/server and peer-to-peer

(Middle) In a client/server LAN, individual microcomputer users, or "clients," share the services of a centralized computer called a *server.* In this case, the server is a file server, which allows users to share files of data and some programs.

(Bottom) In a peer-to-peer LAN, computers share equally with one another without having to rely on a central server.

Client/server LAN

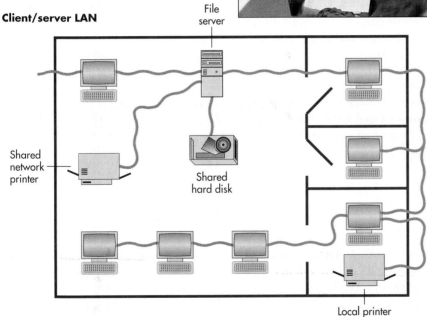

File server

Shared network printer

Shared hard disk

Local printer

Peer-to-peer LAN

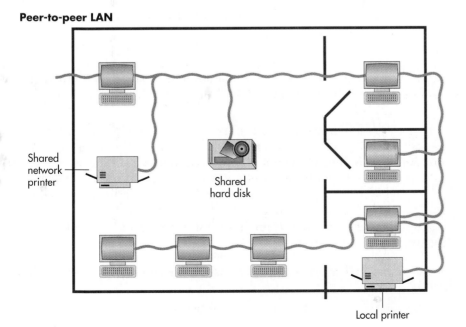

Shared network printer

Shared hard disk

Local printer

Components of a LAN

Local area networks are made up of several standard components.

- **Connection or cabling system:** LANs do not use the telephone network. Instead, they use some other cabling or connection system, either wired or wireless. Wired connections may be twisted-pair wiring, coaxial cable, or fiber-optic cable. Wireless connections may be infrared or radio-wave transmission. Wireless networks are especially useful if computers are portable and are moved often. However, they are subject to interference.

- **Microcomputers with interface cards:** Two or more microcomputers are required, along with network interface cards. A *network interface card*, which is inserted into an expansion slot in a microcomputer, enables the computer to send and receive messages on the LAN.

- **Network operating system:** The network operating system software manages the activity of the network. Depending on the type of network, the operating system software may be stored on the file server or on each microcomputer on the network. Examples of network operating systems are Novell's NetWare and Apple's LocalTalk.

- **Other shared devices:** Printers, fax machines, scanners, storage devices, and other peripherals may be added to the network as necessary and shared by all users.

- **Bridges and gateways:** A LAN may stand alone, but it may also connect to other networks, either similar or different in technology. Hardware and software devices are used as interfaces to make these connections. **A bridge is an interface that enables similar networks to communicate. A gateway is an interface that enables dissimilar networks to communicate,** such as a LAN with a WAN.

Topology of LANs

Networks can be laid out in different ways. The logical layout, or shape, of a network is called a *topology*. The five basic topologies are *star, ring, bus, hybrid,* and *FDDI*. The first three are shown opposite. (■ *See Panel 8.11.*)

SRBH, FDDI

- **Star network:** **A *star network* is one in which all microcomputers and other communications devices are connected to a central server.** Electronic messages are routed through the central hub to their destinations. The central hub monitors the flow of traffic. A PBX system is an example of a star network.

- **Ring network:** **A *ring network* is one in which all microcomputers and other communications devices are connected in a continuous loop.** Electronic messages are passed around the ring until they reach the right destination. There is no central server.

- **Bus network:** **In a *bus network*, all communications devices are connected to a common channel.** There is no central server. Each communications device transmits electronic messages to other devices. If some of those messages collide, the device waits and tries to retransmit again. An example of a bus network is Xerox's Ethernet.

- **Hybrid network:** ***Hybrid networks* are combinations of star, ring, and bus networks.** For example, a small college campus might use a bus network to connect buildings and star and ring networks within certain buildings.

- **FDDI network:** A newer and higher-speed network is the FDDI, short for Fiber Distributed Data Interface. Capable of transmitting 100 megabits per second, **an *FDDI network* uses fiber-optic cable with an adaptation of ring topology.** The FDDI network is being used for such high-tech purposes as electronic imaging, high-resolution graphics, and digital video.

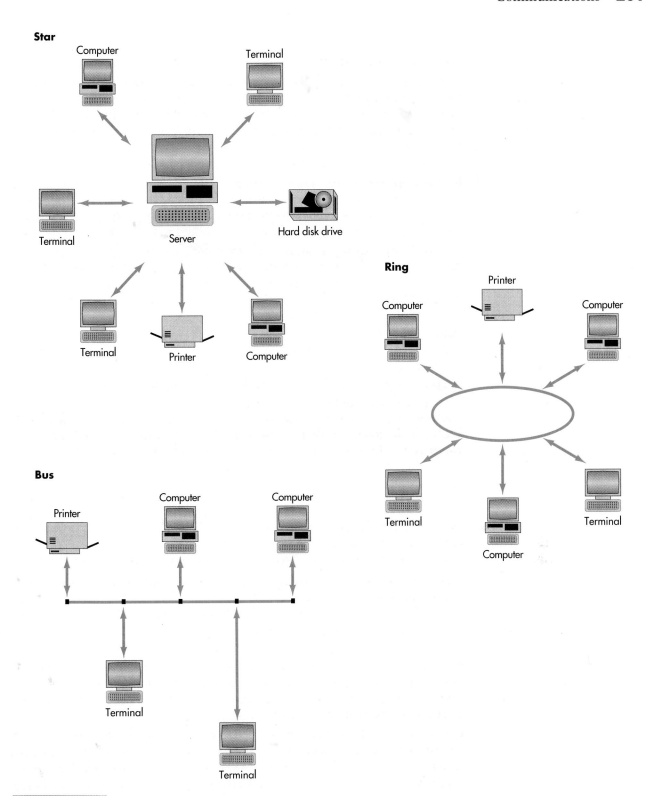

PANEL 8.11

Three LAN topologies: star, ring, bus

(Top) In a star network, all the network's devices are connected to a central server, through which all communications must pass. *(Middle)* In a ring network, the network's devices are connected in a closed loop. If one component fails, the whole system may fail. *(Bottom)* In a bus network, a single channel connects all communications devices.

The Impact of LANs

Sales of mainframes and minicomputers have been falling for some time. This is largely because companies have discovered that LANs can take their place for many functions, and at considerably less expense. This trend is known as *downsizing.* Still, a LAN, like a mainframe, requires a skilled support staff. Moreover, LANs have neither the great storage capacity nor the security that mainframes have, which makes them inappropriate for some applications.

Factors Affecting Data Transmission

Preview & Review: Factors affecting how data is transmitted include the transmission rate (frequency and bandwidth), the line configuration (point-to-point or multipoint), serial versus parallel transmission, the direction of transmission flow (simplex, half-duplex, or full-duplex), transmission mode (asynchronous or synchronous), packet switching, and protocols.

Transmission Rate: Higher Frequency, Wider Bandwidth, More Data

Transmission rate is a function of two variables: frequency and bandwidth.

- **Frequency:** The amount of data that can be transmitted on a channel depends on the wave *frequency*—the cycles of waves per second. Frequency is expressed in hertz: 1 cycle per second equals 1 hertz. The more cycles per second, the more data that can be sent through that channel.
- **Bandwidth:** As mentioned earlier, *bandwidth* is the difference between the highest and lowest frequencies—that is, the range of frequencies. Data may be sent not just on one frequency but on several frequencies within a particular bandwidth, all at the same time. Thus, the greater the bandwidth of a channel, the more frequencies it has available and hence the more data that can be sent through that channel. The rate of speed of data through the channel is expressed in bits per second (bps).

A twisted-pair telephone wire of 4000 hertz might send only 1 kilobyte of data in a second. A coaxial cable of 100 megahertz might send 10 megabytes. And a fiber-optic cable of 200 trillion hertz might send 1 gigabyte.

Line Configurations: Point-to-Point & Multipoint

There are two principal line configurations, or ways of connecting communications lines: point-to-point and multipoint.

- **Point-to-point:** A *point-to-point line* directly connects the sending and receiving devices, such as a terminal with a central computer. This arrangement is appropriate for a private line whose sole purpose is to keep data secure while transmitting it from one device to another.
- **Multipoint:** A *multipoint line* is a single line that interconnects several communications devices to one computer. Often on a multipoint line only one communications device, such as a terminal, can transmit at any given time.

Serial & Parallel Transmission

Data is transmitted in two ways: serially and in parallel.

- **Serial data transmission:** In *serial data transmission,* **bits are transmitted sequentially, one after the other.** This arrangement resembles cars proceeding down a one-lane road.

    Serial transmission is the way most data flows over a twisted-pair telephone line. Serial transmission is found in communications lines, modems, and mice.

- **Parallel data transmission:** In *parallel data transmission,* **bits are transmitted through separate lines simultaneously.** The arrangement resembles cars moving in separate lanes at the same speed on a multilane freeway.

    Parallel lines move information faster than serial lines do, but they are only efficient for up to 15 feet. Thus, parallel lines are used, for example, to transmit data from a computer's CPU to a printer.

Direction of Transmission Flow: Simplex, Half-Duplex, & Full-Duplex

When two computers are in communication, data can flow in three ways: simplex, half-duplex, or full-duplex. These are fancy terms for easily understood processes.

- **Simplex transmission:** In *simplex transmission,* **data can travel in only one direction.** An example is a traditional television broadcast, in which the signal is sent from the transmitter to your TV antenna. There is no return signal. Some computerized data collection devices also work this way (such as seismograph sensors that measure earthquakes).

- **Half-duplex transmission:** In *half-duplex transmission,* **data travels in both directions but only in one direction at a time.** This arrangement resembles traffic on a one-lane bridge; the separate streams of cars must take turns. Half-duplex transmission is seen with CB or marine radios, in which both parties must take turns talking. It is also a common transmission method with microcomputers. When you log onto an electronic bulletin board, you may be using half-duplex transmission.

- **Full-duplex transmission:** In *full-duplex transmission,* **data is transmitted back and forth at the same time.** This arrangement resembles automobile traffic on a two-way street. An example is two people on the telephone talking and listening simultaneously. Full-duplex is used frequently between computers in communications systems.

Transmission Mode: Asynchronous Versus Synchronous

Suppose your computer sends the word CONGRATULATIONS! to someone as bits and bytes over a communications line. How does the receiving equipment know where one byte (or character) ends and another begins? This matter is resolved through either *asynchronous transmission* or *synchronous transmission.* (■ *See Panel 8.12, next page.*)

- **Asynchronous transmission:** This method, used with most microcomputers, is also called *start-stop transmission.* In *asynchronous transmission,* **data is sent one byte (or character) at a time. Each string of bits making up the byte is bracketed, or marked off, with special control bits.** That is, a "start" bit represents the beginning of a character, and a "stop" bit represents its end.

Transmission modes

There are two ways that devices receiving data transmissions can determine the beginnings and ends of strings of bits (bytes, or characters). *(Top)* In asynchronous transmission, each character is preceded by a "start" bit and followed by a "stop" bit. *(Bottom)* In synchronous transmission, messages are sent in blocks, with start and stop patterns of bits, called sync bytes, before and after the blocks. The sync bytes synchronize the timing of the internal clocks between sending and receiving devices.

Asynchronous transmission

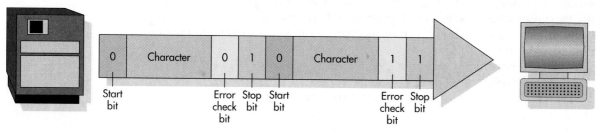

Synchronous transmission

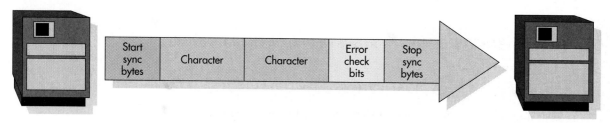

Transmitting only one byte at a time makes this a relatively slow method. As a result, asynchronous transmission is not used when great amounts of data must be sent rapidly. Its advantage is that the data can be transmitted whenever it is convenient for the sender.

- **Synchronous transmission:** Instead of using start and stop bits, **synchronous transmission sends data in blocks. Start and stop bit patterns, called sync bytes, are transmitted at the beginning and end of the blocks.** These start and stop bit patterns synchronize internal clocks in the sending and receiving devices so that they are in time with each other.

This method is rarely used with microcomputers because it is more complicated and more expensive than asynchronous transmission. It also requires careful timing between sending and receiving equipment. It is appropriate for computer systems that need to transmit great quantities of data quickly.

Packet Switching: Getting More Data on a Network

A *packet* is a fixed-length block of data for transmission. The packet also contains instructions about the destination of the packet. **Packet switching is a technique for dividing electronic messages into packets for transmission over a network to their destination through the most expedient route.** The benefit of packet switching is that it can handle high-volume traffic in a network. It also allows more users to share a network, thereby offering cost savings. The method is particularly appropriate for sending messages long distances, such as across the country. Accordingly, it is used in large networks such as Telenet, Tymnet, and AT&T's Accunet.

Here's how packet switching works: A sending computer breaks an electronic message apart into packets. The various packets are sent through a communications network—often by different routes, at different speeds, and sandwiched in between packets from other messages. Once the packets arrive at their destination, the receiving computer reassembles them into proper sequence to complete the message.

Protocols: The Rules of Data Transmission

The word *protocol* is used in the military and in diplomacy to express rules of precedence, rank, manners, and other matters of correctness. (An example would be the protocol for who will precede whom into a formal reception.) Here, however, **a *protocol*, or *communications protocol*, is a set of conventions governing the exchange of data between hardware and/or software components in a communications network.**

Protocols are built into the hardware or software you are using. The protocol in your communications software, for example, will specify how receiving devices will acknowledge sending devices, a matter called *handshaking.* Protocols will also specify the type of electrical connections used, the timing of message exchanges, error-detection techniques, and so on.

Cyberethics: Netiquette, Controversial Material & Censorship, & Privacy Issues Ⓔ

Preview & Review: Users of communications technology must weigh standards of behavior and conduct in three areas: netiquette, controversial material, and privacy.

Communications technology gives us more choices of nearly every sort. Not only does it provide us with different ways of working, thinking, and playing; it also presents us with some different moral choices—determining right actions in the digital and online universe. Next, let us consider three important aspects of "cyberethics"—netiquette, controversial material and censorship, and matters of privacy.

Netiquette

One morning *New Yorker* magazine writer John Seabrook, who had recently published an article about Microsoft chairman Bill Gates, checked into the e-mail on his computer to find the following reaction to his story:

> Listen, you toadying [*deleted*] scumbag . . . remove your head [*three words deleted*] long enough to look around and notice that real reporters don't fawn over their subjects, pretend that their subjects are making some sort of special contact with them, or, worse, curry favor by TELLING their subjects how great the [*deleted*] profile is going to turn out and then brag in print about doing it. . . .

On finishing the message, Seabrook rocked back in his chair. "Whoa," he said aloud to himself. "I got flamed."[36]

A form of speech unique to online communication, *flaming* is writing an online message that uses derogatory, obscene, or inappropriate language. The attack on Seabrook is probably unusual, since most flaming happens when someone violates online manners or "netiquette."

As mentioned, many online discussion groups have a set of FAQ's—frequently asked questions—that newcomers, or "newbies," are expected to become familiar with before joining in any chat forums. Most FAQs offer *netiquette*, or "net etiquette," guides to appropriate behavior while online. The commercial online services also have special online sites where the uninitiated can go to learn how to avoid an embarrassing breach of manners. Examples of netiquette blunders are typing with the CAPS LOCK key on—the Net equivalent of yelling—discussing subjects not appropriate to the forum, repetition of points made earlier, and improper use of the software.[37,38] *Spamming*, or sending unsolicited mail, is especially irksome; a *spam* includes chain letters, advertising, or similar junk mail.

Controversial Material & Censorship

In Saudi Arabia, officials police taboo subjects (sex, religion, politics) on the Internet.[39] In China, Net users must register with the government.[40,41] In Germany, in late 1995, officials persuaded Ohio-based CompuServe to cut off access to 200 sex-related news forums worldwide. (CompuServe reopened them a few months later, offering to provide filtering software to users wanting to block offensive material.)[42,43] In the United States, however, free speech is protected by the First Amendment to the Constitution.

If a U.S. court decides *after* you have spoken that you have defamed or maliciously damaged someone, you may be sued for slander (spoken speech) or libel (written speech) or charged with harassment, but you cannot be stopped beforehand. However, "obscene" material is not constitutionally protected free speech. Obscenity is defined as sexually explicit material that is offensive as measured by "contemporary community standards"—a definition with considerable leeway, depending on localities.

Since computers are simply another way of communicating, there should be no surprise that a lot of people use them to communicate about sex. Yahoo!, the Internet directory company, says that the word "sex" is the most popular search word on the Net.[44] All kinds of online X-rated bulletin boards, chat rooms, and Usenet newsgroups exist. A special problem is with children having access to sexual conversations, downloading hard-core pictures, or encountering odious adults tempting them into a meeting. "Parents should never use an online service as an electronic baby-sitter," says computer columnist Lawrence Magid. People online are not always what they seem to be, he points out, and a message seemingly from a 12-year-old girl could really be from a 30-year-old man. "Children should be warned never to give out personal information," says Magid, "and to tell their parents if they encounter mail or messages that make them uncomfortable."[45]

What can be done about X-rated materials?

• **Filtering software:** Some software developers have discovered a golden opportunity in making programs like SurfWatch and Net Nanny. These filtering programs screen out objectionable matter typically by identifying certain nonapproved keywords in a user's request or comparing the user's request for information against a list of prohibited sites.[46]

• **Browsers with ratings:** Another proposal in the works is browser software that contains built-in ratings for Internet, Usenet, and World Wide Web files. Parents could, for example, choose a browser that has been endorsed by the local school board or the online service provider.[47]

• **The V-chip:** The 1996 Telecommunications Law officially launched the era of the V-chip, a device that will be required equipment in most new television sets within 2 years.[48,49] The *V-chip* allows parents to automatically block out programs that have been labeled as high in violence, sex, or other objectionable material.

The difficulty with any attempts at restricting the flow of information is the basic Cold War design of the Internet itself, with its strategy of offering different roads to the same place. "If access to information on a computer is blocked by one route," writes the *New York Times*'s Peter Lewis, "a moderately skilled computer user can simply tap into another computer by an alternative route." Lewis points out an Internet axiom attributed to an engineer named John Gilmore: "The Internet interprets censorship as damage and routes around it."[50]

Privacy

Privacy **is the right of people not to reveal information about themselves.** Technology, however, puts constant pressure on this right.

A number of people, for example, have been undone by using cellular phones. The location of football star O.J. Simpson, after he was charged with murdering his ex-wife and a friend, was traced by police through his cellular phone signal.[51] Indeed, anyone using a scanner receiving in the 800–900 megahertz range can listen in to cellular phone conversations.[52] The *Electronic Communications Privacy Act* of 1986 makes eavesdropping on private conversations illegal without a court order. However, authorities have no way to catch people using scanners.

Think you're anonymous online when you don't sign your real name? America Online, in response to a legal petition for discovery, turned over the real name, address, and credit card information of a subscriber named Jenny TRR. Attorneys for a resort were considering suing her for defamation for a critical message she posted on an online bulletin board.[53]

Think the boss can't snoop on your e-mail at work? The Electronic Communications Privacy Act allows employers to "intercept" employee communications if one of the parties involved, such as the employer, agrees to the "interception."[54] Indeed, employer snooping seems to be widespread.

Think your medical records are sacrosanct? Actually, private medical information is bought and sold freely by various companies since there is no federal law prohibiting it. (And they simply ignore the patchwork of varying state laws.)[55]

A great many people are concerned about the loss of their right to privacy. Indeed, a 1995 survey found that 80% of the people contacted worried that they had lost "all control" of the personal information being collected and tracked by computers.[56] Although the government is constrained by several laws on acquiring and disseminating information and listening in on private conversations, there are reasons to be alarmed.

Onward

In late 1995, the Federal Communications Commission opened up virgin territory on the electromagnetic spectrum, a section known as *millimeter waves,* those situated above 40 gigahertz.[57] These open up possibilities not only for wireless campuses—so that phones, pagers, and other communications devices could be easily connected—but also for "smart homes," with appliances, heating and cooling, and security controlled by wireless systems. There could also be "smart cars," with radar systems to alert drivers to potential collisions.

Clearly, more roads will be added in cyberspace. We will consider the implications of these developments in the final chapter of this book.

Job Searching on the Internet & World Wide Web

"If you haven't done a job search in a while, you will find many changes in a modern-day, high-quality search for a new position," says Mary Anne Buckman, consultant at Career Directions Inc.[58] Indeed, technological change has so affected the whole field of job hunting that futurists refer to it as a *paradigm shift*. This means that, in one definition, the change is of a magnitude in which the "prevailing structure is radically, rapidly, and unalterably transformed by new circumstances."[59]

Within five years online services will become the most prevalent means of nonlocal hiring and recruitment, says Tom Jackson, a career development expert in Woodstock, New York, who created one of the first computerized job banks. Moreover, he predicts, multimedia resumes and online interviews will become commonplace. "Anyone who doesn't know how to use these services in the next 12 to 18 months will lose out," Jackson says.[60] Even if you never hunt for a job by computer, states Martin Yate, a career consultant, online networking—making friends and exchanging news with others in your field—will become "imperative for your professional survival."[61]

Let's take this further and describe how you can use the Internet and the World Wide Web to help you search for jobs. Online areas of interest for the job seeker include:

- Resources for career advice
- Ways for you to find employers
- Ways for employers to find you

Resources for Career Advice

It's 3 A.M. Still, if you're up at this hour (or indeed at any other time) you can find job-search advice, tips on interviewing and resume writing, and postings of employment opportunities around the world. One means for doing so is through Catapult (*http://www.wm.edu/catapult/catapult.html*), which was developed by Leo J. Charette, director of career services at the College of William and Mary in Williamsburg, Virginia, but has links to other colleges and potential employers.[62] For instance, through Catapult you could access a database located at Hartwick College called Barterbase, which unifies the expertise of 25 colleges in different employment areas. Barterbase offers, with "one-stop shopping," far more information than would probably be available through your own college's career center.

Another route is to use your Web browser to access a directory such as Yahoo! (*http://www.yahoo.com/*) to ob-

tain a list of popular Web sites. In the menu, you can click on Business, then Employment, then Jobs. This will bring up a list of sites that offer career advice, resume postings, job listings, research about specific companies, and other services.[63] (Caution: As might be expected, there is also a fair amount of junk out there: get-rich-quick offers, resume-preparation firms, and other attempts to separate you from your money.)

Advice about careers, occupational trends, employment laws, and job hunting is also available through on-line chat groups and bulletin boards, such as those on the Big Three online services—America Online, CompuServe, and Prodigy.[64,65] For instance, CompuServe offers career-specific discussion groups, such as the PR Marketing Forum. Through these groups you can get tips on job searching, interviewing, and salary negotiations.

Ways for You to Find Employers

As you might expect, companies seeking people with technical backgrounds and technical people seeking employment pioneered the use of cyberspace as a job bazaar. However, as the public's interest in commercial services and the Internet has exploded, the technical orientation of online job exchanges has changed. Now, says one writer, "interspersed among all the ads for programmers on the Internet are openings for English teachers in China, forest rangers in New York, physical therapists in Atlanta, and models in Florida."[66] Most Web sites are free to job seekers, although some may require you to fill out an online registration form.

Some jobs are posted on Usenets by individuals, companies, and universities or colleges, such as computer networking company Cisco Systems of San Jose, California, and the University of Utah in Salt Lake City. Others are posted by professional or other organizations, such as the American Astronomical Society, Jobs Online New Zealand, and Volunteers in Service to America (VISTA).

Some of the principal organizations posting job listings are listed below.[67-72] Among the most established and largest are America's Job Bank, Career Mosaic, Career Path, E-Span, JobTrak, Job Web, and Online Career Center.

- *American Employment Weekly:* This employment tabloid (*http://branch.com/aew/aew/html*) features ads from the Sunday editions of 50 leading American newspapers.

- *America's postedJob Bank:* A joint venture of the New York State Department of Labor and the federal Employment and Training Administration, America's Job Bank (*http://www.ajb.dni.us/index.html*) advertises more than 100,000 jobs of all types. There are links to each state's employment office. More than a quarter of the jobs posted are sales, service, or clerical. Another quarter are managerial, professional, and technical. Other major types are construction, trucking, and manufacturing. The companies listed have their company Web links included.

- *Career Mosaic:* A service run by Bernard Hodes Advertising, Career Mosaic (*http://www.careermosaic.com/*) offers links to nearly 200 major corporations, most of them high-technology companies. One section is aimed at college students and offers tips on resumes and networking. A major strength is the J.O.B.S database, which lets you fill out forms to narrow your search, then presents you with a list of jobs meeting your criteria. A *New York Times* reporter who did this said a search for "writer" turned up 45 job listings, the oldest less than a month old.

- *Career Path:* Career Path (*http://www.careerpath.com/*) is a classified-ad employment listing from six of the country's largest newspapers, which you can search either individually or all at once. The papers are the *Boston Globe,* the *Chicago Tribune,* the *Los Angeles Times,* the *New York Times,* the *San Jose Mercury News,* and the *Washington Post.*

- *Employment Edge:* Containing both job listings and links to other recruiting Web sites, Employment Edge (*http://www.employmentedge.com/employment.edge/*) lists jobs by category (accounting, management, and so on.) It also offers links to sites with interviewing tips and resume help.

- *E-Span:* One of the oldest and biggest services, the E-Span Interactive Employment Network (*http://www.espan.com*) features all-paid ads from employers.

- *FedWorld:* This bulletin board (*http://www.fedworld.gov*) offers job postings from the U.S. Government.

- *Internet Job Locator:* Combining all major job-search engines on one page, the Internet Job Locator (*http://www.joblocator.com/jobs/*) lets you do a search of all of them at once.

- *Job Hunt:* Started by Dr. Dane Spearing, a geologist at Stanford University, the well-organized Job Hunt page (*http://rescomp.stanford.edu/jobs/*) contains a list of more than 200 sites related to online recruiting.

- *JobLinks:* This resource (*http://www.brandeis.edu/hiatt/web_data/Job_Listings.html*) offers job listings for business, government, health, law, science, technology, and other fields.

- *JobTrak:* The nation's leading online job listing service, JobTrak (*http://www.jobtrak.com*) claims to have been used by more than a million students and alumni, with more than 150,000 employers and 300 college career centers posting new jobs daily.

- *JobWeb:* Operated by the National Association of Colleges and Employers, Job Web (*http://www.jobweb.org/*) is a college placement service.

- *The Monster Board:* Not just for computer techies, the Monster Board (*http://www.monster.com*) offers real jobs for real people, although a lot of the companies listed are in the computer industry.

- *NationJob Network:* Despite the name, NationJob Network (*http://www.nationjob.com*) lists job opportunities primarily in the Midwest.

- *Online Career Center:* Based in Indianapolis, Online Career Center (*http://www.occ.com/occ/*) is a nonprofit national recruiting service listing jobs at more than 3000 companies. About 30% of the jobs are nontechnical, with many in sales and marketing and in health care.

- *Workplace:* An employment resource offering staff and administrative positions in colleges and universities, government, and the arts (*http://galaxy.einet.net/galaxy/Community/Workplace.html/*).

The difficulty with searching through these resources is that it can mean wading through thousands of entries in numerous databanks, with many of them not being suitable for or interesting to you. An alternative to trying to find an employer is to have employers find you.

Ways for Employers to Find You

Because of its low (or zero) cost and wide reach, do you have anything to lose by posting your resume on line for prospective employers to view? Certainly you might if the employer happens to be the one you're already working for. In addition, you have to be aware that you lose control over anything broadcast into cyberspace—you're putting your credentials out there for the whole world to see, and you need to be somewhat concerned about who might gain access to them.

Posting your resume with an electronic jobs registry is certainly worth doing if you have a technical background, since technology companies in particular find this an efficient way of screening and hiring. However, it may also benefit people with less-technical backgrounds. Online recruitment "is popular with companies because it pre-screens applicants for at least basic computer skills," says one writer. "Anyone who can master the Internet is likely to know something about word processing, spreadsheets, or database searches, knowledge required in most good jobs these days."[73]

Resumes may be prepared in a traditional manner. The latest variant, however, is to produce a resume with hypertext links and/or clever graphics and multimedia effects, then put it on a Web site to entice employers to chase after you.[74] If you don't know how to do this, there are many companies that—for a fee—can convert your resume to HTML (✓ p. 290) and publish it on their own Web sites. Some of these services can't dress it up with fancy graphics or multimedia, but since complex pages take longer for employers to download anyway, the extra pizzazz is probably not worth the effort. In any case, for you the bottom line is how much you're willing to pay for these services. For instances, OneWayResume (*http://www2.connectnet.com/users/blorincz*) charges $35 to write a resume and nothing to post it on a Web site. Actors can post their resumes and head shots for free on ActorsPavilion *(http://www.ios.com/unisoft/act.html)*. You can post your own resume for free on Intellimatch and the Internet Employment Network. (Further information on preparing a hypertext resume may be found in Scott Grusky, "Winning Resume," *Internet World,* February 1996.)

Some of the principal places for posting your online resume are as follows:

- *E-Span:* Featuring paid ads from employers, the E-Span Interactive Employment Network (*http://www.espan.com*) also allows job seekers to post their resumes.

- *Intellimatch:* A free resume posting service, Intellimatch (*http://www.intellimatch.com*) allows applicants to fill out a structured resume, as well as to search for posted jobs.

- *Internet Employment Network:* This free resume referral service also allows you to search a database of all occupational categories (*http:// garnet.msen.com:70/1/vendor/napa/jobs*).

- *JobTailor Employment Online Service:* This resume posting service is free, but your resume but follow a certain structure (*http://www.jobtailor.com*).

- *123 Resume Distribution Service:* A free service that submits resume information to employers (*http://www.webplaza.com/pages/Careers/123 Careers/123Careers.html*).

- *Online Career Center:* This nonprofit job registry allows job searchers to post their resumes for free (*http://www.occ.com/occ/*).

- *Skill Search:* An online employment service that creates an applicant profile, Skill Search (*http://www.internetis.com/skillsearch/*) works with 60 alumni groups.

Companies are also beginning to replace their campus visits by recruiters with online interviewing. For example, the firm VIEWnet Inc. of Madison, Wisconsin, offers first-round screenings or interviews for summer internships through its teleconferencing "InterVIEW" technology, which allows video signals to be transmitted (at 17 frames per second) via telephone lines.[75]

Suggested Resources

Dixon, Pam, and Sylvia Tiersten. *Be Your Own Headhunter Online: Get the Job You Want Using the Information Superhighway.* New York: Random House, 1995. Explains how to conduct a national or international job search, connect with employers, and use online resources to prepare for interviews.

Glossbrenner, Alfred, and Emily Glossbrenner. *Finding a Job on the Internet.* New York: McGraw-Hill Computing, 1995.

Gonyea, James C. *The On-Line Job Search Companion.* New York: McGraw-Hill, 1995.

Kennedy, Joyce Lain. *Hook Up, Get Hired! The Internet Job Search Revolution.* New York: Wiley, 1995.

SUMMARY

analog cellular phone *(p. 254, LO 3)* Mobile telephone designed primarily for communicating by voice through a system of cells. Calls are directed to cells by a mobile telephone switching office (MTSO). Moving between cells requires that calls be "handed off" by the MTSO between cells.

Cellular phone systems allow callers mobility.

analog signal *(p. 247, LO 2)* Continuous electrical signal in the form of a wave. The wave is called a *carrier wave.* Two characteristics of analog carrier waves that can be altered are frequency and amplitude. Computers cannot process analog signals.

Analog signals are used to convey voices and sounds over wire telephone lines, as well as in radio and TV broadcasting. Computers, however, use digital signals, which must be converted to analog signals in order to be transmitted over telephone wires.

asynchronous transmission *(p. 263, LO 5)* Also called *start-stop transmission;* data is sent one byte (character) at a time. Each string of bits making up the byte is bracketed with special control bits; a "start" bit represents the beginning of a character, and a "stop" bit represents its end.

This method of communications is used with most microcomputers. Its advantage is that data can be transmitted whenever convenient for the sender. Its drawback is that transmitting only one byte at a time makes it a relatively slow method that cannot be used when great amounts of data must be sent rapidly.

bands (bandwidths) *(p. 252, LO 3, 5)* Ranges of frequencies. The bandwidth is the difference between the lowest and highest frequencies transmitted.

Different telecommunications systems use different bandwidths for different purposes, whether cellular phones or network television.

bits per second (bps) *(p. 249, LO 2)* Measurement of data transmission speeds. Modems transmit at 1200 and 2400 bps (slow), 4800 and 9600 bps (moderately fast), and 14,400, 19,200, and 28,800 bps (high-speed).

A 10-page single-spaced letter can be transmitted by a 2400-bps modem in 2½ minutes. It can be transmitted by a 9600-bps modem in 38 seconds and by a 19,200-bps modem in 19 seconds. The faster the modem, the less time online and therefore less expense.

bridge *(p. 260, LO 4)* Interface that enables similar networks to communicate.

Smaller networks (local area networks) can be joined together to create larger networks.

bus network *(p. 260, LO 4)* Type of network in which all communications devices are connected to a common channel, with no central server. Each communications device transmits electronic messages to other devices. If some of those messages collide, the device waits and tries to retransmit again.

The advantage of a bus network is that it may be organized as a client-server or peer-to-peer network. The disadvantage is that extra circuitry and software are needed to avoid collisions between data. Also, if a connection is broken, the entire network may stop working.

cable modem *(p. 250, LO 2)* Modem that connects a PC to a cable-TV system that offers online services, as well as TV.

Cable modems transmit data faster than standard modems.

channel *(p. 251, LO 3)* Also called *links, lines,* or *media;* path over which information travels in a telecommunications system from its source to its destination.

There are many different telecommunications channels, both wired and wireless, some more efficient than others for different purposes.

client/server LAN *(p. 258, LO 4)* Type of local area network (LAN); it consists of requesting microcomputers, called *clients,* and supplying devices that provide a service, called *servers.* The server is a computer that manages shared devices, such as laser printers.

Client/server networks are the most common type of LAN. Compare with *peer-to-peer LAN.*

What It Is / What It Does	**Why It's Important**

coaxial cable *(p. 253, LO 3)* Type of communications channel; commonly called *co-ax,* it consists of insulated copper wire wrapped in a solid or braided metal shield, then in an external cover.

Coaxial cable is much better at resisting noise than twisted-pair wiring. Moreover, it can carry voice and data at a faster rate.

communications *(p. 232, LO 1)* Also called *telecommunications;* the electronic transfer of information from one location to another. Also refers to electromagnetic devices and systems for communicating data.

Communications systems have helped to expand human communication beyond face-to-face meetings to electronic connections called the *global village.*

communications protocol *(p. 265, LO 5)* Set of conventions governing the exchange of data between hardware and/or software components in a communications network. Protocols are built into hardware and software. For example, the protocol in communications software will specify how receiver devices will acknowledge sending devices ("handshaking").

In the past, because not all hardware and software developers subscribed to the same protocols, many kinds of equipment and programs did not work with one another.

communications satellites *(p. 254, LO 3)* Microwave relay stations orbit 22,300 miles above the equator. Because they travel at the same speed as the earth, thus appearing stationary in space, microwave earth stations can beam signals to a fixed location above. The satellite has solar-powered receivers and transmitters (transponders) that receive the signals, amplify them, and retransmit them to another earth station.

An orbiting satellite contains many communications channels and receives both analog and digital signals from ground microwave stations anywhere on earth.

communications software *(p. 249, LO 2)* Software that manages the transmission of data between computers or video display terminals.

Without communications software, computers cannot communicate—with or without modems or other communications equipment.

connectivity *(p. 232, LO 1)* The state of being able to connect devices by communications technology to other devices and sources of information.

Computers offer greater varieties of connectivity than other communications devices such as telephones or radio systems.

dedicated fax machine *(p. 233, LO 1)* Specialized device that does nothing except scan in, send, and receive documents over telephone lines to and from other fax machines.

Fax machines have enabled people to instantly transmit graphics and documents for the price of a phone call.

digital signal *(p. 248, LO 2)* Type of electrical signal that uses on/off or present/absent electrical pulses in discontinuous, or discrete, bursts, rather than a continuous wave.

This two-state kind of signal works perfectly in representing the two-state binary language of 0s and 1s that computers use.

electromagnetic spectrum *(p. 252, LO 3)* All the fields of electrical energy and magnetic energy, which travel in waves. This includes all radio signals, light rays, X-rays, and radioactivity.

The part of the electromagnetic spectrum of particular interest is the area in the middle, which is used for communications purposes. Various frequencies are assigned by the federal government for different purposes.

electronic data interchange (EDI) *(p. 244, LO 1)* System of direct electronic exchange between organizations' computer systems of standard business documents, such as purchase orders, invoices, and shipping documents.

EDI allows the companies involved to do away with standard paper handling and its costs.

electronic mail (e-mail) *(p. 235, LO 1)* System in which computer users, linked by wired or wireless communications lines, may use their keyboards to post messages and to read responses on their display screens.

E-mail allows users to send messages to a single recipient's "mailbox"—a file stored on the computer system—or to multiple users. It is a much faster way of transmitting written messages than traditional mail services.

fax modem *(p. 233, LO 1, 2)* Type of modem installed as a circuit board inside a computer; it exchanges fax messages with another fax machine or fax modem.

The benefit of fax modems is that messages can be transmitted directly from a microcomputer; no paper or scanner is required.

FDDI network *(p. 260, LO 4)* Short for Fiber Distributed Data Interface; a type of local area network that uses fiber-optic cable with a dual counter-rotating ring topology.

The FDDI network is being used for such high-tech purposes as electronic imaging, high-resolution graphics, and digital video.

What It Is / What It Does	Why It's Important

fiber-optic cable *(p. 253, LO 3)* Type of communications channel consisting of hundreds or thousands of thin strands of glass that transmit pulsating beams of light. These strands, each as thin as a human hair, can transmit billions of pulses per second, each "on" pulse representing one bit.

When bundled together, fiber-optic strands in a cable 12 inches thick can support a quarter- to a half-million simultaneous voice conversations. Moreover, unlike electrical signals, light pulses are not affected by random electromagnetic interference in the environment and thus have much lower error rates than telephone wire and cable.

file server *(p. 258, LO 4)* Type of computer used on a local area network (LAN) that acts like a disk drive and stores the programs and data files shared by users of the LAN.

A file server enables users of a LAN to all have access to the same programs and data.

full-duplex transmission *(p. 263, LO 5)* Type of data transmission in which data is transmitted back and forth at the same time, unlike simplex and half-duplex.

Full-duplex is used frequently between computers in communications systems.

gateway *(p. 260, LO 4)* Interface that enables dissimilar networks to communicate with one another.

With a gateway, a local area network may be connected to a larger network, such as a wide area network.

groupware *(p. 244, LO 1)* Software that allows two or more people on a network to work on the same information at the same time.

Groupware has become the glue that ties organizations together, permitting office workers to collaborate with colleagues, suppliers, and customers and to tap into company information through computer networks.

half-duplex transmission *(p. 263, LO 5)* Type of data transmission in which data travels in both directions but only in one direction at a time, as with CB or marine radios; both parties must take turns talking.

Half-duplex is a common transmission method with microcomputers, as when logging onto an electronic bulletin board system.

home page *(p. 243, LO 1)* The first page (main page)—that is, the first screen—seen upon accessing a Web site.

The home page provides a menu or explanation of the topics available on that Web site.

host computer *(p. 256, LO 4)* The central computer that controls a network. On a local area network, the host's functions may be performed by a computer called a *server*.

The host is responsible for managing the entire network.

hybrid network *(p. 260, LO 4)* Type of local area network (LAN) that combines star, ring, and bus networks.

A hybrid network can link different types of LANs. For example, a small college campus might use a bus network to connect buildings and star and ring networks within certain buildings.

Integrated Services Digital Network (ISDN) *(p. 250, LO 2)* A set of international communications standards for transmitting voice, video, and data simultaneously as digital signals over twisted-pair telephone lines.

The main benefit of ISDN is speed. It allows people to send digital data ten times faster than most modems can now deliver on the analog voice network.

Internet *(p. 238, LO 1)* International network composed of approximately 36,000 smaller networks. Created as ARPAnet in 1969 by the U.S. Department of Defense, Internet was designed to share research among military, industry, and university sources and to sustain communication in the event of nuclear attack.

Today the Internet is essentially a self-governing and noncommercial community offering both scholars and the public such features as information gathering, electronic mail, and discussion and newsgroups.

Internet service provider (ISP) *(p. 240, LO 1)* Local or national company that provides unlimited public access to the Internet and the Web for a flat fee.

Unless they are connected to the Internet through an online information service or a direct network connection, microcomputer users need an ISP to connect to the Internet.

intranet *(p. 244, LO 1)* Internal corporate network that uses the infrastructure and standards of the Internet and the World Wide Web.

Intranets can connect all types of computers.

local area network (LAN) *(p. 258, LO 4)* A network consisting of a communications link, network operating system, microcomputers or workstations, servers, and other shared hardware such as printers or storage devices. LANs are of two principal types: client/server and peer-to-peer.

LANs have replaced mainframes and minicomputers for many functions and are considerably less expensive. However, LANs have neither the great storage capacity nor the security of mainframes.

local network *(p. 256, LO 4)* Privately owned communications network that serves users within a confined geographical area. The range is usually within a mile.

Local networks are of two types: private branch exchanges (PBXs) and local area networks (LANs).

What It Is / What It Does	**Why It's Important**
metropolitan area network (MAN) *(p. 256, LO 4)* Communications network covering a geographic area the size of a city or suburb. Cellular phone systems are often MANs.	The purpose of a MAN is often to bypass telephone companies when accessing long-distance services.
microwave systems *(p. 254, LO 3)* Communications systems that transmit voice and data through the atmosphere as super-high-frequency radio waves. Microwaves are the electromagnetic waves that vibrate at 1 billion hertz per second or higher.	Microwave frequencies are used to transmit messages between ground-based earth stations and satellite communications systems. More than half of today's telephone system uses microwave transmission.
modem *(p. 248, LO 2)* Short for *modulater/demodulater*. A device that converts digital signals into a representation of analog form (modulation) to send over phone lines; a receiving modem then converts the analog signal back to a digital signal (demodulation).	A modem enables users to transmit data from one computer to another by using standard telephone lines instead of special communications lines such as fiber optic or cable.
network (communications network) *(p. 255, LO 4)* System of interconnected computers, telephones, or other communications devices that can communicate with one another.	Networks allow users to share applications and data.
online information service *(p. 236, LO 1)* Company that provides access to databases and electronic meeting places to subscribers equipped with telephone-linked microcomputers—for example Prodigy, CompuServe, and America Online.	Online information services offer a wealth of services, from electronic mail to home shopping to videogames to enormous research facilities to discussion groups.
packet *(p. 264, LO 5)* Fixed-length block of data for transmission. The packet also contains instructions about the destination of the packet.	By creating data in the form of packets, a transmission system can deliver the data more efficiently and economically, as in packet switching.
packet switching *(p. 264, LO 5)* Technique for dividing electronic messages into packets—fixed-length blocks of data—for transmission over a network to their destination through the most expedient route. A sending computer breaks an electronic message apart into packets, which are sent through a communications network—via different routes and speeds—to a receiving computer, which reassembles them into proper sequence to complete the message.	The benefit of packet switching is that it can handle high-volume traffic in a network. It also allows more users to share a network, thereby offering cost savings.
parallel data transmission *(p. 263, LO 5)* Method of transmitting data in which bits are sent through separate lines simultaneously.	Unlike serial lines, parallel lines move information fast, but they are efficient for only up to 15 feet. Thus, parallel lines are used, for example, to transmit data from a computer's CPU to a printer.
peer-to-peer LAN *(p. 258, LO 4)* Type of local area network (LAN); all microcomputers on the network communicate directly with one another without relying on a server.	Peer-to-peer networks are less expensive than client/server networks and work effectively for up to 25 computers. Thus, they are appropriate for networking in small groups.
privacy *(p. 267, LO 6)* Right of people not to reveal information about themselves.	Computer technology and electronic databases have made it more difficult for people to protect their privacy.
private branch exchange (PBX) *(p. 258, LO 4)* Private or leased telephone switching system that connects telephone extensions inhouse as well as to the outside telephone system.	Newer PBXs can handle not only analog telephones but also digital equipment, including computers.
ring network *(p. 260, LO 4)* Type of local area network (LAN) in which all communications devices are connected in a continuous loop and messages are passed around the ring until they reach the right destination. There is no central server.	The advantage of a ring network is that messages flow in only one direction and so there is no danger of collisions. The disadvantage is that if a connection is broken, the entire network stops working.
serial data transmission *(p. 263, LO 5)* Method of data transmission in which bits are sent sequentially, one after the other, through one line.	Serial transmission is found in communications lines, modems, and mice.
server *(p. 256, LO 4)* Computer shared by several users in a network.	With servers, users on a LAN can share several devices, as well as data.

simplex transmission *(p. 263, LO 5)* Type of transmission in which data can travel in only one direction; there is no return signal.

Some computerized data collection devices, such as seismograph sensors that measure earthquakes, use simplex transmission.

star network *(p. 260, LO 4)* Type of local area network (LAN) in which all microcomputers and other communications devices are connected to a central hub, such as a file server. Electronic messages are routed through the central hub to their destinations. The central hub monitors the flow of traffic.

The advantage of a star network is that the hub prevents collisions between messages. Moreover, if a connection is broken between any communications device and the hub, the rest of the devices on the network will continue operating.

synchronous transmission *(p. 264, LO 5)* Type of transmission in which data is sent in blocks. Start and stop bit patterns, called sync bytes, are transmitted at the beginning and end of the blocks. These start and end bit patterns synchronize internal clocks in the sending and receiving devices so that they are in time with each other.

Synchronous transmission is rarely used with microcomputers because it is more complicated and more expensive than asynchronous transmission. It is appropriate for computer systems that need to transmit great quantities of data quickly.

telecommuting *(p. 245, LO 1)* Way of working at home and communicating ("commuting") with the office by phone, fax, and computer.

Telecommuting can help ease traffic and the stress of commuting by car and extend employment opportunities to more people, such as those who need or want to stay at home.

twisted-pair wire *(p. 252, LO 3)* Type of communications channel consisting of two strands of insulated copper wire, twisted around each other in pairs.

Twisted-pair wire has been the most common channel or medium used for telephone systems. It is relatively slow and does not protect well against electrical interference.

URL (Uniform Resource Location) *(p. 243, LO 1)* Address that points to a specific resource on the Web.

Addresses are necessary to distinguish among Web sites.

videoconferencing *(p. 236, LO 1)* Also called *teleconferencing;* form of conferencing using video cameras and monitors that allow people at different locations to see, hear, and talk with one another.

Videoconferencing may be done from a special videoconference room or handled with equipment rolled on casters from room to room.

virtual office *(p. 245, LO 1)* A nonpermanent and mobile office run with computer and communications technology.

Employees work not in a central office but from their homes, cars, and customers' offices. They use pocket pagers, portable computers, fax machines, and various phone and network services to conduct business.

voice mail *(p. 233, LO 1)* System in which incoming voice messages are stored in a recipient's "voice mailbox" in digitized form. The system converts the digitized versions back to voice messages when they are retrieved. With voice mail, callers can direct calls within an office using buttons on their touch-tone phone.

Voice mail enables callers to deliver the same message to many people, to forward calls, to save or erase messages, and to dictate replies. The main benefit is that voice mail helps eliminate "telephone tag."

Web browser *(p. 243, LO 1)* Internet software used to browse through multimedia Websites.

See World Wide Web.

Web site *(p. 243, LO 1)* File stored on a computer as part of the World Wide Web.

Each Web site focuses on a particular topic. The information on a site is stored on "pages." The starting page is called the *home page.*

wide area network (WAN) *(p. 256, LO 4)* Type of communications network that covers a wide geographical area, such as a state or a country.

Wide area networks provide worldwide communications systems.

workgroup computing *(p. 244, LO 1)* Also called *collaborative computing;* technology that enables teams of coworkers to use networks of microcomputers to share information and cooperate on projects. Workgroup computing is made possible not only by networks and microcomputers but also by groupware.

Workgroup computing permits office workers to collaborate with colleagues, suppliers, and customers and to tap into company information through computer networks.

World Wide Web *(p. 242, LO 1)* Interconnected system of sites of the Internet that store information in multimedia form.

Web software allows users to view information that includes not just text but graphics, animation, video, and sound.

EXERCISES

(Selected answers appear at the back of the book.)

Short-Answer Questions

1. What is the definition of *connectivity*?
2. List three methods you can use to connect your PC to the Internet.
3. Why is workgroup computing significant?
4. Why would you buy an internal modem rather than an external one?
5. Give the definition of *bandwidth*.
6. What is a virtual office?
7. What is the difference between asynchronous and synchronous transmission?
8. What could you use the Internet for?
9. What advantages are gained by using networks?

Fill-in-the-Blank Questions

1. Two types of fax hardware are ~~*dedicated*~~ and ~~*fax modems*~~.
2. The Internet is an international ~~*database network*~~ made up of *36,000 smaller networks*.
3. A nonpermanent and mobile office run with computer and communications technology is called a(n) *virtual* office.
4. Workgroup computing uses a type of software called *database*.
5. In an analog signal, the number of times a wave repeats during a specific time interval is called its ~~*bandwidth*~~ *frequency* and the height of a wave within a given period of time is called its ~~*frequency*~~ *amplitude*.
6. With _____, the user works at home but can communicate, using hardware and software, with the office.
7. When using the World Wide Web, *browser* software helps you to get the information you want by clicking on underlined words or pictures on the screen.

Multiple-Choice Questions

1. Which of the following allows different computers on different networks to communicate?
   a. e-mail
   b. BBS
   c. LAN
   d. TCP/IP
   e. all of the above
2. Which of the following *isn't* a principal feature of the Internet?
   a. e-mail
   b. file transfer
   c. LANs
   d. information services
   e. all of the above
3. Which of the following would you use if you wanted to copy files for free from a remote computer?
   a. World Wide Web
   b. FTP
   c. Usenet
   d. none of the above
4. Which of the following provides information in multimedia form?
   a. World Wide Web
   b. FTP
   c. Usenet
   d. none of the above
5. A communications network that covers a wide geographical area, such as a state or country, is called a(n)_____ network.
   a. wide area network (WAN)
   b. metropolitan area network (MAN)
   c. local area network (LAN)
   d. privatized network (PN)
   e. none of the above
6. A(n) _____ can be either a PBX or a LAN.
   a. MAN
   b. local network
   c. WAN
   d. Internet-based network
   e. none of the above
7. A network in which all communications devices are connected to a common channel is called a _____ network.
   a. star
   b. ring
   c. bus
   d. hybrid
   e. all of the above

8. A communications _____ is used in a communications network to govern the exchange of data between hardware and/or software.
   a. protocol
   b. multiplexer
   c. channel
   d. information service
   e. none of the above

True/False Questions

T F 1. Before you can communicate over the phone lines, your computer requires communications software.

T **F** 2. A metropolitan area network (MAN) typically covers a greater area than a wide area network (WAN).

T F 3. Coaxial cable is better able to resist noise than twisted-pair wiring.

T F 4. The term *download* refers to retrieving a file from another computer and copying it onto your computer.

T F 5. A private branch exchange (PBX) typically connects telephone extensions in a city or metropolitan area.

T F 6. Local area networks may be client/server or peer-to-peer.

T F 7. A file server is a computer that stores programs and data files for users on a LAN.

T **F** 8. A gateway is an interface that enables similar networks to communicate.

Projects/Critical-Thinking Questions

1. Are the computers at your school or at work connected to a network? If so, what are the characteristics of the network? What advantages does the network provide in terms of hardware and software support? What types of computers are connected to the network (microcomputers, minicomputers, and/or mainframes)? Specifically, what software/hardware allows the network to function?

2. "Distance learning" or "distance education" uses electronic links to extend college campuses to people who otherwise would not be able to take college courses. In one variant, college instructors using such systems are able to lecture "live" to students in distant locations. Is your school involved in distance learning? If so, research the system's components and uses. What hardware does it use? Software? Communications media?

3. You need to purchase a computer to use at home to perform business-related tasks. You want to be able to communicate with the network at school or work so that you can use its software and access its data. Include the following in a report:
   - A description of the hardware and software used at school or work.
   - A description of the types of tasks you will want to perform at home.
   - The name of the computer you would buy. (Include a detailed description of the computer, such as the RAM capacity and secondary storage capacity.)
   - The communications hardware/software you would need to purchase.
   - A cost estimate.

4. In the world today, what country or group of individuals do you think could benefit the most from using communications technology? Why isn't communications technology used now and what impact do you think this has? How would communications technology be beneficial? What are the barriers to implementing communications technology? Explore the answers to these questions in a two-page report.

net Using the Internet

Objective: *In this exercise we describe how to subscribe to a newsgroup and reply to articles using Netscape Navigator.*

Before you continue: *We assume you have access to the Internet through your university, business, or commercial service provider and to the Web browser tool named Netscape Navigator 2.0 or 2.01. Additionally, we assume you know how to connect to the Internet and then load Netscape Navigator. If necessary, ask your instructor or system administrator for assistance.*

1. Make sure you have started Netscape. The home page for Netscape Navigator should appear on your screen.

2. To subscribe to a newsgroup that discusses issues relating to future computer technology (news:comp.society.futures), perform the following steps:
   CLICK: Open button ( ).
   TYPE: `news:comp.society.futures`
   PRESS: [Enter]
   CLICK: *Subscribe box, located next to the "news:comp.society" newsgroup name.* (*Note:* You can easily unsubscribe to a newsgroup by clicking in the newsgroup's associated check box.)

3. To display the messages in the "comp.society. futures" newsgroup:
   DOUBLE-CLICK: *"comp.society.futures" in the newsgroup list*

4. The newsgroup's articles are listed in the right pane. Click an article you're interested in.

5. After reading an article, you may want to use the Re:News button ( ), which sends a copy of your message to the newsgroup for all subscribers to read, or the Re:Both button ( ), which sends a copy of your message to the newsgroup and an e-mail copy to the sender of the message. A dialog box will appear into which you can type your message. The message dialog box will look different depending on your computer system. On a Windows 95-based system, the following dialog box appeared after clicking the Re:News button ( ) to respond to an article related to how the Internet can change your life.

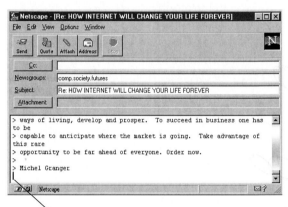

Begin typing your message here.

6. Now that you've seen the general process of replying to a newsgroup article, let's exit from the dialog box without posting a message.

7. Display Netscape's home page and then exit Netscape.

CHAPTER 9

Systems
Development, Programming, & Languages

Concepts You Should Know

After reading this chapter, you should be able to:

1. Discuss the six phases of the systems development life cycle.
2. Explain several systems analysis tools and design tools.
3. Compare four strategies for converting to a new system.
4. Explain what a program is.
5. Explain the five steps in programming.
6. Explain what compilers and interpreters do.
7. Identify a few of the major programming languages used today.

Organizations can make mistakes, and big organizations can make really big mistakes.

California's state Department of Motor Vehicles' databases needed to be modernized, and in 1988 Tandem Computers said it could do the job. "The fact that the DMV's database system, designed around an old IBM-based platform, and Tandem's new system were as different as night and day seemed insignificant at the time to the experts involved," said one writer investigating the project later.[1] The massive drivers' license database, containing the driving records of more than 30 million people, first had to be "scrubbed" of all information that couldn't be translated into the language used by Tandem Computers. One such scrub yielded 600,000 errors. Then the DMV had to translate all its IBM programs into the Tandem language. "Worse, DMV really didn't know how its current IBM applications worked anymore," said the writer, "because they'd been custom-made decades before by long-departed programmers and rewritten many times since." Eventually the project became a staggering $44 million loss to California's taxpayers.

In Denver, airport officials weren't trying to upgrade an old system but to do something completely new. At the heart of the Denver International Airport was supposed to be a high-tech baggage system. This was intended to whisk bags between terminals and at speeds that would mean that passengers would practically never have to wait for their luggage. As the system failed test after test, airport officials eventually decided they had to *build a manual baggage system*—at an additional cost of $50 million. Spending the money on old technology, it developed, was cheaper than continuing to spend millions paying interest on construction bonds for a nonoperating airport.[2]

Both these examples show how important planning is, especially when an organization is trying to launch a new kind of system. How can we try to avoid such mistakes? By employing the principles of systems analysis and design and of software programming, the subjects of this chapter.

Systems Development: The Six Phases of Systems Analysis & Design

Preview & Review: Knowledge of systems analysis and design helps you explain your present job, improve personal productivity, and lessen risk of a project's failure. The initiative for suggesting analysis and possibly change in an information system may come from users, managers, or technical staff.

The six phases of systems analysis and design are known as the *systems development life cycle (SDLC)*.

Why Know About Systems Analysis & Design?

You may not have to wrestle with problems on the scale of motor-vehicle departments and airports. That's a job for computer professionals. You're mainly interested in using computers and communications to increase your own productivity. Why, then, do you need to know anything about systems analysis and design?

In many types of jobs, you may find your department or your job the focus of a study by a systems analyst, discussed shortly. Knowing how analysis and design works will help you better explain how your job works and what goals your department is supposed to achieve. In progressive companies, management is always interested in employees' suggestions for improving productivity. In some cases, employee input is required.

The Purpose of a System

A *system* **is defined as a collection of related components that interact to perform a task in order to accomplish a goal.** A system may not work very well, but it is nevertheless a system. The point of systems analysis and design is to ascertain how a computer-and-communications system works and then take steps to make it better.

An organization's computer-based information system consists of hardware, software, people, procedures, and data, as well as communications setups. These work together to provide management with information for running the organization.

Getting the Project Going: How It Starts, Who's Involved

All it takes is a single individual who believes that something badly needs changing to get a systems development project rolling. An employee may influence a supervisor. A customer or supplier may get the attention of someone in higher management. Top management on its own may decide to take a look at a system that seems to be inefficient. A steering committee may be formed to decide which of many possible projects should be worked on.

Participants in the project are of three types:

- **Users:** The system under discussion should *always* be developed in consultation with users, whether floor sweepers, research scientists, or customers. Indeed, inadequate user involvement in analysis and design can be a major cause of system failure.

- **Management:** Managers within the organization should also be consulted about the system.

- **Technical staff:** Members of the company's information systems department, consisting of systems analysts and software programmers, need to be involved. For one thing, they may well have to carry out and execute the project. Even if they don't, they may have to work with outsiders contracted to do the job.

Complex projects will require a systems analyst. **A *systems analyst* is an information specialist who performs systems analysis, design, and implementation.** His or her job is to study the information and communications needs of an organization and determine what changes are required to deliver better information to people who need it, when they need it. "Better" information means information that is accurate, timely, and useful. The systems analyst achieves this goal through the problem-solving method of systems analysis and design.

The Six Phases of Systems Analysis & Design

Systems analysis and design is a six-phase problem-solving procedure for examining an information system and improving it. The six phases make up what is called the *systems development life cycle.* **The *systems development life cycle (SDLC)* is defined as the step-by-step process that many organizations follow during systems analysis and design.**

Whether applied to a Fortune 500 company or a three-person engineering business, the phases in systems analysis and design are the same.

1. **Preliminary investigation:** Conduct preliminary analysis, propose alternative solutions, and describe the costs and benefits of each solution. Submit a preliminary plan with recommendations.

2. **Systems analysis:** Gather data, analyze the data, and make a written report.
3. **Systems design:** Do a preliminary design and then a detailed design, and write a report.
4. **Systems development:** Acquire the hardware and software and test the system.
5. **Systems implementation:** Convert the hardware, software, and files to the new system and train the users.
6. **Systems maintenance:** Audit the system, and evaluate it periodically.

Phases often overlap, and a new one may start before the old one is finished. After the first four phases, management must decide whether to proceed to the next phase. *User input and review is a critical part of each phase.*

The First Phase: Conduct a Preliminary Investigation

The objective of Phase 1, *preliminary investigation*, is to conduct a preliminary analysis, propose alternative solutions, describe costs and benefits, and submit a preliminary plan with recommendations.

During preliminary investigation, you need to find out what the organization's objectives are and the nature and scope of the problems under study. Even if a problem pertains only to a small segment of the organization, you cannot study it in isolation. You need to find out what the objectives of the organization itself are. Then you need to see how the problem being studied fits in with them.

In delving into the organization's objectives and the specific problems, you may have already discovered some solutions. Other possible solutions can come from interviewing people inside the organization, clients or customers affected by it, suppliers, and consultants. You can also study what competitors are doing. With this data, you then have three choices. You can leave the system as is, improve it, or develop a new system.

Whichever of the three preceding alternatives is chosen, it will have costs and benefits, and you need to indicate what these are. Costs may depend on benefits, which may offer savings. There are all kinds of benefits that may be derived. A process may be speeded up, streamlined through elimination of unnecessary steps, or combined with other processes. Input errors or redundant output may be reduced. Systems and subsystems may be better integrated. Users may be happier with the system. Customers or suppliers may interact better with the system. Security may be improved. Costs may be cut.

Now you need to wrap up all your findings in a written report. The readers of this report will be the executives (probably top managers) who are in a position to decide in which direction to proceed—make no changes, change a little, or change a lot. You should describe the potential solutions, costs, and benefits and indicate your recommendations.

The Second Phase: Do an Analysis of the System

The objective of Phase 2, *systems analysis*, is to gather data, analyze the data, and write a report.

In this second phase of the SDLC, you will follow the course that management has indicated after having read your Phase 1 report. We are assuming that they have directed you to perform Phase 2—to do a careful analysis or study of the existing system to understand how the new system you proposed would differ. This analysis will also consider how people's positions and tasks will have to change if the new system is put into effect. During this phase you will gather data by reviewing written documents, interviewing employees and managers, developing questionnaires, and observing people and processes at work.

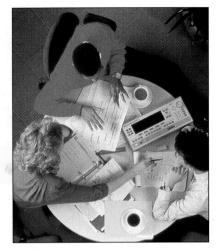

Once the data is gathered, you need to come to grips with it and analyze it. To do this, systems analysts use a variety of analytical tools, or modeling tools. *Modeling tools* enable a systems analyst to present graphic, or pictorial, representations of a system. Some of these tools involve creating flowcharts and diagrams on paper.

Once you have completed the analysis, you need to document this phase. This report to management should have three parts. First, it should explain how the existing system works. Second, it should explain the problems with the existing system. Finally, it should describe the requirements for the new system and make recommendations about what to do next.

At this point, not a lot of money will have been spent on the systems analysis and design project. If the costs of going forward seem to be prohibitive, this is a good time for the managers reading the report to call a halt. Otherwise, you will be called on to move to Phase 3.

The Third Phase: Design the System

The objective of Phase 3, *systems design*, is to do a preliminary design and then a detail design, and write a report. In this third phase of the SDLC, you will essentially create a "rough draft" and then a "detail draft" of the proposed information system.

A *preliminary design* describes the general functional capabilities of a proposed information system. It reviews the system requirements and then considers major components of the system. Usually several alternative systems (called *candidates*) are considered, and the costs and the benefits of each are evaluated.

Three tools may be used in the preliminary design:

- **Prototyping tools:** *Prototyping* refers to building a working model or experimental version of all or part of a system so that it can be quickly tested and evaluated.

  A *prototype* is a limited working system developed to test out design concepts. A prototype allows users to find out immediately how a change in the system might benefit them. Prototypes are built with *prototyping tools*. These are special software packages that can be used to design screen displays.

- **CASE tools:** CASE tools are another type of software tool. **CASE (**stands for **computer-aided software engineering) tools are software that provides computer-automated means of designing and changing systems.** There are many packages of such specialized software. (Sample names: Application Development Workbench, BACHMAN/Analyst, Excelerator, HyperAnalyst, Information Engineering Facility, PacBASE, System Architect.) CASE tools may be used at almost any stage of the systems development life cycle, not just design.

- **Project management software:** As we described in Chapter 2 (✓ p. 68), *project management software* consists of programs used to plan, schedule, and control the people, costs, and resources required to complete a project on time.

A *detail design* describes how a proposed information system will deliver the general capabilities described in the preliminary design. The detail design usually considers the following parts of the system, in this order: *output requirements, input requirements, storage requirements, processing requirements,* and *system controls and backup.*

- **Output requirements:** What do you want the system to produce? That is the first requirement to determine. In this first step, the systems analyst determines what media the output will be—whether hardcopy and/or softcopy. He or she will also design the appearance or format of the output, such as headings, columns, menu, and the like.

- **Input requirements:** Once you know the output, you can determine the inputs. Here, too, you must define the type of input, such as keyboard or source data entry (✓ p. 139). You must determine in what form data will be input and how it will be checked for accuracy. You also need to figure what volume of data the system can be allowed to take in.

- **Storage requirements:** Using the data dictionary (✓ p. 218) as a guide, you need to define the files and databases in the information system. How will the files be organized? What kind of storage devices will be used? How will they interface with other storage devices inside and outside of the organization? What will be the volume of database activity?

- **Processing requirements:** What kind of computer or computers will be used to handle the processing? What kind of operating system will be used? Will the computer or computers be tied to others in a network? Exactly what operations will be performed on the input data to achieve the desired output information?

- **System controls and backup:** Finally, you need to think about matters of security, privacy, and data accuracy. You need to prevent unauthorized users from breaking into the system, for example, and snooping in people's private files. You need to have auditing procedures and set up specifications for testing the new system (Phase 4). You need to institute automatic ways of backing up information and storing it elsewhere in case the system fails or is destroyed.

All the work of the preliminary and detail designs will end up in a large, detailed report. When you hand over this report to senior management, you will probably also make some sort of presentation or speech.

The Fourth Phase: Develop the System

In Phase 4, *systems development,* the systems analyst or others in the organization acquire the software, acquire the hardware, and then test the system.

During the design stage, the systems analyst may have had to address what is called the "make-or-buy" decision, but that decision certainly cannot be avoided now. In the *make-or-buy decision,* you decide whether you have to create a program—have it custom-written—or buy it, meaning simply purchase an existing software package. Sometimes programmers decide they can buy an existing program and modify it rather than write it from scratch.

If you decide to create a new program, then the question is whether to use the organization's own staff programmers or hire outside contract programmers. Whichever way you go, the task could take many months.

Once the software has been chosen, the hardware to run it must be acquired or upgraded. It's possible your new system will not require obtaining any new hardware. It's also possible that the new hardware will cost millions of dollars and involve many items: microcomputers, minicomputers, mainframes, monitors, modems, and many other devices. The organization may find it's better to lease rather than to buy some equipment, especially since chip capability doubles about every 18 months.

With the software and hardware acquired, you can now start testing the system. Testing is usually done in stages called *unit testing*; then *system testing* is done.

- **Unit testing:** In *unit testing*, individual parts of the program are tested, using test (made-up) data. If the program is written as a collaborative effort by multiple programmers, each part of the program is tested separately.

- **System testing:** In *system testing*, the parts are linked together, and test data is used to see if the parts work together. At this point, actual organization data may also be used to test the system. The system is also tested with erroneous and massive amounts of data to see if it can be made to fail ("crash").

At the end of this long process, the organization will have a workable information system, one ready for the implementation phase.

The Fifth Phase: Implement the System

Phase 5, *systems implementation*, consists of converting the hardware, software, and files to the new system and training the users. Whether the new information system involves a few handheld computers, an elaborate telecommunications network, or expensive mainframes, the fifth phase will involve some close coordination to make the system not just workable but successful.

Conversion, the process of converting from an old information system to a new one, involves converting hardware, software, and files.

Hardware conversion may be as simple as taking away an old PC and plunking a new one down in its place. Or it may involve acquiring new buildings and putting in elaborate wiring, climate-control, and security systems.

Software conversion means making sure the applications that worked on the old equipment also work on the new.

File conversion means converting the old files to new ones without loss of accuracy. For example, can the paper contents from the manila folders in the personnel department be input to the system with a scanner? Or do they have to be keyed in manually, with the consequent risk of errors being introduced?

There are four strategies for handling conversion: *direct, parallel, phased,* and *pilot*.

- **Direct approach:** *Direct implementation* means the user simply stops using the old system and starts using the new one. The risk of this method should be evident: What if the new system doesn't work? If the old system has truly been discontinued, there is nothing to fall back on.

- **Parallel approach:** *Parallel implementation* means that the old and new systems are operated side by side until the new system has shown it is reliable, at which time the old system is discontinued. Obviously there are benefits in taking this cautious approach. If the new system fails, the organization can switch back to the old one. The difficulty of this method is the expense of paying for the equipment and people to keep two systems going at the same time.

- **Phased approach:** *Phased implementation* means that parts of the new system are phased in separately—either at different times (parallel) or all at once in groups (direct).

- **Pilot approach:** *Pilot implementation* means that the entire system is tried out but only by some users. Once the reliability has been proved, the system is implemented with the rest of the intended users. The pilot approach still has its risks, since *all* of the users in a particular group are taken off the old system. However, the risks are confined to only a small part of the organization.

In general, the phased and pilot approaches are the most favored methods. Phased is best for large organizations in which people are performing different jobs. Pilot is best for organizations in which all people are performing the same task (such as order takers at a direct-mail house).

Training users in the use of a new system is done with a variety of tools. They run from documentation (instruction manuals) to videotapes to live classes to one-on-one, side-by-side teacher-student training. Sometimes training is conducted by the organization's own staff; at other times it is contracted out.

The Sixth Phase: Maintain the System

Phase 6, *systems maintenance*, adjusts and improves the system by having system audits and periodic evaluations and by making changes based on new conditions. Even with the conversion accomplished and the users trained, the system won't just run itself. The information system must be monitored to ensure that it is successful. Maintenance includes not only keeping the machinery running but also updating and upgrading the system to keep pace with new products, services, customers, government regulations, and other requirements.

In the fourth phase of the SDLC, the software for the new system was acquired. (■ *See Panel 9.1.*) Whether this software was custom-developed or purchased off the shelf, it had to be written—created— by programmers. The next section covers the topic of programming.

Programming: A Five-Step Procedure

Preview & Review: Programming is a five-step procedure for producing a program—a list of instructions—for the computer. Programmers use special programming languages to create a program's code.

To see how programming works, consider what a program is. **A *program* is a list of instructions that the computer must follow in order to process data into information.** The instructions consist of *statements* written in a programming language, such as BASIC. Examples of programs are those that do word processing, desktop publishing, or payroll processing.

Programming, **also called *software engineering,* is a five-step process for creating that list of instructions.** Only one of those steps (the step called *coding*) consists of sitting at the keyboard typing words into a computer.

The five steps are as follows.

1. Define the problem—include needed output, input, processing requirements.
2. Design a solution—use modeling tools to chart the program.
3. Code the program—use a programming language's syntax, or rules, to write the program.
4. Test the program—get rid of any logic errors, or "bugs," in the program ("debug" it).
5. Document the program—include written instructions for users, explanation of the program, and operating requirements.

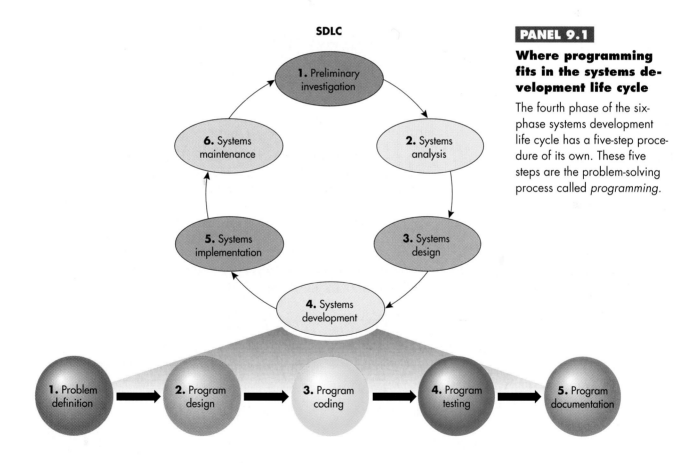

SDLC

PANEL 9.1

Where programming fits in the systems development life cycle

The fourth phase of the six-phase systems development life cycle has a five-step procedure of its own. These five steps are the problem-solving process called *programming.*

Coding is what many people think of when they think of programming, although it is only one of the five steps. Coding consists of translating the logic requirements of a planned program into a programming language—the letters, numbers, and symbols that make up the program.

A *programming language* is a set of rules that tells the computer what operations to do. Examples of well-known programming languages are BASIC, COBOL, Pascal, FORTRAN, and C. (■ *See Panel 9.2.*) Not all languages are appropriate for all uses. Thus, the language needs to be chosen based on such considerations as what purpose the program is designed to serve and what languages are already being used in the organization or field you are in.

Programming languages are also called *high-level languages.* For the computer to be able to "understand" them they must be translated into the low-level language called *machine language.* **Machine language is the basic language of the computer, representing data as 1s and 0s.** (■ *See Panel 9.3.*) Machine language programs vary from computer to computer; that is, they are *machine dependent.*

A high-level language allows users to write in a familiar notation, rather than numbers or abbreviations. Most high-level languages are not machine dependent—they can be used on more than one kind of computer. The translator (✓ p. 117) for high-level languages is, depending on the language, either a *compiler* or an *interpreter.*

- Compiler—execute later: **A *compiler* is a language translator that converts the *entire* program of a high-level language into machine language before the computer executes the program.** The high-level language is called the *source code.* The compiler translates it into machine language, which in this case is called the *object code.* The significance of this distinction is that the object code *can be saved.* Thus, it can be executed later rather than run right away. (■ *See Panel 9.4.*)

    Examples of high-level languages using compilers are COBOL, FORTRAN, and Pascal.

PANEL 9.2

Some common programming languages*

Language Name	Sample Code Fragment	Use
FORTRAN (FORmula TRANslator)	IF (XINVO.GT.500.00) THEN	Widely used for mathematical, scientific, and engineering programs.
COBOL (COmmon Business Oriented Language)	IF INVOICE=AMT>500	Most frequently used for business applications.
BASIC (Beginner's All-purpose Symbolic Instruction Code)	IF INV.AMT A>500	Used by nonprofessional and beginning programmers.
Pascal (named after Blaise Pascal)	if INVOICEAMOUNT>500.00 then	Used for teaching purposes at schools.
C	if (invoice_amount>500.00)	Used by many professional programmers.

*These are but a few of *many* languages.

- *Interpreter—execute immediately:* **An *interpreter* is a language translator that converts each high-level language statement into machine language and executes it immediately, statement by statement.** No object code is saved, as with the compiler. However, with an interpreter, processing seems to take less time.

An example of a high-level language using an interpreter is BASIC.

Object-Oriented & Visual Programming

Two new developments have made programming easier—*object-oriented programming* and *visual programming.*

Object-oriented programming (OOP)* is a programming method that combines data and instructions for processing that data into a self-sufficient "object" that can be used in other programs. The important thing here is the object. **An *object* is a block of preassembled programming code that is a self-contained module. The module contains, or encapsulates, both (1) a chunk of data and (2) the processing instructions that may be called on to be performed on that data.**

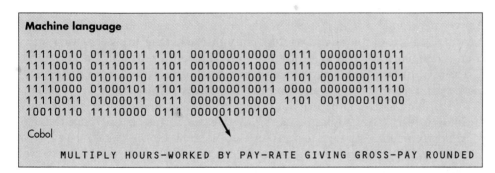

```
Machine language

11110010 01110011 1101 001000010000 0111 000000101011
11110010 01110011 1101 001000011000 0111 000000101111
11111100 01010010 1101 001000010010 1101 001000011101
11110000 01000101 1101 001000010011 0000 000000111110
11110011 01000011 0111 000001010000 1101 001000010100
10010110 11110000 0111 000001010100

Cobol

          MULTIPLY HOURS-WORKED BY PAY-RATE GIVING GROSS-PAY ROUNDED
```

PANEL 9.3

Low-level and high-level languages.

(Top) Machine language is all binary 0s and 1s—very difficult for people to work with. *(Bottom)* COBOL, a third-generation language, uses English words that can be understood by people.

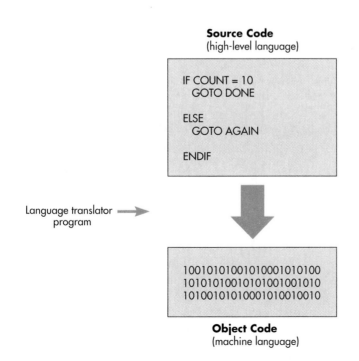

Source Code
(high-level language)

```
IF COUNT = 10
    GOTO DONE

ELSE
    GOTO AGAIN

ENDIF
```

Language translator program →

Object Code
(machine language)

```
1001010100101010001010100
1010101001010100100101010
101001010101000101001001010010
```

PANEL 9.4

Compiler

This language translator converts the high-level language (source code) into machine language (object code) before the computer can execute the program.

Once you've written a block of program code (that computes overtime pay, for example), it can be reused in any number of programs. Thus, unlike with traditional programming, with OOP you don't have to start from scratch—that is, reinvent the wheel—each time. Some examples of object-oriented programming languages are Smalltalk, C++, Turbo Pascal, and Hypertalk.

Visual programming **is a method of creating programs in which the programmer makes connections between objects by drawing, pointing, and clicking on diagrams and icons.**

The goal of visual programming is to make programming easier for programmers and more accessible to nonprogrammers. It does so by borrowing the object orientation of OOP languages but exercising it in a graphical or visual way. Visual programming enables users to think more about solving the problem than about handling the programming language.

With one example, ObjectVision (from Borland), the user doesn't employ a programming language but simply connects icons and diagrams on screen. Visual BASIC offers another visual environment for program construction, allowing you to build various components using buttons, scroll bars, and menus.

Internet Programming

As we mentioned in Chapter 8, the Internet connects thousands of data and information sites around the world. Many of these sites are text-based only; that is, the user sees no graphics, animation, or video, and hears no sound. To build multimedia sites on the World Wide Web, people use some recently developed programming languages: HTML, VRML, and Java.

- **HTML:** *HTML (hypertext markup language)* is a type of programming language that embeds simple commands within standard ASCII (✓ p. 116) text documents to provide an integrated, two-dimensional display of text and graphics. In other words, a document created in any word processor and stored in ASCII format can become a Web page with the addition of a few HTML commands. One of the main features of HTML is its ability to insert hypertext (✓ p. 69) links into a document. Hypertext links enable you to display another Web document simply by clicking on a link area on the current screen. One document may contain links to many other related documents. The related documents may be on the same server (✓ p. 258) as the first document, or they may be on a computer halfway around the world. A link may be a word, a group of words, or a picture.

- **VRML:** *VRML (virtual reality markup language)* is a type of programming language used to create three-dimensional Web pages. VRML (rhymes with "thermal") is not an extension of HTML; thus, HTML Web browsers cannot interpret it. That is, users need a VRML Web browser to receive VRML Web pages. If they are not on a large computer system, they also need a high-end microcomputer such as a Power Macintosh or a Pentium-based PC. Like HTML, VRML is a document-centered ASCII language. Unlike HTML, it tells the computer how to create 3-D worlds. VRML pages can also be linked to other VRML pages.

- **Java:** Java is a new programming language from Sun Microsystems. Basically, *Java* is a way to create any conceivable type of software applications that will work on the Internet. If the use of Java becomes widespread, the Web will be transformed from the information-delivering medium it is today into a completely interactive computing environment. You will be able to treat the Web as a giant hard disk loaded with a never-ending supply of software applications. However, Java is not compatible with most existing microprocessors, such as those from Intel and Motorola. For this reason, users need to use a small "interpreter" program to be able to use Java applications (called "applets"). They also need a recent-version operating system and a Java-capable browser in order to view Java special effects. In the meantime, Sun is working on developing special Java chips designed to run Java software directly.

Onward

What do object-oriented, visual, and Internet programming imply for the future? Will tomorrow's programmer look less like a writer typing out words and more like an electrician wiring together circuit components, as one magazine suggests?[3] What does this mean for the five-step programming model we have described?

Some institutions are now teaching only object-oriented design techniques, which allow the design and ongoing improvement of working program models. Here programming stages overlap, and users repeatedly flow through analysis, design, coding, and testing stages. Thus, users can test out new parts of programs and even entire programs as they go along. They need not wait until the end of the process to find out if what they said they wanted is what they really wanted.

This new approach to programming is not yet in place in business, but in a few years it may be. If you're interested in being able to communicate with programmers in the future, or in becoming one yourself, the new approaches are worth your attention.

SUMMARY

What It Is / What It Does	Why It's Important

compiler *(p. 288, LO 6)* Language translator that converts the entire program of a high-level language (called the *source code*) into machine language (called the *object code*) for execution later. Examples of compiler languages: COBOL, FORTRAN, Pascal.

Unlike other language translators (assemblers and interpreters), with a compiler program the object code can be saved and executed later rather than run right away. The advantage of a compiler is that, once the object code has been obtained, the program executes faster.

computer-aided software engineering (CASE) tools *(p. 283, LO 2)* Software that provides computer-automated means of designing and changing systems.

CASE tools may be used in almost any phase of the SDLC, not just design. So-called *front-end CASE tools* are used during the first three phases—preliminary analysis, systems analysis, systems design—to help with the early analysis and design. So-called *back-end CASE tools* are used during two later phases—systems development and implementation—to help in coding and testing, for instance.

interpreter *(p. 289, LO 6)* Language translator that converts each high-level language statement into machine language and executes it immediately, statement by statement. An example of a high-level language using an interpreter is BASIC.

Unlike with the language translator called the compiler, no object code is saved. The advantage of an interpreter is that programs are easier to develop.

machine language *(p. 288, LO 6)* Lowest level of programming language; the language of the computer, representing data as 1s and 0s. Most machine language programs vary from computer to computer—they are machine-dependent.

Machine language, which corresponds to the on and off electrical states of the computer, is not convenient for people to use. High-level languages were developed to make programming easier.

object *(p. 289, LO 7)* In object-oriented programming, block of preassembled programming code that is a self-contained module. The module contains (encapsulates) both (1) a chunk of data and (2) the processing instructions that may be called on to be performed on that data.

The object can be reused and interchanged among programs, thus making the programming process easier, more flexible and efficient, and faster.

object-oriented programming (OOP) *(p. 289, LO 7)* Programming method in which data and the instructions for processing that data are combined into a self-sufficient object—piece of software. Examples of OOP languages: Smalltalk, C++, Turbo Pascal, Hypertalk.

Objects can be reused and interchanged among programs, producing greater flexibility and efficiency than is possible with traditional programming methods.

preliminary investigation *(p. 282, LO 1)* Phase 1 of the SDLC; the purpose is to conduct a preliminary analysis (determine the organization's objectives, determine the nature and scope of the problem), propose alternative solutions (leave the system as is, improve the efficiency of the system, or develop a new system), describe costs and benefits, and submit a preliminary plan with recommendations.

The preliminary investigation lays the groundwork for the other phases of the SDLC.

program *(p. 287, LO 4)* List of instructions the computer follows to process data into information. The instructions consist of statements written in a programming language (for example, BASIC).

Without programs, data could not be processed into information by a computer.

What It Is / What It Does	Why It's Important
programming *(p. 287, LO 4)* Five-step process for creating software instructions: (1) define the problem; (2) design a solution; (3) write (code) the program; (4) test (debug) the program; (5) document the program.	Programming is one step in the systems development life cycle.
programming language *(p. 288, LO 4)* Set of words and symbols that allow programmers to tell the computer what operations to follow. Some common programming languages are COBOL, FORTRAN, BASIC, Pascal, and C.	Not all programming languages are appropriate for all uses. Thus, a language must be chosen to suit the purpose of the program and to be compatible with other languages being used by users.
system *(p. 281, LO 1)* Collection of related components that interact to perform a task in order to accomplish a goal.	The point of systems analysis and design is to ascertain how a system works and then take steps to make it better.
systems analysis *(p. 282, LO 1)* Phase 2 of the SDLC; the purpose is to gather data (using written documents, interviews, questionnaires, observation, and sampling), analyze the data, and write a report.	The results of systems analysis will determine whether the system should be redesigned.
systems analyst *(p. 281, LO 1)* Information specialist who performs systems analysis, design, and implementation.	The systems analyst studies the information and communications needs of an organization to determine how to deliver information that is more accurate, timely, and useful. The systems analyst achieves this goal through the problem-solving method of systems analysis and design.
systems design *(p. 283, LO 1)* Phase 3 of the SDLC; the purpose is to do a preliminary design and then a detail design, and write a report.	Systems design is one of the most crucial phases of the SDLC.
systems development *(p. 284, 1)* Phase 4 of the SDLC; hardware and software for the new system are acquired and tested. The fourth phase begins once management has accepted the report containing the design and has approved the way to development.	This phase may involve the organization in investing substantial time and money.
systems development life cycle (SDLC) *(p. 281, LO 1)* Six-phase process that many organizations follow during systems analysis and design: (1) *preliminary investigation;* (2) *systems analysis;* (3) *systems design;* (4) *systems development;* (5) *systems implementation;* (6) *systems maintenance.* Phases often overlap, and a new one may start before the old one is finished. After the first four phases, management must decide whether to proceed to the next phase. User input and review is a critical part of each phase.	The SDLC is a comprehensive tool for solving organizational problems, particularly those relating to the flow of computer-based information.
systems implementation *(p. 285, LO 1,3)* Phase 5 of the SDLC; consists of converting the hardware, software, and files to the new system and training the users.	This phase is important because it involves putting design ideas into operation.
systems maintenance *(p. 286, LO 1)* Phase 6 of the SDLC; consists of keeping the system working by having system audits and periodic evaluations.	This phase is important for keeping a new system operational and useful.
visual programming *(p. 290, LO 7)* Method of creating programs; the programmer makes connections between objects by drawing, pointing, and clicking on diagrams and icons. Programming is made easier because the object orientation of object-oriented programming is used in a graphical or visual way.	Visual programming enables users to think more about the problem solving than about handling the programming language.

(Selected answers appear at the back of the book.)

Short-Answer Questions

1. What is the purpose of the systems development life cycle?
2. Why is it important for users to understand the principles of the SDLC?
3. How would you define *programming language*?
4. What are the five basic steps of program development?
5. What is the basic difference between a compiler and an interpreter?

Fill-in-the-Blank Questions

1. _____ involves building a model or experimental version of all or part of a system so that it can be quickly tested and evaluated.
2. A(n) _____ is a block of pre-assembled programming code that forms a self-contained module that can be used in creating many different programs.
3. Writing a computer program is called _____.
4. To _____ a program means to detect, locate, and remove all errors in a computer program.
5. In _____ programming, programmers make connections between objects by drawing, pointing, and clicking on diagrams and icons.

Multiple-Choice Questions

1. In which phase of the SDLC are software and hardware obtained?
   a. preliminary investigation
   b. system analysis
   c. systems design
   d. systems development
   e. systems implementation
2. In which phase of the SDLC is data gathered and analyzed?
   a. preliminary investigation
   b. system analysis
   c. systems design
   d. systems development
   e. systems implementation
3. The rules of a programming language are called _____.
   a. documentation
   b. grammar
   c. syntax
   d. flowcharts
4. Which of the following languages was developed to help students learn programming?
   a. FORTRAN
   b. COBOL
   c. BASIC
   d. Pascal
   e. none of the above
5. To convert a high-level language statement into machine language and execute it immediately, you need a(n):
   a. converter
   b. report generator
   c. interpreter
   d. application generator
   e. none of the above

True/False Questions

T F 1. Users are never involved in systems development.

T F 2. System testing is performed in the fourth phase of systems development.

T F 3. Pascal is a high-level programming language.

T F 4. Customized software can be purchased off the shelf of a computer store.

T F 5. Most, but not all, programs have to be converted into machine language before your computer can execute them.

Projects/Critical-Thinking Questions

1. Interview a student majoring in computer science who plans to become a systems analyst. Why is this person interested in this field? What does he or she hope to accomplish in it? What courses must be taken to satisfy the requirements for becoming an analyst?
2. Scan the employment ads in a few major newspapers and professional journals. What programming languages are most in demand? For what types of jobs?
3. Interview several students who are majoring in computer science and are studying to become computer programmers. What languages do they plan to master? Why? What kinds of jobs do these people expect to get? What kinds of future developments do they anticipate in the field of software programming?

Society & the Digital Age
Challenges & Promises

Concepts You Should Know

After reading this chapter, you should be able to:

1. Describe computers-and-communications security issues—accidents, hazards, crime, viruses—and security safeguards.

2. Discuss some related quality-of-life issues—environment, mental health, the workplace.

3. Explain the implications of information technology for employment and the future.

4. Define and explain a few types of artificial intelligence.

5. Discuss the promises of the Information Superhighway.

here's a fundamental shift going on in society right now," says Jane Metcalfe. ". . . [T]he digital revolution is whipping through our lives like a Bengali typhoon."

Metcalfe, co-founder of the popular information-technology magazine *Wired*, says she is "profoundly optimistic about how we can use technology to change things."[1] Not everyone thinks the future is so rosy. Unemployment, says technology critic Jeremy Rifkin gloomily, "can be expected to climb steadily and inexorably over the next four decades as the global economy makes the transition to the Information Age."[2] Who's right? In this final chapter, we consider the challenges and promises of the Digital Age.

The Information Superhighway: The Challenge of Creating a Blueprint

Preview & Review: The Information Superhighway envisions using wired and wireless capabilities of telephones and networked computers with cable-TV.

As we said in Chapter 1, the *Information Superhighway* (I-way) is a vision or a metaphor. It envisions a fusion of wired and wireless capabilities of telephones and networked computers with cable-TV's capacity to transmit hundreds of programs. When complete, the I-way would supposedly give us, for example, video on-demand, multimedia, fast data exchange, teleconferencing, distance learning, enormous research databases, and teleshopping services.

What shape will the Information Superhighway take? Some government officials hope it will follow a somewhat orderly model, such as that envisioned in the National Information Infrastructure.

As envisioned by various U.S. government officials, the **National Information Infrastructure (NII) would include today's existing networks and technologies as well as technologies yet to be deployed. Its services would be delivered by telecommunications companies, cable-television companies, and the Internet.** Applications would be varied—education, health care, information access, electronic commerce, and entertainment.[3] (■ *See Panel 10.1, next page.*)

Who would put the pieces of the NII together? The present national policy is to let private industry do it, with the government trying to ensure fair competition among phone and cable companies and compatibility among various technological systems.[4] In addition, it strives for open access to people of all income levels.

However, no matter what developments occur in computer and communications technology, these systems will remain vulnerable to certain threats, as we describe next.

Security Issues: Challenges to Computers & Communications Systems

Preview & Review: Information technology can be disabled by a number of occurrences. It may be harmed by people; by procedural, and software errors; by electromechanical problems; by viruses; and by "dirty data." It may be threatened by natural hazards and by civil strife and terrorism.

PANEL 10.1

Toward a National Information Infrastructure

Some services that supporters of the NII hope might be offered.

Commerce

With inexpensive access charges, small companies could afford to act like big ones. Boundaries would be erased between company departments, suppliers, and customers. Designs for new products could be tested and exchanged with factories in remote locations. With information flowing faster, goods could be sent to market faster and inventories kept low.

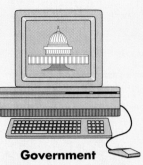

Government

An information highway could extend electronic democracy through electronic voting, allow interactive local-government meetings between electors and elected, and help deliver government services such as administering Social Security forms.

Education

"Virtual" classrooms and distance learning would replace lecture halls and scheduled class times. Students could take video field trips to distant places and get information from remote museums and libraries (such as the Library of Congress).

Home Services

Consumers would be able to receive movies on demand, home shopping, and videogames; do electronic bill paying; and tap into libraries and schools.

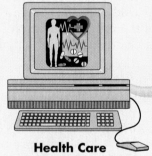

Health Care

Through telemedicine, health-care providers and researchers could share medical images, patient records, and research and perform long-distance patient examinations. Interactive, multimedia materials directed to the public would outline health-care options.

Information

Government records, patents, contracts, legal documents, and satellite maps could be made available to the public online. Libraries could also be digitized, with documents available for downloading.

Mobile Communications

Users with handheld personal communicators or personal digital assistants would be able to send and receive voice, fax, text, and video messages anywhere.

Security issues go right to the heart of the workability of computer and communications systems. Here we discuss several threats to computers and communications systems:

Errors & Accidents

ROBOT SENT TO DISARM BOMB GOES WILD IN SAN FRANCISCO, read the headline.[5] Evidently, a hazardous-duty police robot started spinning out of control when officers tried to get it to grasp a pipe bomb. Fortunately, it was shut off before any damage could be done. Most computer glitches are not so spectacular, although they can be almost as important.

In general, errors and accidents in computer systems may be classified as people errors, procedural errors, software errors, electromechanical problems, and "dirty data" problems.

- **People errors:** Recall that one part of a computer system is the people who manage it or run it (✓ p. 9). For instance, Brian McConnell of Roanoke, Virginia, found that he couldn't get past a bank's automated telephone system to talk to a real person. This was not the fault of the system so much as of the people at the bank. McConnell, president of a software firm, thereupon wrote a program that automatically phoned eight different numbers at the bank. People picking up the phone heard the recording, "This is an automated customer complaint. To hear a live complaint, press. . . ."[6] Quite often, what may seem to be "the computer's fault" is human indifference or bad management.

- **Procedural errors:** Some spectacular computer failures have occurred because someone didn't follow procedures. Consider the 2½-hour shutdown of Nasdaq, the nation's second largest stock market. Nasdaq is so automated that it likes to call itself "the stock market for the next 100 years." In July 1994, Nasdaq was closed down by an effort, ironically, to make the computer system more user-friendly. Technicians were phasing in new software, adding technical improvements a day at a time. A few days into this process, the technicians tried to add more features to the software, flooding the data-storage capability of the computer system. The result was a delay in opening the stock market that shortened the trading day.[7]

- **Software errors:** We are forever hearing about "software glitches" or "software bugs." A **software bug** is an error in a program that causes it not to work properly.

  An example of a somewhat small error was when a school employee in Newark, New Jersey, made a mistake in coding the school system's master scheduling program. When 1000 students and 90 teachers showed up for the start of school at Central High School, half the students had incomplete or no schedules for classes. Some classrooms had no teachers while others had four instead of one.[8]

- **Electromechanical problems:** Mechanical systems, such as printers, and electrical systems, such as circuit boards, don't always work. They may be incorrectly constructed, get dirty or overheated, wear out, or become damaged in some other way. Power failures (brownouts and blackouts) can shut a system down. Power surges can burn out equipment.

- **"Dirty data" problems:** When keyboarding a research paper, you undoubtedly make a few typing errors. So do all the data-entry people around the world who feed a continual stream of raw data into computer systems. A lot of problems are caused by this kind of "dirty data." **Dirty data is data that is incomplete, outdated, or otherwise inaccurate.**

Natural & Other Hazards

Some disasters can do more than lead to temporary system downtime, they can wreck the entire system. Examples are natural hazards, and civil strife and terrorism.

- **Natural hazards:** Whatever is harmful to property (and people) is harmful to computers and communications systems. This certainly includes natural disasters: fires, floods, earthquakes, tornadoes, hurricanes, blizzards, and the like. If they inflict damage over a wide area, as have some Florida and Hawaii hurricanes, natural hazards can disable all the electronic systems we take for granted. Without power and communications connections, automated teller machines, credit-card verifiers, and bank computers are useless.

- **Civil strife and terrorism:** We may take comfort in the fact that wars and insurrections seem to take place in other parts of the world. Yet we are not immune to civil unrest, such as the so-called Rodney King riots that wracked Los Angeles in 1992. Nor are we immune, apparently, to acts of terrorism, such as the February 1993 bombing of New York's World Trade Center. In the latter case, companies found themselves frantically having to move equipment to new offices and reestablishing their computer networks. The Pentagon itself (which has 650,000 terminals and workstations, 100 WANs, and 10,000 LANs) is taking steps to reduce its own systems' vulnerability to intruders.[9]

Crimes Against Computers & Communications

An *information-technology crime* can be one of two types. It can be an illegal act perpetrated against computers or telecommunications. Or it can be the use of computers or telecommunications to accomplish an illegal act. Here we discuss the first type.

Crimes against technology include theft—of hardware, of software, of computer time, of cable or telephone services, of information. Other illegal acts are crimes of malice and destruction. Some examples are as follows:

- **Theft of hardware:** Stealing of hardware can range from shoplifting an accessory in a computer store to removing a laptop or cellular phone from someone's car. Professional criminals may steal shipments of microprocessor chips off a loading dock or even pry cash machines out of shopping-center walls. Portable computer thefts rose 39% from 1994 to 1995.[10] Popular sites for laptop larceny are airports and hotels.[11,12]

- **Theft of software:** Stealing software can take the form of physically making off with someone's diskettes, but it is more likely to be copying of programs. Software makers secretly prowl electronic bulletin boards in search of purloined products, then try to get a court order to shut down the bulletin boards.[13] They also look for companies that "softlift"—buying one copy of a program and making copies for as many computers as they have.

  Many pirates are reported by co-workers or fellow students to the "software police," the Software Publishers Association. The SPA has a toll-free number (800-388-7478) on which anyone can report illegal copying, to initiate antipiracy actions. In mid-1994, two New England college students were indicted for allegedly using the Internet to encourage the exchange of copyrighted software.[14]

Another type of software theft involves selling copies or counterfeits of well-known software programs. These pirates often operate in China, Taiwan, Mexico, Russia, and various parts of Asia and Latin America. In some countries, more than 90% of U.S. microcomputer software in use is thought to be illegally copied.[15]

- **Theft of time and services:** The theft of computer time is more common than you might think. Probably the biggest use of it is people using their employer's computer time to play games. Some people also may run sideline businesses.

  For years "phone phreaks" have bedeviled the telephone companies. They have found ways to get into company voice-mail systems, then use an extension to make long-distance calls at the company's expense.[16] In addition, they have also found ways to tap into cellular phone networks.

- **Theft of information:** "Information thieves" have been caught infiltrating the files of the Social Security Administration, stealing confidential personal records and selling the information.[17] Thieves have also broken into computers of the major credit bureaus and have stolen credit information. They have then used the information to charge purchases or have resold it to other people. On college campuses, thieves have snooped on or stolen private information such as grades.

- **Crimes of malice and destruction:** Sometimes criminals are more interested in abusing or vandalizing computers and telecommunications systems than in profiting from them. For example, a student at a Wisconsin campus deliberately and repeatedly shut down a university computer system, destroying final projects for dozens of students. A judge sentenced him to a year's probation, and he left the campus.[18]

(E) Crimes Using Computers & Communications

Just as a car can be used to assist in a crime, so can a computer or communications system. For example, four college students on New York's Long Island who met via the Internet used a specialized computer program to steal credit-card numbers, then, according to police, went on a one-year, $100,000 shopping spree. When arrested, they were charged with grand larceny, forgery, and scheming to defraud.[19]

In addition, investment fraud has come to cyberspace. Many people now use online services to manage their stock portfolios through brokerages hooked into the services. Scam artists have followed, offering nonexistent investment deals and phony solicitations and manipulating stock prices.[20-22]

Worms & Viruses

Worms and viruses are forms of high-tech maliciousness. **A *worm* is a program that copies itself repeatedly into memory or onto a disk drive until no more space is left.** An example is the worm program unleashed by a student at Cornell University that traveled through an e-mail network and shut down thousands of computers around the country.

A *virus* is a "deviant" program that attaches itself to computer systems and destroys or corrupts data. (■ *See Panel 10.2.*) Viruses are passed in two ways:

- **By diskette:** The first way is via an infected diskette, such as one you might get from a friend or a repair person. It's also possible to get a virus from a sales demo disk or even (in 3% of cases) from a shrink-wrapped commercial disk.

- *Boot-sector virus:* The boot sector is that part of the system software containing most of the instructions for booting, or powering up, the system. The boot-sector virus replaces these boot instructions with some of its own. Once the system is turned on, the virus is loaded into main memory before the operating system. From there it is in a position to infect other files.

    Any diskette that is used in the drive of the computer then becomes infected. When that diskette is moved to another computer, the contagion continues.

SAM Update Virus Definitions [?]
Phone Settings
Dial Phone Number: [   ]
[X] If Busy, Redial [5] Times
Every [10] Seconds
Modem Settings
Modem: [Hayes-Compatible Modem ▼]
Baud: [9600 ▼] Dial: [Tone ▼]
Modem Port   Printer Port
[Schedule] [Cancel] [**Dial**]

- *File virus:* File viruses attach themselves to executable files—those that actually begin a program. (In DOS these files have the extensions .com and .exe.) When the program is run, the virus starts working, trying to get into main memory and infecting other files.
- *Logic bomb:* Logic bombs, or simply *bombs,* differ from other viruses in that they are set to go off at a certain date and time. A disgruntled programmer for a defense contractor created a bomb in a program that was supposed to go off two months after he left. Designed to erase an inventory tracking system, the bomb was discovered only by chance.
- *Trojan horse:* The Trojan horse covertly places illegal, destructive instructions in the middle of a legitimate program, such as a computer game. Once you run the program, the Trojan horse goes to work, doing its damage while you are blissfully unaware.
- *Polymorphic virus:* A polymorphic virus, of which there are several kinds, can mutate and change form just as human viruses can. These are especially troublesome because they can change their profile, making existing antiviral technology ineffective.
- *Macro virus:* These viruses take advantage of a procedure in which miniature programs, known as macros, are embedded inside common data files, such as those created by e-mail or spreadsheets, which are sent over computer networks. Such documents are typically ignored by antivirus software. The Word Concept virus is an example of this type.

- **By network:** The second way is via a network, as from e-mail or an electronic bulletin board. This is why, with all the freebie games and other software available online, you should use virus-scanning software to check downloaded files before you open them.

The virus usually attaches itself to your hard disk. It might then display annoying messages ("Your PC is stoned—legalize marijuana") or cause Ping-Pong balls to bounce around your screen and knock away text. More seriously, it might add garbage to or erase your files or destroy your system software. It may evade your detection and create havoc elsewhere.

A variety of virus-fighting programs are available at stores, although you should be sure to specify the viruses you want to protect against. **Antivirus software scans a computer's hard disk, diskettes, and main memory to detect viruses and, sometimes, to destroy them.** We described some antivirus programs in Chapter 3 (✓ p. 98). (A detailed list of antivirus software can be found on the World Wide Web at *http://www.ncsa.com.* Up-to-date material may also be found at *FTP://mcafee.com/pub/antivirus.*)

Computer Criminals

What kind of people are perpetrating most of the information-technology crime? Over 80% may be employees, and the rest are outside users, hackers and crackers, and professional criminals.

- **Employees:** Says Michigan State University criminal justice professor David Carter, who surveyed companies about computer crime, "Seventy-five to 80% of everything happens from inside."[23] Most common frauds, Carter found, involved credit cards, telecommunications, employees' personal use of computers, unauthorized access to confidential files, and unlawful copying of copyrighted or licensed software.

  Workers may use information technology for personal profit or steal hardware or information to sell. They may also use it to seek revenge for real or imagined wrongs, such as being passed over for promotion. Sometimes they may use the technology simply to demonstrate to themselves that they have power over people.

- **Outside users:** Suppliers and clients may also gain access to a company's information technology and use it to commit crimes. With both, this becomes more possible as electronic connections such as Electronic Data Interchange (✓ p. 245) systems become more commonplace.

- **Hackers and crackers:** *Hackers* **are people who gain unauthorized access to computer or telecommunications systems for the challenge or even the principle of it.** For example, Eric Corley, publisher of a magazine called *2600: The Hackers' Quarterly*, believes that hackers are merely engaging in "healthy exploration." In fact, by breaking into corporate computer systems and revealing their flaws, he says, they are performing a favor and a public service. Such unauthorized entries show the corporations involved the leaks in their security systems.[24] Indeed, in late 1995, Netscape launched its so-called Bugs Bounty program, offering a cash reward to the first hacker to identify a "significant" security flaw in its latest Web browser software (✓ p. 64).[25]

  Crackers **also gain unauthorized access to information technology but do so for malicious purposes.** (Some observers think the term hacker covers malicious intent, also.) Crackers attempt to break into computers and deliberately obtain information for financial gain, shut down hardware, pirate software, or destroy data.

  The tolerance for "benign explorers"—hackers—has waned. Most communications systems administrators view any kind of unauthorized access as a threat, and they pursue the offenders vigorously. Educators try to point out to students that universities can't provide an education for everybody if hacking continues.[26] The most flagrant cases of hacking are met with federal prosecution.

- **Professional criminals:** Members of organized crime rings don't just steal information technology. They also use it the way that legal businesses do—as a business tool, but for illegal purposes. For instance, databases can be used to keep track of illegal gambling debts and stolen goods. Not surprisingly, the old-fashioned illegal bookmaking operation has gone high-tech, with bookies using computers and fax machines in place of betting slips and paper tally sheets.[27] In addition, telecommunications can be used to transfer funds illegally.

As information-technology crime has become more sophisticated, so have the people charged with preventing it and disciplining its outlaws. Campus administrators are no longer being quite as easy on offenders and are turning them over to police. Industry organizations such as the Software Publishers Association are going after software pirates large and small. (Commercial software piracy is now a felony, punishable by up to five years in prison and fines of up to $250,000 for anyone convicted of stealing at least 10 copies of a program, or more than $2500 worth of software.) Police departments as far apart as Medford, Massachusetts, and San Jose, California, now have police patrolling a "cyber beat." That is, they cruise online bulletin boards looking for pirated software, stolen trade secrets, child molesters, and child pornography.[28]

Security: The Challenge of Safeguarding Computers & Communications

Preview & Review: Information technology requires vigilance in security. Four areas of concern are identification and access, encryption, protection of software and data, and disaster-recovery planning.

The ongoing dilemma of the Digital Age is balancing convenience against security. **Security is a system of safeguards for protecting information technology against disasters, systems failure, and unauthorized access that can result in damage or loss.** We consider four components of security.

Identification & Access

Are you who you say you are? The computer wants to know.

There are three ways a computer system can verify that you have legitimate right of access. Some security systems use a mix of these techniques. The systems try to authenticate your identity by determining (1) what you have, (2) what you know, or (3) who you are.

- **What you have—cards, keys, signatures, badges:** Credit cards, debit cards, and cash-machine cards all have magnetic strips or built-in computer chips that identify you to the machine. Many require you to display your signature, which someone may compare as you sign their paperwork. Computer rooms are always kept locked, requiring a key. Many people also keep a lock on their personal computers. A computer room may also be guarded by security officers, who may need to see an authorized signature or a badge with your photograph before letting you in.

    Of course, credit cards, keys, and badges can be lost or stolen. Signatures can be forged. Badges can be counterfeited.

- **What you know—PINs, passwords, and digital signatures:** To gain access to your bank account through an automated teller machine (ATM), you key in your PIN. A **PIN**, or **personal identification number, is the security number known only to you that is required to access the system.** Telephone credit cards also use a PIN. If you carry either an ATM or phone card, never carry the PIN written down elsewhere in your wallet (even disguised).

    A **password is a special word, code, or symbol that is required to access a computer system.** Passwords, however, are one of the weakest security links, says AT&T security expert Steven Bellovin. Passwords can be guessed, forgotten, or stolen.

- **Who you are—physical traits:** Some forms of identification can't be easily faked—such as your physical traits. Biometrics tries to use these in security devices. **Biometrics is the science of measuring individual body characteristics.**

    For example, before a number of University of Georgia students can use the all-you-can-eat plan at the campus cafeteria, they must have their hands read. As one writer describes the system, "a camera automatically compares the shape of a student's hand with an image of the same hand pulled from the magnetic strip of an ID card. If the patterns match, the cafeteria turnstile automatically clicks open. If not, the would-be moocher eats elsewhere."[29]

Besides hand prints, other biological characteristics read by biometric devices are fingerprints (computerized "finger imaging"), voices, the blood vessels in the back of the eyeball, the lips, and even one's entire face.[30-32]

Some computer security systems have a "call-back" provision. In a *call-back system*, the user calls the computer system, punches in the password and hangs up. The computer then calls back a certain preauthorized number. This measure will block anyone who has somehow got hold of a password but is calling from an unauthorized telephone.

Encryption

PGP (for Pretty Good Privacy) is a computer program written for encrypting computer messages—putting them into secret code. **Encryption, or enciphering, is the altering of data so that it is not usable unless the changes are undone.** PGP is so good that it is practically unbreakable; even government experts can't crack it.[33]

Encryption is clearly useful for some organizations, especially those concerned with trade secrets, military matters, and other sensitive data. However, from the standpoint of our society, encryption is a two-edged sword. For instance, police in Sacramento, California, found that PGP blocked them from reading the computer diary of a convicted child molester and finding links to a suspected child pornography ring. Should the government be allowed to read the coded e-mail of its citizens? What about its being blocked from surveillance of overseas terrorists, drug dealers, and other enemies?

Protection of Software & Data

Organizations go to tremendous lengths to protect their programs and data. As might be expected, this includes educating employees about making backup disks, protecting against viruses, and so on.

Other security procedures include the following:

- **Control of access:** Access to online files is restricted only to those who have a legitimate right to access—because they need them to do their jobs. Many organizations have a transaction log that notes all accesses or attempted accesses to data.

- **Audit controls:** Many networks have audit controls, which track the programs and servers used, the files opened, and so on. This creates an audit trail, a record of how a transaction was handled from input through processing and output.

- **People controls:** Because people are the greatest threat to a computer system, security precautions begin with the screening of job applicants. That is, resumes are checked to see if people did what they said they did. Another control is to separate employee functions, so that people are not allowed to wander freely into areas not essential to their jobs. Manual and automated controls—input controls, processing controls, and output controls—are used to check that data is handled accurately and completely during the processing cycle. Printouts, printer ribbons, and other waste that may yield passwords and trade secrets to outsiders are disposed of through shredders or locked trash barrels.

Disaster-Recovery Plans

A *disaster-recovery plan* **is a method of restoring information processing operations that have been halted by destruction or accident.** "Among the countless lessons that computer users have absorbed in the hours, days, and weeks after the [New York] World Trade Center bombing," wrote one reporter, "the most enduring may be the need to have a disaster-recovery plan. The second most enduring lesson may be this: Even a well-practiced plan will quickly reveal its flaws."[34]

Mainframe computer systems are operated in separate departments by professionals, who tend to have disaster plans. Mainframes are usually backed up. However, many personal computers, and even entire local area networks, are not backed up. The consequences of this lapse can be great. It has been reported that, on average, a company loses as much as 3% of its gross sales within eight days of a sustained computer outage. In addition, the average company struck by a computer outage lasting more than 10 days never fully recovers.[35]

A disaster-recovery plan is more than a big fire-drill. It includes a list of all business functions and the hardware, software, data, and people to support those functions, as well as arrangements for alternate locations. The disaster-recovery plan includes ways for backing up and storing programs and data in another location, ways of alerting necessary personnel, and training for those personnel.

Quality-of-Life Challenges: The Environment, Mental Health, & the Workplace

Preview & Review: Information technology can create problems for the environment, people's mental health, and the workplace.

Earlier in this book, we pointed out some of the worrisome effects of technology on intellectual property rights and truth in art and journalism (✓ pp. 44, 174), on health matters and ergonomics (✓ p. 153), and on privacy (✓ pp. 221, 265). Here are some other quality-of-life issues.

Environmental Problems

"This county will do peachy fine without computers," says Micki Haverland, who has lived in rural Hancock County, Tennessee, for 20 years.[36] Telecommunications could bring jobs to an area that badly needs them, but several people moved there precisely because they like things the way they are—pristine rivers, unspoiled forests, and mountain views.

But it isn't just people in rural areas who are concerned. Suburbanites in Idaho and Utah, for example, worry that lofty metal poles topped by cellular-transmitting equipment will be eyesores that will destroy views and property values.[37] City dwellers everywhere are concerned that the federal government's 1996 decision to deregulate the telecommunications industry will lead to a rat's nest of roof antennas, satellite dishes, and above-ground transmission stations.[38] As a result, telecommunications companies are now experimenting with hiding transmitters in the "foliage" of fake trees made of metal.[39]

Political scientist James Snider of Northwestern University points out that the problems of the cities could expand well beyond the cities, if telecommuting triggers a massive movement of people to rural areas. "If all Americans succeed in getting their dream homes with several acres of land," he writes, "the forests and open lands across the entire continental United States will be destroyed" as they become carved up with subdivisions and roads.[40]

Mental-Health Problems: Isolation, Gambling, Net-Addiction, Stress

Many people say that improved computers and communications systems will liberate us from the traditional constraints of time and place. But, from a mental-health standpoint, will being wired together really set us free? Consider:

- **Isolation:** Automation allows us to go days without actually speaking with or touching another person, from buying gas to playing games. Even the friendships we make online in cyberspace, some believe, "are likely to be trivial, short lived, and disposable—junk friends." Says one writer, "We may be overwhelmed by a continuous static of information and casual acquaintance, so that finding true soul mates will be even harder than it is today."[41]

- **Gambling:** Gambling is already widespread in North America, but information technology could make it almost unavoidable. Although gambling by wire is illegal in the U.S., all kinds of moves are afoot to get around that. For example, host computers for Internet casinos and sports books are being set up in Caribbean tax havens, and satellites, decoders, and remote-control devices are being used so TV viewers can do racetrack wagering from home.[42-44]

- **Net-addiction:** Don't let this happen to you: "A student e-mails friends, browses the World Wide Web, blows off homework, botches exams, flunks out of school."[45] This is the downward spiral of the "Net addict," often a college student—because schools give students no-cost/low-cost linkage to the Internet—though it can be anyone. Some become addicted (although some mental-health professionals feel "addiction" is too strong a word) to chat groups, some to online pornography, some simply to escape from real life.[46,47] Indeed, sometimes the computer replaces one's spouse or boyfriend/girlfriend in the user's affections. In one instance, a man sued his wife for divorce for having an "online affair" with a partner who called himself The Weasel.[48-51]

- **Stress:** In a 1995 survey of 2802 American PC users, three-quarters of the respondents (whose ages ranged from children to retirees) said personal computers had increased their job satisfaction and were a key to success and learning. However, many found PCs stressful: 59% admitted getting angry at them within the previous year. And 41% said they thought computers have reduced job opportunities rather than increased them.[52]

E Workplace Problems

First the mainframe computer, then the desktop stand-alone PC, and lately the networked computer were all brought into the workplace for one reason only: to improve productivity. How is it working out? Let's consider two aspects: the misuse of technology and information overload.

- **Misuse of technology:** "For all their power," says an economics writer, "computers may be costing U.S. companies tens of billions of dollars a year in downtime, maintenance and training costs, useless game playing, and information overload."[53]

Consider games. Employees may look busy, staring into their computer screens with brows crinkled. But often they're just hard at work playing Doom or surfing the Net. Workers with Internet access average 10 hours a week online.[54] However, fully 23% of computer game players use their office PCs for their fun, according to one survey.[55] A study of employee online use at one major company concluded that the average worker wastes 1½ hours each day.[56]

Another reason for so much wasted time is all the fussing that employees do with hardware and software. A 1992 study estimated that microcomputer users waste 5 billion hours a year waiting for programs to run, checking computer output for accuracy, helping co-workers use their applications, organizing cluttered disk storage, and calling for technical support.[57]

Many companies don't even know what kind of microcomputers they have, who's running them, or where they are. The corporate customer of one computer consultant, for instance, swore it had 700 PCs and 15 users per printer. An audit showed it had 1200 PCs with one printer each.[58]

- **Information overload:** "My boss basically said, 'Carry this pager seven days a week, 24 hours a day, or find another job,'" says the chief architect for a New Jersey school system. (He complied, but pointedly notes that the pager's "batteries run out all the time.")[59] "It used to be considered a status symbol to carry a laptop computer on a plane," says futurist Paul Saffo. "Now anyone who has one is clearly a working dweeb who can't get the time to relax. Carrying one means you're on someone's electronic leash."[60]

The new technology is definitely a two-edged sword. Cellular phones, pagers, fax machines, and modems may unleash employees from the office, but these employees tend to work longer hours under more severe deadline pressure than do their tethered counterparts who stay at the office, according to one study.[61]

What does being overwhelmed with information do to you, besides inducing stress and burnout? One result is that because we have so many choices to entice and confuse us, we become more averse to making decisions. "The volume of information available is so great that I think people generally are suffering from a lack of meaning in their lives," says Neil Postman, chair of the Department of Culture and Communication at New York University. "People are just adrift in the sea of information, and they don't know what the information is about or why they need it."[62]

People and businesses are beginning to realize the importance of coming to grips with these problems. Some companies are employing GameCop, a software program that catches unsuspecting employees playing computer games on company time.[63] Some are installing asset-management software that tells them how many PCs are on their networks and what they run. Some are imposing strict hardware and software standards to reduce the number of different products they support.[64] To avoid information overload, some people—those who have a choice—no longer carry cell phones or even look at their e-mail. Others are installing so-called Bozo filters, software that screens out trivial e-mail messages and cellular calls and assigns priorities to the remaining files.[65] Still others are beginning to employ programs called intelligent agents to help them make decisions, as we discuss shortly.

But the real change may come as people realize that they need not be tied to the technological world in order to be themselves, that solitude is a scarce resource, and that seeking serenity means streamlining the clutter and reaching for simpler things.[66]

E Economic Challenges: Employment & the Haves/Have-Nots

Preview & Review: Many people worry that jobs are being eliminated by the effects of information technology. They also worry that it is widening the gap between the haves and have-nots.

In recent times a number of books (such as Clifford Stoll's *Silicon Snake Oil*, Stephen Talbott's *The Future Does Not Compute*, and Mark Slouka's *War of the Worlds*) have appeared that try to provide a counterpoint to the hype and overselling of information technology to which we have long been exposed. Some of these strike a sensible balance, but some make the alarming case that technological progress is actually no progress at all—indeed, it is a curse. The two biggest charges (which are related) are, first, that information technology is killing jobs, and second, that it is widening the gap between the rich and the poor.

- **Technology, the job killer?** There's probably no question that technological advances play an ambiguous role in social progress. But is it true, as Jeremy Rifkin says in *The End of Work*, that intelligent machines are replacing humans in countless tasks, "forcing millions of blue-collar and white-collar workers into temporary, contingent, and part-time employment and, worse, unemployment"?[67]

  This is too large a question to be fully considered in this book. Many factors are responsible for the decline in economic growth, the downsizing of companies, the rise of unemployment among many sectors of society. The U.S. economy is undergoing powerful structural changes, brought on not only by the widespread diffusion of technology but also by the growth of international trade, the shift from manufacturing to service employment, the weakening of labor unions, more rapid immigration, and other factors.[68,69]

  Many economists seem to agree that the boom times of economic growth that the United States enjoyed in the 1950s and 1960s won't return until there is more public investment and more personal saving instead of spending—savings that could be used for machinery and other tools of a thriving economy. Investment in recent years has been concentrated in computers. Surprisingly, however, as one economics writer points out, "so far computers have not yielded the rapid growth in production that came from investments in railroads, autos, highways, electric power, and aircraft—all huge outlays, involving government as well as the private sector, that changed the way Americans lived and worked."[70]

- **Gap between rich and poor:** "In the long run," says Stanford University economist Paul Krugman, "improvements in technology are good for almost everyone. . . . Unfortunately, what is true in the long run need not be true over shorter periods."[71] We are now, he believes, living through one of those difficult periods in which technology doesn't produce widely shared economic gains but instead widens the gap between those who have the right skills and those who don't.

A U.S. Department of Commerce survey of "information have-nots" reveals that about 20% of the poorest households in the U.S. do not have telephones. Moreover, only a fraction of those poor homes that do have phones will be able to afford the information technology that most economists agree is the key to a comfortable future.[72] The richer the family, the more likely it is to have and use a computer.

Schooling—especially college—makes a great difference. Every year of formal schooling after high school adds 5–15% to annual earnings later in life.[73] Being well educated is only part of it, however; one should also be technologically literate. Employees who are skilled at technology "earn roughly 10–15% higher pay," according to the chief economist for the U.S. Labor Department.[74]

Advocates of information access for all find hope in the promises of NII proponents for "universal service" and the wiring of every school to the Net. But this won't happen automatically. Ultimately we must become concerned with the effects of growing economic disparities on our social and political health. "Computer technology is the most powerful and the most flexible technology ever developed," says Terry Bynum, chair of the American Philosophical Association's Committee on Philosophy and Computing. "Even though it's called a technical revolution, at heart it's a social and ethical revolution because it changes everything we value."[75]

Now that we've considered the challenges, let us discuss some of the promises of information technology not described so far. They include artificial intelligence and the promises of the Information Superhighway.

The Promise of Artificial Intelligence

Preview & Review: Artificial intelligence (AI) is a research and applications discipline that develops systems that mimic the intelligence and behavior of human beings.

You're having trouble with your new software program. You call the customer "help desk" at the software maker. Do you get a busy signal or get put on hold to listen to music (or, worse, advertising) for several minutes? Technical support lines are often swamped, and waiting is commonplace. Or, to deal with your software difficulty, do you find yourself dealing with . . . other software?

This event is not unlikely. Programs that can walk you through a problem and help solve it are called *expert systems*. As the name suggests, these are systems that are imbued with knowledge by a human expert.[76] Expert systems, as we will discuss, are one of the most useful applications of an area known as artificial intelligence.

Artificial intelligence (AI) is a group of related technologies that attempt to develop machines to emulate human-like qualities, such as learning, reasoning, communicating, seeing, and hearing. Among other sophisticated innovations, the field of artificial intelligence includes:

* Robotics
* Expert systems
* Natural language processing

Robotics

Robotics is a field that attempts to develop machines that can perform work normally done by people. The machines themselves, of course, are called robots. A **robot is an automatic device that performs functions ordinarily ascribed to human beings or that operates with what appears to be almost human intelligence.** Dante II, for instance, is an eight-legged, 10-foot-high, satellite-linked robot used by scientists to explore the inside of Mount Spurr, an active volcano in Alaska.[77] Robots may be controlled from afar, as in an experiment at the University of Southern California in which Internet users thousands of miles away were invited to manipulate a robotic arm to uncover objects in a sandbox.[78]

Robots that resemble R2D2 in the movie *Star Wars* are some way from being realized. Such "personal robots" as have been developed are like B.O.B. (for Brains On Board), a device sold by Visual Machines. B.O.B. speaks prerecorded phrases and maneuvers around objects by using ultrasonic sound.

Expert Systems

We mentioned one example of an expert system in the "help desk" software. An **expert system is an interactive computer program that helps users solve problems that would otherwise require the assistance of a human expert.**

Such programs simulate the reasoning process of experts in certain well-defined areas. That is, professionals called *knowledge engineers* interview the expert or experts and determine the rules and knowledge that must go into the program. Programs incorporate not only surface knowledge ("textbook knowledge") but also deep knowledge ("tricks of the trade"). Expert systems exist in many areas. For example, MYCIN helps diagnose infectious diseases. PROSPECTOR assesses geological data to locate mineral deposits. DENDRAL identifies chemical compounds. Home-Safe-Home evaluates the residential environment of an elderly person. Business Insight helps businesses find the best strategies for marketing a product. REBES (Residential Burglary Expert System) helps detectives investigate crime scenes.

Natural Language Processing

Natural languages are ordinary human languages, such as English. **Natural language processing is the study of ways for computers to recognize and understand ordinary human language, whether in spoken or written form.**

Think how challenging it is to make a computer translate English into another language. In one instance, the English sentence "The spirit is willing, but the flesh is weak" came out in Russian as "The wine is agreeable, but the meat is spoiled." The problem with human language is that it is often ambiguous and often interpreted differently by different listeners.

Today you can buy a handheld computer that will translate a number of English sentences—principally travelers' phrases ("May I see a menu, please?")—into another language. This trick is similar to teaching an English-speaking child to sing "Frère Jacques." More complex is the work being done by AI scientists trying to discover ways to endow the computer with an "understanding" of how human language works. This means working with ideas about the instinctual instructions or genetic code that babies are born with for understanding language.

Still, some natural-language systems are already in use. Intellect is a product that uses a limited English vocabulary to help users orally query databases. LUNAR, developed to help analyze moon rocks, answers questions about geology from an extensive database. Verbex, used by the U.S. Postal Service, lets mail sorters read aloud an incomplete address and then replies with the correct zip code.

Artificial Intelligence, the Turing Test, & AI Ethics

How can we know when we have reached the point where computers have achieved human intelligence? How will you always know, say, when you're on the phone, whether you're talking to a human being or to a computer? Clearly, with the strides made in the fields of artificial intelligence and artificial life, this question is no longer just academic.

Interestingly, this matter was addressed back in 1950 by Alan Turing, an English mathematician and computer pioneer. Turing predicted that by the end of the century computers would be able to mimic human thinking and converse so naturally that their communications would be indistinguishable from a person's. Out of these observations came the Turing test. The *Turing test* is a test or game for determining whether a computer is considered to possess "intelligence" or "self-awareness."

In the Turing test, a human judge converses by means of a computer terminal with two entities hidden in another location. One entity is a person typing on a keyboard. The other is a software program. As the judge types in and receives messages on the terminal, he or she must decide whether the entity is human. In this test, intelligence and the ability to think is demonstrated by the computer's success in fooling the judge. (■ *See Panel 12.3.*)

Judith Anne Gunther participated as one of eight judges in the third annual Loebner Prize Competition, which is based on Turing's ideas.[79] (There have been two other competitions since.) The "conversations" are restricted to predetermined topics, such as baseball. This is because today's best programs have neither the databases nor the syntactical ability to handle an unlimited number of subjects. Conversations with each entity are limited to 15 minutes. At the end of the contest, the program that fools the judges most is the one that wins.

PANEL 12.3

The Turing test

In this test, a human judge converses via computer terminal with two entities in another location. One is a human being, the other a computer program. The task of the judge is to figure out whether the "intelligence" he or she is conversing with is human or not. Often the computer just gives itself away by being illogical, as in this "conversation."

"**Computer:** As a psychoanalyst, I am familiar with discussing such marital troubles with you as the patient on the couch. If you don't have a bad marriage just pretend that you do. Please start with 'My wife . . .' or 'My husband . . .'.
Judge: Ok, I'll pretend I have a bad marriage. Let's say the reason is because he never picks up his dirty underwear. Is this marriage irreparable?
Computer: Well, if you think I'm hopeless, you can always just swear at me."

—Judith Anne Gunther, "An Encounter with A.I.," *Popular Science*

Gunther found that she wasn't fooled by any of the computer programs. The winning program, for example, relied as much on deflection and wit as it did on responding logically and conversationally. (For example, to a judge trying to discuss a federally funded program, the computer said: "You want logic? I'll give you logic: shut up, shut up, shut up, shut up, shut up, now go away! How's that for logic?") However, Gunther *was* fooled by one of the five humans, a real person discussing abortion. "He was so uncommunicative," wrote Gunther, "that I pegged him for a computer."

Behind everything to do with artificial intelligence and artificial life—just as it underlies everything we do—is the whole matter of ethics. In his book *Ethics in Modeling,* William A. Wallace, professor of decision sciences at Rensselaer Polytechnic Institute, points out that many users are not aware that computer software, such as expert systems, is often subtly shaped by the ethical judgments and assumptions of the people who create it.[80] In one instance, he points out, a bank had to modify its loan-evaluation software after it discovered that it tended to reject some applications because it unduly emphasized old age as a negative factor. Another expert system, used by health maintenance organizations (HMOs), instructs doctors on when they should opt for expensive medical procedures, such as magnetic resonance imaging tests. HMOs like the systems because they help control expenses, but critics are concerned that doctors will have to base decisions not on the best medicine but simply on "satisfactory" medicine combined with cost cutting.[81] Clearly, there is no such thing as completely "value-free" technology. Human beings build it, use it, and have to live with the results.

Promises of the Information Superhighway

Preview & Review: The Information Superhighway promises great benefits in the areas of education and information, health, commerce and electronic money, entertainment, and government and electronic democracy.

"Where can I find the on-ramp to the information highway?" When a video crew asked this single question of several passers-by on the streets of New York City, the answers showed how vague people are about the concept. ("Take a left on Houston Street, and keep going straight," one man said. "Ask Reynaldo, the doorman," said another.)[82] Companies, too, are searching for the on-ramp to the Information Highway or Superhighway, the catch phrase for the convergence of computer, telephone, and television technologies that is supposed to deliver text, video, and sound to the home screen. Ultimately, the notion is that you would be able to hook up your "information appliance"—whether desktop model or mobile personal communicator—and access numerous services. Here business, government, and educators all see different priorities and needs. Nevertheless, potential services include the following.

Education & Information

The government is interested in reforming education, and technology can assist that effort. Presently the United States has more computers in its classrooms than other countries, but not all the machines are new and not all teachers are computer-literate. Of course, the poorer the school district, the less likely it is to have an integrated computer-and-communications system. President Clinton proposed a "high-tech barn-raising," a government-industry collaboration to put every school in the nation on the Internet by the year 2000.[83]

Computers can be used to create "virtual" classrooms not limited by scheduled class time. Some institutions (Stanford, MIT) are replacing the lecture hall with forms of learning featuring multimedia programs, workstations, and television courses at remote sites. The Information Superhighway could be used to enable students to take video field trips to distant places and to pull information from remote museums and libraries.[84–86]

Of particular interest is distance learning, or the "virtual university." *Distance learning* is the use of computer and/or video networks to teach courses to students outside the conventional classroom. At present, distance learning is largely outside the mainstream of campus life. That is, it concentrates principally on part-time students, those who cannot easily travel to campus, those interested in noncredit classes, or those seeking special courses in business or engineering. However, part-timers presently make up about 45% of all college enrollments. This, says one writer, is "a group for whom 'anytime, anywhere' education holds special appeal."[87]

Health

Another goal for the Information Superhighway is to improve health care. The government is calling for an expansion of "telemedicine." *Telemedicine* is the use of telecommunications to link health-care providers and researchers, enabling them to share medical images, patient records, and research. Of particular interest would be the use of networks for "teleradiology" (the exchange of X-rays, CAT scans, and the like), so that specialists could easily confer. Telemedicine would also allow long-distance patient examinations, using video cameras and, perhaps, virtual-reality kinds of gloves that would transmit and receive tactile sensations.[88,89]

Commerce & Electronic Money

Businesses clearly see the Information Superhighway as a way to enhance productivity and competitiveness. However, the changes will probably go well beyond this.

The thrust of the original 19th-century Industrial Revolution was separation—to break work up into its component parts to permit mass production. The effect of computer networks in the Digital Revolution, however, is unification—to erase boundaries between company departments, suppliers, and customers.[90]

Indeed, the parts of a company can now as easily be global as down the hall from one another. Thus, designs for a new product can be tested and exchanged with factories in remote locations. With information flowing faster, goods can be sent to market faster and inventories kept reduced. Says an officer of the Internet Society, "Increasingly you have people in a wide variety of professions collaborating in diverse ways in other places. The whole notion of 'the organization' becomes a blurry boundary around a set of people and information systems and enterprises."[91]

The electronic mall, in which people make purchases online, is already here. Record companies, for instance, are making sound and videos of new albums available on Web sites, which you can sample and then order as a cassette or CD.[92] Banks in cyberspace are allowing users to adopt avatars or personas of themselves in three-dimensional virtual space on the World Wide Web, where they can query bank tellers and officers and make transactions.[93] Wal-Mart Stores and Microsoft have developed a joint online shopping venture that allows shoppers to browse online and buy merchandise.[94]

Cybercash, or e-cash, will change the future of money. Whether it takes the form of smart cards or of electronic blips online, cybercash will probably begin to displace (though not completely supplant) checks and paper currency. This would change the nature of how money is regulated as well as the way we spend and sell.[95,96]

Entertainment

Among the future entertainment offerings could be movies on-demand, videogames, and gaming ("telegambling"). *Video on-demand* would allow viewers to browse through a menu of hundreds of movies, select one, and start it when they wanted. This definition is for true video on-demand, which is like having a complete video library in your house. (An alternative, simpler form could consist of running the same movie on multiple channels, with staggered starting times.) True video on-demand will require a server (✓ p. 372), a storage system with the power of a supercomputer that would deliver movies and other data to thousands of customers at once.

Government & Electronic Democracy

Will information technology help the democratic process? There seem to be two parts to this. The first is its use as a campaign tool, which may, in fact, skew the democratic process in some ways. The second is its use in governing and in delivering government services.

Santa Monica, California, established a computer system called Public Electronic Network (PEN), which residents may hook into free of charge. PEN gives Santa Monica residents access to city council agendas, staff reports, public safety tips, and the public library's online catalog. Citizens may also enter into electronic conferences on topics both political and non-political; this has been by far the most popular attraction.[97]

PEN could be the basis for wider forms of electronic democracy. For example, electronic voting might raise the percentage of people who vote. Interactive local-government meetings could enable constituents and town council members to discuss proposals.

The Information Superhighway could also deliver federal services and benefits. In 1994, the government unveiled a program in which Social Security pensioners and other recipients of federal aid without bank accounts could use an automated-teller-machine card to walk up to any ATM and withdraw the funds due them.[98]

Onward

The four principal information technologies—computer networks, imaging technology, massive data storage, and artificial intelligence—will affect probably 90% of the workforce by 2010.[99] Clearly, information technology is driving the new world of jobs, leisure, and services, and nothing is going to stop it.

Where will you be in all this? People pursuing careers find the rules are changing very rapidly. Up-to-date skills are becoming ever more crucial. Job descriptions of all kinds are metamorphosing, and even familiar jobs are becoming more demanding. Today, experts advise, you need to prepare to continually upgrade your skills, prepare for specialization, and prepare to market yourself. In a world of breakneck change, you can still thrive. The most critical knowledge, however, may turn out to be self-knowledge.

SUMMARY

antivirus software *(p. 301, LO 1)* Program that scans a computer's hard disk, diskettes, and main memory to detect viruses and, sometimes, to destroy them.

Computer users must find out what kind of antivirus software to install in their systems to protect them against damage or shut-down.

artificial intelligence (AI) *(p. 309, LO 4)* Group of related technologies that attempt to develop machines to emulate human-like qualities, such as learning, reasoning, communicating, seeing, and hearing.

AI is important for enabling machines to do things formerly possible only with human effort.

biometrics *(p. 303, LO 1)* Science of measuring individual body characteristics.

Biometrics is used in some computer security systems—for example, to verify individuals' fingerprints before allowing access.

crackers *(p. 302, LO 1)* People who gain unauthorized access to information technology for malicious purposes.

Crackers attempt to break into computers and deliberately obtain information for financial gain, shut down hardware, pirate software, or destroy data.

dirty data *(p. 298, LO 1)* Data that is incomplete, outdated, or otherwise inaccurate.

Dirty data can cause information to be inaccurate.

disaster-recovery plan *(p. 305, LO 1)* Method of restoring information processing operations that have been halted by destruction or accident.

A disaster-recovery plan is important if a company desires to resume its computer and business operations in short order.

encryption *(p. 304, LO 1)* Also called *enciphering;* the altering of data so that it is not usable unless the changes are undone.

Encryption is useful for users transmitting trade or military secrets or other sensitive data.

expert system *(p. 310, LO 4)* Interactive computer program that helps users solve problems that would otherwise require the assistance of a human expert.

Expert systems allow users to solve problems without assistance of a human expert; they incorporate both surface knowledge ("textbook knowledge") and deep knowledge ("tricks of the trade").

hackers *(p. 302, LO 1)* People who gain unauthorized access to computer or telecommunications systems for the challenge or even the principle of it.

The acts of hackers create problems not only for the institutions that are victims of break-ins but also for ordinary users of the systems.

information-technology crime *(p. 299, LO 1)* Crime of two types: an illegal act perpetrated against computers or telecommunications; or the use of computers or telecommunications to accomplish an illegal act.

Information-technology crimes cost billions of dollars every year.

National Information Infrastructure (NII) *(p. 296, LO 5)* The Information Superhighway; services will be delivered via networks and technologies of several information providers—the telecommunications networks, cable-TV networks, and the Internet—to users' "information appliances."

Services could include education, health, information, commerce, government, home services, and mobile communications.

natural language processing *(p. 310, LO 4)* Study of ways for computers to recognize and understand human language, whether in spoken or written form.

Natural language processing could further reduce the barriers to human/computer communications.

What It Is / What It Does	Why It's Important
password *(p. 303, LO 1)* Special word, code, or symbol that is required to access a computer system.	One of the weakest links in computer security, passwords can be guessed, forgotten, or stolen.
PIN (personal identification number) *(p. 303, LO 1)* Security number known only to you that is required to access the system.	PINs are required to access many computer systems and automated teller machines and to charge telephone calls.
robot *(p. 310, LO 4)* Automatic device that performs functions ordinarily ascribed to human beings or that operates with what appears to be almost human intelligence.	Robots are performing more and more functions in business and the professions.
security *(p. 303, LO 1)* System of safeguards for protecting information technology against disasters, systems failure, and unauthorized access that can result in damage or loss.	With proper security, organizations and individuals can minimize losses caused to information technology from disasters, system failures, and unauthorized access.
software bug *(p. 298, LO 1)* Error in a program caused by incorrect use of the programming language or faulty logic.	An error in a program will cause it not to work properly.
virus *(p. 300, LO 1)* Deviant program that attaches itself to computer systems and destroys or corrupts data.	Viruses can cause users to lose data or files or even shut down entire computer systems.
worm *(p. 300, LO 1)* Program that copies itself repeatedly into memory or onto a disk drive until no more space is left.	Like viruses, worms can shut down a user's computer system.

((Selected answers appear at the back of the book.)

Short-Answer Questions

1. How would you define *information-technology crime*?

2. What is the Turing test used for?

3. What is the difference between a hacker and a cracker?

4. How would you define *information superhighway*?

Fill-in-the-Blank Questions

1. A program that contains a(n) _____ won't function properly.

2. The purpose of_____ is to scan a computer's disk devices and memory to detect viruses and, sometimes, to destroy them.

3. Data that is incomplete, outdated, or otherwise inaccurate is referred to as _____.

4. _____ is a system of safeguards for protecting information technology.

5. List four areas in which the Information Superhighway promises great benefits.
   a. _____
   b. _____
   c. _____
   d. _____

6. A(n) _____ is a special word, code, or symbol that the user must provide before obtaining access to a computer system.

7. So that information processing operations can be restored after destruction or accident, a company should adopt a(n) _____.

8. When telecommunications technology is used to link healthcare providers and researchers, the field is referred to as _____.

Multiple-Choice Questions

1. Which of the following is an example of an information-technology crime?
   a. theft of hardware
   b. theft of software
   c. theft of time and services
   d. theft of information
   e. all of the above

2. Which of the following copies itself into memory or a disk drive until no more space is left?
   a. virus
   b. worm
   c. dirty data
   d. antiviral agent
   e. none of the above

3. Which of the following destroys or corrupts data?
   a. virus
   b. worm
   c. dirty data
   d. antiviral agent
   e. none of the above

4. Which of the following is an interactive computer program that helps users solve problems?
   a. robot
   b. perception system
   c. expert system
   d. natural language
   e. all of the above

5. Which of the following can disable information technology?
   a. people
   b. software errors
   c. dirty data
   d. crackers
   e. all of the above

6. Which of the following groups perpetrate over 80% of information technology crime?
   a. hackers
   b. crackers
   c. professional criminals
   d. employees
   e. none of the above

True/False Questions

T F   1. Information technology has been associated with people's mental health.

T F   2. Distance learning involves the use of a computer and/or a video network.

T F   3. A virus is a combination of hardware and software.

T F   4. Viruses can be passed to another computer by a diskette or through a network.

T F   5. Encrypted data isn't directly usable.

Projects/Critical-Thinking Questions

1. Write a few paragraphs about the challenges and obstacles facing the completion of the Information Superhighway.

2. In addition to *2600: The Hacker's Quarterly*, where do hackers find new information about their field? Are support groups available? In what ways do hackers help companies? Does a hacker underground exist? Research your answers using current computer periodicals and/or the Internet.

3. Aside from helping you do all of your chores, what do you think is a good application for a robot? For example, a snake-like robot was developed in GMD's Institute for System Design Technology that is ideal for inspection tasks that deal with tight tubes. Would your robot be hard to develop? Is it likely that it will be developed someday? If so, when? By whom? Is anything similar to your robot under development today?

4. Assuming you have a microcomputer in your home that includes a modem, what security threats, if any, should you be concerned with? List as many ways as you can think of to ensure that your computer is protected.

5. Explore the National Information Infrastructure (NII) in more detail. Create an executive report describing the objectives for the NII, its guiding principles, and its agenda for action. Does the NII exist today or is it a plan for the future? Or both? Research your answers using current periodicals and/or the Internet.

Answers

Following are answers to the odd-numbered fill-in-the-blank, multiple-choice, and true/false questions.

Chapter 1
Fill-in-the-Blank Questions
1. computer   3. end-user   5. analog, digital   7. storage hardware   9. telecommuting

Multiple-Choice Questions
1. c   3. b   5. d   7. d   9. c

True/False Questions
1. t   3. f   5. t   7. t   9. f

Chapter 2
Fill-in-the-Blank Questions
1. applications software   3. electronic spreadsheet   5. dialog box   7. integrated   9. browser

Multiple-Choice Questions
1. d   3. b   5. a   7. c

True/False Questions
1. f   3. t   5. t   7. t

Chapter 3
Fill-in-the-Blank Questions
1. operating systems, utility programs, language translators
3. format   5. utility programs   7. supervisor   9. plug and play

Multiple-Choice Questions
1. b   3. c   5. c   7. d   9. b

True/False Questions
1. f   3. t   5. f   7. f   9. t

Chapter 4
Fill-in-the-Blank Questions
1. bus line, or bus   3. system clock   5. ASCII   7. silicon

Multiple-Choice Questions
1. a   3. d   5. b

True/False Questions
1. t   3. f   5. f   7. t   9. t

Chapter 5
Fill-in-the-Blank Questions
1. Source data entry system (source data automation)   3. sensor
5. scanner   7. dumb

Multiple-Choice Questions
1. a   3. b   5. a   7. b

True/False Questions
1. f   3. t   5. t   7. t

Chapter 6
Fill-in-the-Blank Questions
1. resolution   3. hardcopy, softcopy   5. plotters   7. videoconferencing   9. robot

Multiple-Choice Questions
1. c   3. d   5. d   7. c   9. a

True/False Questions
1. t   3. f   5. f   7. t   9. t

Chapter 7
Fill-in-the-Blank Questions
1. 3½-inch, 5¼-inch   3. access speed   5. compression   7. bits, bytes, fields, records, files

Multiple-Choice Questions
1. d   3. d   5. b   7. e

True/False Questions
1. t   3. f   5. t   7. f   9. t   11. t

Chapter 8
Fill-in-the-Blank Questions
1. dedicated fax machine, fax modem   3. virtual   5. frequency, amplitude   7. browser

Multiple-Choice Questions
1. d   3. b   5. a   7. c

True/False Questions
1. t   3. t   5. t   7. t

Chapter 9
Fill-in-the-Blank Questions
1. prototyping   3. coding (programming)   5. visual

Multiple-Choice Questions
1. d   3. c   5. c

True/False Questions
1. f   3. t   5. f

Chapter 10
Fill-in-the-Blank Questions
1. software bug   3. dirty data   5. [answers will vary]
7. disaster recovery plan

Multiple-Choice Questions
1. e   3. a   5. e

True/False Questions
1. t   3. f   5. t

Notes

Chapter 1

1. Thomas A. Stewart, "The Information Age in Charts," *Fortune,* April 4, 1994, pp. 75–79.
2. Donald Spencer, *Webster's New World Dictionary of Computer Terms,* 4th ed. (New York: Prentice Hall, 1992), p. 206.
3. We are grateful to Prof. John Durham for contributing these ideas.
4. Sandra D. Atchison, "The Care and Feeding of 'Lone Eagles,'" *Business Week,* November 15, 1993, p. 58.
5. Link Resources, cited in Carol Kleiman, "At-Home Employees Multiplying," *San Jose Mercury News,* November 12, 1995, pp. 1PC–2PC.
6. Susan N. Futterman, "Quick-Hit Research of a Potential Employer," *CompuServe Magazine,* September 1993, p. 36.
7. Lee Gomes, "Hollow Dreams," *San Jose Mercury News,* November 12, 1995, pp. 1D, 3D.
8. Bloomberg Business News, "Motorola Links PC to Cable," *San Francisco Chronicle,* November 30, 1995, p. D1.
9. Bloomberg Business News, "New Technology Ties Internet, TV to PCs," *San Francisco Chronicle,* October 24, 1995, p. C3.
10. Catherine Arnst, Paul M. Eng, Richard Brandt, and Peter Burrows, "The Information Appliance," *Business Week,* November 22, 1993, pp. 98–110.
11. Tom Forester and Perry Morrison, *Computer Ethics: Cautionary Tales and Ethical Dilemmas in Computing* (Cambridge, MA: The MIT Press, 1990), pp. 1–2.

Chapter 2

1. John Markoff, "A Free and Simple Computer Link," *New York Times,* December 8, 1993, p. C1.
2. Alan Deutschman, "Mac vs. Windows: Who Cares?" *Fortune,* October 4, 1993, p. 114.
3. Barbara Kantrowitz, Andrew Cohen, and Melinda Lieu, "My Info Is NOT Your Info," *Newsweek,* July 18, 1994, p. 54.
4. Teresa Riordan, "Writing Copyright Law for an Information Age," *New York Times,* July 7, 1994, pp. C1, C5.
5. David L. Wheeler, "Computer Networks Are Said to Offer New Opportunities for Plagiarists," *Chronicle of Higher Education,* June 30, 1993, pp. A17, A19.
6. Denise K. Magner, "Verdict in a Plagiarism Case," *Chronicle of Higher Education,* January 5, 1994, pp. A17, A20.
7. Robert Tomsho, "As Sampling Revolutionizes Recording, Debate Grows Over Aesthetics, Copyrights," *Wall Street Journal,* November 5, 1990, p. B1.
8. Andy Ihnatko, "Right-Protected Software," *MacUser,* March 1993, pp. 29–30.
9. Rick Tetzeli, "Videogames: Serious Fun," *Fortune,* December 27, 1993, pp. 110–116.
10. Steve G. Steinberg, "Back in Your Court, Software Designers," *Los Angeles Times,* December 7, 1995, pp. D2, D11.
11. Tetzeli, 1993.
12. Herb Brody, "Video Games That Teach?" *Technology Review,* November/December 1993, pp. 50–57.
13. Jay Sivin-Kachala, Interactive Educational Systems Design, quoted in Nicole Carroll, "How Computers Can Help Low-Achieving Students," *USA Today,* November 20, 1995, p. 5D.
14. Charles Bermant, "Databases for Everyone," *San Francisco Examiner,* February 28, 1993, p. E-16.
15. Lohr, 1993, p. C3.
16. David Kirkpatrick, "Groupware Goes Boom," *Fortune,* December 27, 1993, p. 100.
17. Kirkpatrick, 1993, pp. 99–106.
18. Gary McWilliams, "Lotus 'Notes' Get a Lot of Notice," *Business Week,* March 29, 1993, pp. 84–85.
19. Kirkpatrick, 1993, pp. 100–101.
20. Edward Rothstein, "Between the Dream and the Reality Lies the Shadow. Or Is It the Interface?" *New York Times,* December 11, 1995, p. C3.
21. Margaret Trejo, quoted in Richard Atcheson, "A Woman for *Lear's*," *Lear's,* November 1993, p. 87.
22. Bernie Ward, "Computer Chic," *Sky,* April 1993, pp. 84–90.
23. Alan Freedman, *The Computer Glossary,* 6th ed. (New York: AMACOM, 1993), pp. 196–97.

Chapter 3

1. Tim Eckles, quoted in Jim Carlton, "Computer Firms Try to Make PCs Less Scary," *Wall Street Journal,* May 6, 1994, pp. B1, B7.

2. Carlton, 1994.
3. Peter H. Lewis, "Champion of MS-DOS, Admirer of Windows," *New York Times,* April 4, 1993, sec. 3, p. 11.
4. BIS, cited in David Kirkpatrick, "Mac vs. Windows," *Fortune,* June 20, 1993, sec 3, p. 8.
5. Peter H. Lewis, "A Strong New OS/2, with an Uncertain Future," *New York Times,* June 20, 1993, sec. 3, p. 8.
6. Dan Gillmor, "Best 32-bit Software Is Yet to Hit Market," *San Jose Mercury News,* August 27, 1995, p. 4E.
7. Lewis, 1993.
8. Robert Frankenberg, quoted in Buck, 1995.
9. Fisher, September 18, 1995.
10. Tosca Moon Lee, "Utility Software: Your PC's Life Preserver," *PC Novice,* March 1993, pp. 68–73.
11. "A New Model for Personal Computing," 1995.
12. Lee Gomes, "Hollow Dreams," *San Jose Mercury News,* November 12, 1995, pp. 1D, 3D.
13. Gomes, November, 1995.
14. Jennings, December, 1995.
15. Mark Fleming [letter] and Mike McGowan [letter], "Present at the Creation of the Net," *Business Week,* December 25, 1995, p. 12.

Chapter 4

1. Hank Roberts, quoted in "Talking About Portables," *Wall Street Journal,* November 16, 1992, p. R19.
2. Hadas Dembo, "The Way Things Were," *Wall Street Journal,* November 16, 1992, pp. R16–R17.
3. Michael S. Malone, "The Tiniest Transformer," *San Jose Mercury News,* September 10, 1995, pp. 1D–2D; excerpted from *The Microprocessor: A Biography* (New York: Telos/Springer Verlag, 1995).
4. Laurence Hooper, "No Compromises," *Wall Street Journal,* November 16, 1992, p. R8.
5. Dataquest Inc., reported in "Personal Computer Market Seen Doubling in 5 Years," *Wall Street Journal,* June 29, 1995, p. B13.
6. Brian Nadel, "Power to the PC," *PC Magazine,* April 26, 1994, pp. 114–183.
7. Nadel, 1994, p. 116.
8. Dan Gillmor, "Old Computer Will Mean a Lot to Those in Need," *San Jose Mercury News,* December 24, 1995, p. 1F.
9. Kyle Pope, "Changing Work Habits Fuel Popularity of Notebooks," *Wall Street Journal,* November 11, 1993, pp. B1, B6.
10. Phillip Robinson, "Access to Web Varies in Price, Ease, Speed," *San Jose Mercury News,* September 24, 1995, p. 2E.
11. Tom Abate, "Internet Overload," *San Francisco Examiner,* December 24, 1995, pp. D-1, D-4, D-6.
12. Mike Langberg, "Impatient Web Users Bog Down, Tune Out, Log Off," *San Jose Mercury News,* December 24, 1995, pp. 1A, 14A.
13. Alfred and Emily Glossbrenner, *Finding a Job on the Internet* (New York: McGraw-Hill, 1995).
14. Alan Phelps, "The World Wide Web: Your Mouse's License to Netcruise," *PC Novice,* June 1995, pp. 79–81.
15. Peter H. Lewis, "The Mac's High Road to a Spot on the Internet," *New York Times,* December 5, 1995, p. B10.
16. Robinson, September, 1995.

Chapter 5

1. Survey by The Wirthlin Group, Washington, D.C., reported in: Reuters, "It's Always 12:00 for VCR Incompetents," *San Francisco Chronicle,* February 2, 1994, p. A4.
2. David Gelernter, quoted in Associated Press, "Bombing Victim Says He's Lucky to Be Alive," *San Francisco Chronicle,* January 28, 1994, p. A15.
3. Frederick Williams, *Technology and Communication Behavior* (Belmont, CA: Wadsworth, 1987), p. 30.
4. David Berquel, quoted in Mary Geraghty, "Pen-Based Computer Seen as Tool to Ease Burden of Note Taking," *Chronicle of Higher Education,* November 9, 1994, p. A22.
5. Michael M. Phillips, "Voice Recognition Systems Work, but Only If You Speak American," *Wall Street Journal,* February 7, 1995, p. B1.
6. Diana Hembree and Ricardo Sandoval, "The Lady and the Dragon," *Columbia Journalism Review,* August 1991, pp. 44–45.
7. A. Hopkins, "The Social Recognition of Repetition Strain Injuries: An Australian/American Comparison," *Social Science and Medicine,* 1990, *30,* pp. 365–372.
8. Edward Felsenthal, "An Epidemic or a Fad? The Debate Heats Up Over Repetitive Stress," *Wall Street Journal,* July 14, 1994, pp. A1, A4.

9. Bart Ziegler and Peter Coy, "The Cellular Cancer Risk: How Real Is It?" *Business Week,* February 8, 1993, pp. 94–95.
10. David Kirkpatrick, "Do Cellular Phones Cause Cancer?" *Fortune,* March 8, 1993, pp. 82–89.
11. James Barron, "Stanley Adelman, 72, Repairer of Literary World's Typewriters," *New York Times,* December 1, 1995, p. C19.
12. Jonathan Auerbach, "Smith Corona Seeks Protection of Chapter 11," *Wall Street Journal,* July 6, 1995, p. B6.
13. Tom Foremski, "Conquering the Computer Connection," *San Francisco Examiner,* February 12, 1995, pp. D-5, D-7.
14. Federico Faggin, "A Call for Humanized Computing," *San Francisco Examiner,* August 20, 1995, pp. B-5, B-6.

Chapter 6

1. David L. Wheeler, "Recreating the Human Voice," *Chronicle of Higher Education,* January 19, 1996, pp. A8–A9.
2. Larry Armstrong, "The Price Is Right for Printers," *Business Week,* November 6, 1995, pp. 132–134.
3. Phillip Robinson, "Color Comes In," *San Jose Mercury News,* October 1, 1995, pp. 1F, 6F.
4. Liz Spayd, "Taming the Paper Jungle," *San Francisco Chronicle, Sunday Punch,* December 19, 1993, p. 2; reprinted from *The Washington Post.*
5. Clifford Nass, quoted in Spayd, 1993.
6. Lawrence J. Magid, "ProShare Lets Pair Video-edit Document," *San Jose Mercury News,* February 13, 1994, p. 5E.
7. Edmund L. Andrews, "Quest for Sharper TV Is Likely to Produce More TV Instead," *New York Times,* July 10, 1995, pp. C1, C8.
8. Paul M. Eng, "Virtual Buses for Novice Drivers," *Business Week,* January 29, 1996, p. 68D.
9. Matthew L. Wald, "FAA Will Use a Simulator to Train Controllers on Breakdowns," *New York Times,* September 24, 1995, sec. 1, p. 19.
10. Belinda Thurston, "Virtual Reality Lessons Hit Home," *USA Today,* April 11, 1995, p. 6D.
11. Laurie Flynn, "Virtual Reality and Virtual Spaces Find a Niche in Real Medicine," *New York Times,* June 5, 1995, p. C3.
12. Donna Horowitz, "Virtual Reality Takes Acrophones to Greater Heights," *San Francisco Examiner,* March 26, 1995, pp. C-1, C-6.
13. Daniel Goleman, "In Virtual Reality, Phobias Cease to Exist," *San Francisco Chronicle,* September 2, 1995, p. A7; reprinted from *The New York Times.*
14. John Malyon, *Whole Earth Review,* Spring 1992, pp. 80–84.
15. John Holusha, "Down on the Farm with R2D2," *New York Times,* October 7, 1995, pp. 21, 24.
16. Gene Bylinsky, "High-Tech Help for the Housekeeper," *Fortune,* November 2, 1992, p. 117.
17. Tahree Lane, "Robots Can Fill Jobs Nobody Wants," *San Francisco Examiner,* April 25, 1993, p. E-3.
18. Stephen Baker, "A Surgeon Whose Hands Never Shake," *Business Week,* October 4, 1993, pp. 111–114.
19. Jack Cheevers, "Real-Life 'Robocops' Gain Popularity," *San Francisco Chronicle,* August 13, 1994, p. A9; reprinted from *Los Angeles Times.*
20. Holusha, 1995.
21. William Safire, "Art vs. Artifice," *New York Times,* January 3, 1994, p. A11.
22. Hans Fantel, "Sinatra's 'Duets': Music Recording or Wizardry?" *New York Times,* January 1, 1994, p. 13.
23. Cover, *Newsweek,* June 27, 1994.
24. Cover, *Time,* June 27, 1994.
25. Cox News Service, "Computers Manipulate Old Photographs," *San Jose Mercury News,* February 19, 1995, p. 5H.
26. Jonathan Alter, "When Photographs Lie," *Newsweek,* July 30, 1990, pp. 44–45.
27. Fred Ritchin, quoted in Alter, 1990.
28. Robert Zemeckis, cited in Laurence Hooper, "Digital Hollywood: How Computers Are Remaking Movie Making," *Rolling Stone,* August 11, 1994, pp. 55–58, 75.
29. Woody Hochswender, "When Seeing Cannot Be Believing," *New York Times,* June 23, 1992, pp. B1, B3.
30. "Tying Up Your Computer," *PC Novice,* May 1992, p. 9.
31. Marty Jerome, "Killer Color Notebooks," *PC Computing,* January 1994, pp. 142–177.
32. Lori Beckmann Johnson, "Power Protection: Saving Your Computer from Surges, Spikes, and Sags," *PC Novice,* May 1992, pp. 72–73.

33. Jennifer Larson, "Maintaining Your Diskette Collection," *PC Novice,* February 1994, pp. 36–37.
34. Robert Schmidt, "Comfortable Keyboards," *PC Novice,* October 1993, pp. 44–46.

Chapter 7

1. Richard L. Hudson, "Europeans No Longer Scoff at Interactive Multimedia," *Wall Street Journal,* March 2, 1994, p. B6.
2. "Gargantua's 'Lossless' Compression," *The Australian,* March 22, 1994, p. 32; reprinted from *The Economist.*
3. Peter Coy, "Invasion of the Data Shrinkers," *Business Week,* February 14, 1994, pp. 115–116.
4. Coy, 1994.
5. Phillip Robinson, "Between a Floppy and a Hard Drive," *San Jose Mercury News,* October 8, 1995, pp. 1E, 6E.
6. "Team of 3 Companies in Computer Industry Improves Disk Drives," *Wall Street Journal,* May 9, 1995, p. B7.
7. Scott Rosenberg, "Rock 'n' ROM," *San Francisco Examiner,* May 1, 1994, pp. B-13, B-14.
8. Rick Sammon, Associated Press, "Learn Photography via CD-ROMs, Videotapes," *Chicago Tribune,* September 15, 1995, sec. 7, p. 78.
9. Catherine Arnst, Paul M. Eng, and Richard Brandt, "Multimedia: Joyful and Triumphant?" *Business Week,* December 6, 1993, pp. 167–169.
10. Donn Menn, "More Than Music Rock 'n' ROM," *Multimedia World,* August 1995, pp. 58–63.
11. Mike Langberg, "The Minuses of CD Plus," *San Jose Mercury News,* November 12, 1995, pp. 1E, 7E.
12. Edward Baig, "Music to Your Ears—and Eyes," *Business Week,* January 22, 1996, p. 99.
13. Jeffrey Gordon Angus and Carla Thornton, "Do-It-Yourself CD-ROMs," *PC World,* January 1996, pp. 173–175.
14. L. R. Shannon, "Help for Picturing Pictures on Screen," *New York Times,* August 1, 1995, p. B7.
15. L. R. Shannon, "Will Putting All Your Snapshots on a Disk Replace the Photo Album?" *New York Times,* November 16, 1993, p. B7.
16. Peter H. Lewis, "Besides Storing 1,000 Words, Why Not Store a Picture Too?" *New York Times,* October 11, 1992, sec. 3, p. 8.
17. Stephen Manes, "If DVD Is the Next Medium, When Will Future Arrive?" *New York Times,* January 16, 1996, p. B8.
18. Mike Langberg, "The Dawning of the DVD Age," *San Jose Mercury News,* January 14, 1996, pp. 1F, 4F.
19. Phillip Robinson, "With a Tape Drive, Backing Up Isn't Quite So Hard to Do," *San Jose Mercury News,* October 16, 1995, pp. 1F, 6F.
20. Peter H. Lewis, "Revering Redundancy," *San Jose Mercury News,* February 13, 1994, p. 1E; reprinted from *The New York Times.*
21. Amanda Kell, "Modern Monks—Holy, High Tech," *San Francisco Chronicle,* January 9, 1995, p. B3.
22. Jeffrey R. Young, "Modern-Day Monastery," *Chronicle of Higher Education,* January 19, 1996, pp. A21–A22.
23. Richard Lamm, quoted in Christopher J. Feola, "The Nexis Nightmare," *American Journalism Review,* July/August 1994, pp. 39–42.
24. Feola, 1994.
25. Penny Williams, "Database Dangers," *Quill,* July/August 1994, pp. 37–38.
26. Lynn Davis, quoted in Williams, 1994.
27. Associated Press, "Many Companies Are Willing to Give a Cat a Little Credit," *San Francisco Chronicle,* January 8, 1994, p. C1.
28. John R. Emshwiller, "Firms Find Profits Searching Databases," *Wall Street Journal,* January 25, 1993, pp. B1, B2.

Chapter 8

1. Thomas A. Stewart, "The Information Age in Charts," *Fortune,* April 4, 1994, pp. 75–79.
2. Tom Mandel, in "Talking About Portables," *Wall Street Journal,* November 16, 1992, p. R18.
3. Blanton Fortson, in "Talking About Portables," *Wall Street Journal,* November 16, 1992, pp. R18–R19.
4. Jacob M. Schlesinger, "Get Smart," *Wall Street Journal,* October 21, 1991, p. R18.
5. Peter H. Lewis, "The Good, the Bad and the Truly Ugly Faces of Electronic Mail," *New York Times,* September 6, 1994, p. B7.
6. Robert Rossney, "E-Mail's Best Asset—Time to Think," *San Francisco Chronicle,* October 5, 1995, p. E7.
7. Mike Branigan, "The Cost of Using an Online Service," *PC Novice,* January 1992, pp. 65–71.
8. Tammi Wark, "Online Service Subscribers," *USA Today,* Feburary 9, 1996, p. 4B.
9. Jared Sandberg and Bart Ziegler, "Internet's Popularity Threatens to Swamp the On-line Services," *Wall Street Journal,* January 18, 1996, pp. A1, A8.
10. Jesse Kornbluth, "The Truth About the Web," *San Francisco Chronicle,* January 23, 1996, p. C4.
11. Schmit, 1996.
12. David L. Wilson, "Internet@home," *Chronicle of Higher Education,* June 16, 1995, pp. A20, A22.
13. Forrester Research Inc., cited in Eng, 1996.
14. David Einstein, "What They Want Is E-mail," *San Francisco Chronicle,* February 20, 1996, pp. B1, B6.
15. Elizabeth P. Crowe, "The News on Usenet," *Bay Area Computer Currents,* August 8–21, 1995, pp. 94–95.
16. Trip Gabriel, "The Meteoric Rise of Web Site Designers," *New York Times,* February 12, 1996, pp. C1, C5.
17. David Plotnikoff, "The Mercury News Family Guide to Cyberspace," *San Jose Mercury News,* February 11, 1996, pp. 3H–6H.
18. Gabriel, 1996.
19. Peter H. Lewis, "Home Pages Never Die; You Must Kill Them," *New York Times,* January 2, 1996, p. C15.
20. Laurie Flynn, "Companies Use Web Hoping to Save Millions," *New York Times,* July 17, 1995, p. C5.
21. Alison L. Sprout, "The Internet Inside Your Company," *Fortune,* November 27, 1995, pp. 161–168.
22. Alvin Toffler, quoted in Marianne Roberts, "Computers Replace Commuters," *PC Novice,* September 1992, p. 27.
23. American Information User 1994 survey, cited in Sherri Merl, "Resisting the Call to Telecommute," *New York Times,* October 22, 1995, sec. 3, p. 14.
24. Jonathan Marshall, "Eliminating the Permanent Office," *San Francisco Chronicle,* March 10, 1994, pp. D1, D3.
25. Alison L. Sprout, "Moving Into the Virtual Office," *Fortune,* May 2, 1994, p. 103.
26. Neil Gross, Peter Burrows, and Robert D. Hof, "Internet Lite: Who Needs a PC?" *Business Week,* November 13, 1995, pp. 102-103.
27. Jeff Pelline, "Oracle's 'Magic Box,'" *San Francisco Chronicle,* January 18, 1996, p. B1.
28. Don Clark, "Oracle Is Demonstrating Network Computer Today," *Wall Street Journal,* February 26, 1996, p. A3.
29. Lee Gomes, "Stripped-down Computing," *San Jose Mercury News,* February 27, 1996, pp. 1C, 2C.
30. Denise Caruso, "On-line Browsing Got You Down? Don't Get Mad, Get Cable," *New York Times,* June 5, 1995, p. C3.
31. Lawrence J. Magid, "ISDN Lines Boost Speed Limit on Net," *San Jose Mercury News,* May 7, 1995, pp. 1E, 6E; reprinted from *Los Angeles Times.*
32. Michael J. Himowitz, "Agony, Ecstasy, and ISDN," *Fortune,* February 19, 1996, p. 107.
33. Grant Balkema, quoted in Mark Robichaux, "Cable Modems Are Tested and Found to Be Addictive," *Wall Street Journal,* December 27, 1995, pp. 13, 17.
34. Peter Coy, "The Big Daddy of Data Haulers?" *Business Week,* January 29, 1996, pp. 74–76.
35. Lucien Rhodes, "The Race for More Bandwidth," *Wired,* January 1996, pp. 140–145.
36. John Seabrook, "My First Flame," *The New Yorker,* June 6, 1994, pp. 70–79.
37. Jared Sandberg, "Up in Flames," *Wall Street Journal,* November 15, 1993, p. R12.
38. Ramon G. McLeod, "Netiquette—Cyberspace's Cryptic Social Code," *San Francisco Chronicle,* March 6, 1996, pp. A1, A10.
39. Faiza S. Ambah, "An Intruder in the Kingdom," *Business Week,* August 21, 1995, p. 40.
40. Joseph Kahn, Kathy Chen, and Marcus W. Brauchli, "Beijing Seeks to Build Version of the Internet that Can Be Censored," *Wall Street Journal,* January 31, 1996, pp. A1, A14.
41. Associated Press, "China Tells Internet Users to Register with the Police," *San Francisco Chronicle,* February 16, 1996, p. A15.
42. John Markoff, "On-line Service Blocks Access to Topics Called Pornographic," *New York Times,* December 29, 1995, pp. A1, C4.
43. Peter H. Lewis, "On-line Service Ending Its Ban of Sexual Materials on Internet," *New York Times,* February 14, 1996, pp. A1, C2.
44. Yahoo!, cited in Del Jones, "Cyber-porn Poses Workplace Threat," *USA Today,* November 27, 1995, p. 1B.
45. Lawrence J. Magid, "Be Wary, Stay Safe in the On-line World," *San Jose Mercury News,* May 15, 1994, p. 1F.
46. David Einstein, "SurfWatch Strikes Gold as Internet Guardian," *San Francisco Chronicle,* March 7, 1996, pp. D1, D2.
47. Peter H. Lewis, "Limiting a Medium without Boundaries," *New York Times,* January 15, 1996, pp. C1, C4.
48. Richard Zoglin, "Chips Ahoy," *Time,* February 19, 1996, pp. 58–61.
49. Frank Rich, "The Idiot Chip," *New York Times,* February 10, 1996, p. 15.
50. Lewis, "Limiting a Medium without Boundaries," 1996.
51. Mike Mills, "'Cell' Phones Betraying Their Owners," *San Jose Mercury News,* June 26, 1994, p. 14A.
52. Jonathan Marshall, "Why Crime, Cellular Phones Don't Mix," *San Francisco Chronicle,* June 21, 1994, pp. D1, D3.
53. Constance Johnson, "Anonymity On-Line? It Depends Who's Asking," *Wall Street Journal,* November 24, 1995, pp. B1, B3.
54. Carol Kleiman, "The Boss May Be Listening," *San Jose Mercury News,* February 25, 1996, pp. 1PC, 2PC.
55. Gina Kolata, "When Patients' Records Are Commodities for Sale," *New York Times,* November 15, 1995, pp. A1, B7.
56. Survey by Equifax and Louis Harris & Associates, cited in Bruce Horovitz, "80% Fear Loss of Privacy to Computers," *USA Today,* October 31, 1995, p. 1A.
57. Associated Press, "Companies Looking to Harness Airwaves," *San Francisco Chronicle,* January 30, 1996, p. A4.
58. Mary Anne Buckman, quoted in Carol Kleiman, "Tailor Your Resume for Inclusion in a Company Database," *San Jose Mercury News,* April 14, 1996, pp. PC1–PC2.
59. David Borchard, "Planning for Career and Life," *The Futurist,* January–February 1995, pp. 8–12.
60. Tom Jackson, quoted in Jonathan Marshall, "Surfing the Internet Can Land You a Job," *San Francisco Chronicle,* July 17, 1995, pp. D1, D3.
61. Martin Yate, quoted in Kathleen Murray, "Plug In. Log On. Find a Job," *New York Times,* January 2, 1994, sec. 3, p. 23.
62. Lisa Levenson, "High-Tech Job Searching," *Chronicle of Higher Education,* July 14, 1995, pp. A16–A17.
63. "Online Job Search Resources," *San Francisco Chronicle,* July 17, 1995, p. D3.
64. Hal Lancaster, "How to Pick Up Job Tips Strewn Along the Infobahn," *Wall Street Journal,* June 27, 1995, p. B1.
65. Jane Easter Bahls, "Courting Your Career," *CompuServe Magazine,* November 1995, pp. 24–26.
66. Marshall, 1995.
67. Stephen C. Miller, "High-Tech Job Hunting, Even for Low-Tech Positions," *New York Times,* February 19, 1996, p. C6.
68. Cynthia Chin-Lee and Comet, "Surfing the Web for Your Next Job," *High Technology Careers Magazine,* August/September 1995, p. 94.
69. Scott Grusky, "Winning Resume," *Internet World,* February 1996, pp. 58–64.
70. Marshall Loeb, "Getting Hired by Getting Wired," *Fortune,* November 13, 1995, p. 252.
71. Marshall, 1995.
72. Levenson, 1995.
73. Marshall, 1995.
74. Grusky, 1996.
75. Levenson, 1995.

Chapter 9

1. Gary Webb, "Potholes, Not 'Smooth Transition,' Mark Project," *San Jose Mercury News,* July 3, 1994, p. 18A.
2. Dirk Johnson, "Denver May Open Airport in Spite of Glitches," *New York Times,* July 17, 1994, p. A12.
3. K. Kendall and J. Kendall, *Systems Analysis and Design* (Englewood Cliffs, NJ: Prentice Hall, 1992), p. 39.
4. Peter D. Varhol, "Visual Programming's Many Faces," *Byte,* July 1994, pp. 187–188.

Chapter 10

1. Jane Metcalfe, quoted in Laurie Flynn, "Tracking High-Tech Culture," *New York Times,* July 10, 1994, sec. 3, p. 10.
2. Jeremy Rifkin, "The End of Work," *San Jose Mercury News,* May 21, 1995, pp. 1C, 4C; adapted from his book *The End of Work: The Decline of the Global Labor Force and the Dawn of the Post Market Era* (Los Angeles: Jeremy P. Tarcher/Putnam, 1995).
3. Patricia Schnaidt, "The Electronic Superhighway," *LAN Magazine,* October 1993, pp. 6–8.
4. Al Gore, reported in "Toward a Free Market in Telecommunications," *Wall Street Journal,* April 19, 1994, p. A18.
5. Associated Press, "Robot Sent to Disarm Bomb Goes Wild in San Francisco," *New York Times,* August 28, 1993, p. 7.
6. "Frustrated Bank Customer Lets His Computer Make Complaint," *Los Angeles Times,* October 20, 1993, p. A28.
7. Arthur M. Louis, "Nasdaq's Computer Crashes," *San Francisco Chronicle,* July 16, 1994, pp. D1, D3.
8. Joseph F. Sullivan, "A Computer Glitch Causes Bumpy Start in a Newark School," *New York Times,* September 18, 1991, p. A25.
9. John J. Fialka, "Pentagon Studies Art of 'Information Warfare,' to Reduce Its Systems' Vulnerability to Hackers," *Wall Street Journal,* July 3, 1995, p. A10.
10. Safeware and The Insurance Agency, reported in "Laptop Thefts," *USA Today,* February 6, 1996, p. 9B.
11. Davis Stipp, "Laptop Larceny Is Taking Off at Airports," *Fortune,* February 5, 1996, pp. 30–31.
12. "Laptop Larceny on Rise at Airports and Hotels," *USA Today,* February 6, 1996, p. 9B.
13. G. Pascal Zachary, "Software Firms Keep Eye on Bulletin Boards," *Wall Street Journal,* November 11, 1991, p. B1.
14. Thomas J. DeLoughry, "2 Students Are Arrested for Software Piracy," *Chronicle of Higher Education,* April 20, 1994, p. A32.
15. Suzanne P. Weisband and Seymour E. Goodman, "Subduing Software Pirates," *Technology Review,* October 1993, pp. 31–33.
16. John J. Keller, "Hackers Open Voice-Mail Door to Others' Phone Lines," *Wall Street Journal,* March 15, 1991, pp. B1, B3.
17. William Barnhill, "'Privacy Invaders,'" *AARP Bulletin,* May 1992, pp. 1, 10.
18. David L. Wilson, "Gate Crashers," *Chronicle of Higher Education,* October 20, 1993, pp. A22–A23.

19. John T. McQuiston, "4 College Students Charged with Theft Via Computer," *New York Times,* March 18, 1995, p. 38.

20. David Einstein, "Crooks Swindle On-Line Investors," *San Francisco Chronicle,* July 1, 1994, pp. B1, B2.

21. Gary Weiss, "The Hustlers Queue Up on the Net," *Business Week,* November 20, 1995, pp. 146–148.

22. Neil Roland, "Scams Abound on the Net," *San Francisco Chronicle,* November 1, 1995, p. B2.

23. David Carter, quoted in Associated Press, "Computer Crime Usually Inside Job," *USA Today,* October 25, 1995, p. 1B.

24. Eric Corley, cited in Kenneth R. Clark, "Hacker Says It's Harmless, Bellcore Calls It Data Rape," *San Francisco Examiner,* September 13, 1992, p. B-9; reprinted from Chicago Tribune.

25. Philip Elmer-Dewitt, "Bugs Bounty," *Time,* October 23, 1995, p. 86.

26. Wilson, 1993.

27. Selwyn Raab, "New York Bookies Go Computer Age but Wind up Being Raided Anyhow," *New York Times,* August 25, 1995, p. A8.

28. Timothy Ziegler, "Elite Unit Tracks Computer Crime," *San Francisco Chronicle,* May 27, 1994, pp. A1, A17.

29. Eugene Carlson, "Some Forms of Identification Can't Be Handily Faked," *Wall Street Journal,* September 14, 1993, p. B2.

30. Kimberly J. McLarin, "Fingerprinting, Without the Ink, Is Introduced in New York City as a Bar to Welfare Fraud," *New York Times,* July 13, 1995, p. A16.

31. Dana Milbank, "Measuring and Cataloging Body Parts May Help to Weed Out Welfare Cheats," *Wall Street Journal,* December 4, 1995, pp. B1, B6.

32. Evan Perez, "Security System Uses Your Face as the Password," *San Francisco Chronicle,* January 5, 1996, p. C1.

33. William M. Bulkeley, "Popularity Overseas of Encryption Code Has the U.S. Worried," *Wall Street Journal,* April 28, 1994, pp. A1, A7.

34. John Holusha, "The Painful Lessons of Disruption," *New York Times,* March 17, 1993, pp. C1, C5.

35. The Enterprise Technology Center, cited in "Disaster Avoidance and Recovery Is Growing Business Priority," special advertising supplement in *LAN Magazine,* November 1992, p. SS3.

36. Micki Haverland, quoted in Fred R. Bleakley, "Rural County Balks at Joining Global Village," *Wall Street Journal,* January 4, 1996, pp. B1, B2.

37. David Ensunsa, "Proposed Cell-Phone Pole Faces Challenge," *Idaho Statesman,* June 23, 1995, p. 4B.

38. Mary Anne Ostrom, "Coming Distractions," *San Jose Mercury News,* February 18, 1996, pp. 1A, 24A.

39. Andrew Kupfer, "The Trouble with Cellular," *Fortune,* November 13, 1995, pp. 179–188.

40. James H. Snider, "The Information Superhighway as Environmental Menace," *The Futurist,* March–April 1995, pp. 16–21.

41. Andrew Kupfer, "Alone Together," *Fortune,* March 20, 1995, pp. 94–104.

42. William M. Bulkeley, "Electronics Is Bringing Gambling into Homes, Restaurants, and Planes," *Wall Street Journal,* August 16, 1995, pp. A1, A5.

43. Linda Kanamine, "Despite Legal Issues, Virtual Dice Are Rolling," *USA Today,* November 17, 1995, p. 1A, 2A.

44. James Sterngold, "Imagine the Internet as Electronic Casino," *New York Times,* October 22, 1995, sec. 4, p. 3.

45. Marco R. della Cava, "Are Heavy Users Hooked or Just On-line Fanatics?" *USA Today,* January 16, 1996, pp. 1A, 2A.

46. Kenneth Howe, "Diary of an AOL Addict," *San Francisco Chronicle,* April 5, 1995, pp. D1, D3.

47. Kendall Hamilton and Claudia Kalb, "They Log On, but They Can't Log Off," *Newsweek,* December 18, 1995, pp. 60–61.

48. Nanci Hellmich, "When the Computer Replaces the Spouse Bit by Bit," *USA Today,* February 14, 1995, p. 8D.

49. Laura Evenson, "Losing a Mate to the Internet," *San Francisco Chronicle,* August 9, 1995, pp. A1, A13.

50. Associated Press, "Husband Accuses Wife of Having Online Affair," *San Francisco Chronicle,* February 2, 1996, p. A3.

51. Karen S. Peterson and Leslie Miller, "Cyberflings Are Heating Up the Internet," *USA Today,* February 6, 1996, pp. 1D, 2D.

52. Survey by Microsoft Corporation, reported in Don Clark and Kyle Pope, "Poll Finds Americans Like Using PCs but May Find Them to Be Stressful," *Wall Street Journal,* April 10, 1995, p. B3.

53. Jonathan Marshall, "Some Say High-Tech Boom Is Actually a Bust," *San Francisco Chronicle,* July 10, 1995, pp. A1, A4.

54. Yahoo!/Jupiter Communications survey, reported in Del Jones, "On-line Surfing Costs Firms Time and Money," *USA Today,* December 8, 1995, pp. 1A, 2A.

55. Coleman & Associates survey, reported in Julie Tilsner, "Meet the New Office Party Pooper," *Business Week,* January 29, 1996, p. 6.

56. Webster Network Strategies survey, reported in Jones, 1995.

57. STB Accounting Systems 1992 survey, reported in Jones, 1995.

58. Ira Sager and Gary McWilliams, "Do You Know Where Your PCs Are?" *Business Week,* March 6, 1995, pp. 73–74.

59. Alex Markels, "Words of Advice for Vacation-Bound Workers: Get Lost," *Wall Street Journal,* July 3, 1995, pp. B1, B5.

60. Paul Saffo, quoted in Laura Evenson, "Pulling the Plug," *San Francisco Chronicle,* December 18, 1994, "Sunday" section, p. 53.

61. Daniel Yankelovich Group report, cited in Barbara Presley Noble, "Electronic Liberation or Entrapment," *New York Times,* June 15, 1994, p. C4.

62. Neil Postman, quoted in Evenson, 1994.

63. Mike Snider, "Keeping PC Play Out of the Office," *USA Today,* January 26, 1995, p. 3D.

64. Sager & McWilliams, 1995.

65. Matthew Rothschild, "When You're Gagging on E-Mail," *Forbes,* June 6, 1994, pp. S25, S26.

66. Evenson, 1994.

67. Jeremy Rifkin, "Technology's Curse: Fewer Jobs, Fewer Buyers," *San Francisco Examiner,* December 3, 1995, p. C-19.

68. Michael J. Mandel, "Economic Anxiety," *Business Week,* March 11, 1996, pp. 50–56.

69. Bob Herbert, "A Job Myth Downsized," *New York Times,* March 8, 1996, p. A19.

70. Louis Uchitelle, "It's a Slow-Growth Economy, Stupid," *New York Times,* March 17, 1996, sec. 4, pp. 1, 5.

71. Paul Krugman, "Long-Term Riches, Short-Term Pain," *New York Times,* September 25, 1994, sec. 3, p. 9.

72. Department of Commerce survey, cited in "The Information 'Have Nots'" [editorial], *New York Times,* September 5, 1995, p. A12.

73. Beth Belton, "Degree-based Earnings Gap Grows Quickly," *USA Today,* February 16, 1996, p. 1B.

74. Alan Krueger, quoted in LynNell Hancock, Pat Wingert, Patricia King, Debra Rosenberg, and Allison Samuels, "The Haves and the Have-Nots," *Newsweek,* February 27, 1995, pp. 50–52.

75. Terry Bynum, quoted in Lawrence Hardy, "Tapping into New Ethical Quandaries," *USA Today,* August 1, 1995, p. 6D.

76. Lawrence M. Fisher, "Trouble with the Software? Ask Other Software," *New York Times,* July 10, 1994, sec. 3, p. 10.

77. Charles Petit, "8-Legged Robot to Crawl into Volcano," *San Francisco Chronicle,* July 28, 1994, p. A3.

78. David L. Wilson, "On-Line Treasure Hunt," *Chronicle of Higher Education,* March 17, 1995, pp. A19–A20.

79. Judith Anne Gunther, "An Encounter With A.I.," *Popular Science,* June 1994, pp. 90–93.

80. William A. Wallace, Ethics in Modeling (*New York: Elsevier Science, Inc.,* 1994).

81. Laura Johannes, "Meet the Doctor: A Computer That Knows a Few Things," *Wall Street Journal,* December 18, 1995, p. B1.

82. Steve Lohr, "The Elusive Information Highway," *New York Times,* September 23, 1994, pp. C1, C2.

83. Susan Yoachum and Edward Epstein, "Clinton Goal—Internet in Every School," San Francisco Chronicle, September 22, 1995, pp. A1, A19.

84. Bill Workman, "Media Revolution in Stanford Future," *San Francisco Chronicle,* February 16, 1994, p. A12.

85. Edward Barrett, "Collaboration in the Electronic Classroom," *Technology Review,* February/March 1993, pp. 51–55.

86. Louis Freedberg, "A Plan to Make Books Obsolete," *San Francisco Chronicle,* July 15, 1993, pp. A1, A15.

87. Robert L. Johnson, "Extending the Reach of 'Virtual' Classrooms," *Chronicle of Higher Education,* July 6, 1994, pp. A19–A23.

88. Ronald Smothers, "New Video Technology Lets Doctors Examine Patients Many Miles Away," *New York Times,* September 16, 1992, p. B6.

89. John Eckhouse and Ken Siegmann, "A Medical Version of the Superhighway," *San Francisco Chronicle,* January 19, 1994, p. B4.

91. Myron Magnet, "Who's Winning the Information Revolution," *Fortune,* November 30, 1992, pp. 110–117.

92. Tony Rutkowski, quoted in Schnaidt, 1993.

93. Bruce Haring, "Record Companies, Retailers Set Up Shop in Cyberspace," *USA Today,* May 10, 1995, p. 5D.

94. Jared Sandberg, "Virtual Bank Branches Are Coming Via Visa and Worlds Inc.," *Wall Street Journal,* August 9, 1995, p. B3.

95. Ellen Neuborne, "Wal-Mart Going On Line Via Microsoft," *USA Today,* January 29, 1996, p. 1B.

96. Steven Levy, "The End of Money?" *Newsweek,* October 30, 1995, pp. 62–65.

97. Kelley Holland and Amy Cortese, "The Future of Money," *Business Week,* June 12, 1995, pp. 66–78.

98. Pamela Varley, "Electronic Democracy," *Technology Review,* November/December 1991, pp. 43–51.

99. "Federal, State Aid to Go On-line," *San Jose Mercury News,* June 1, 1994, pp. 1A, 12A; reprinted from Los Angeles Times.

Sources & Credits

Page numbers for Panels and README boxes are indicated in **boldface.**

Text and Art Sources

26 Data from Link Resources Corp. in Chart from USA Today Snapshots, *USA Today,* November 19, 1993. **45** Jeff Pelline, "Software Pirates Loot," *San Francisco Chronicle,* May 14, 1996, p. C1, data source Glencoe Engineering Inc.; **64** Adapted from *New York Times,* August 14, 1995, p. C5. **74** By Richard Williams as it appeared in *Globe & Mail,* April 10, 1993, p. B13. **91** Copyright 1993 by Consumers Union of U.S., Inc., Yonkers, NY 10703-1057. Excerpted by permission from *Consumer Reports,* September 1993, p. 570. **92** *Fortune,* October 4, 1993, p. 108. © 1993 Time Inc. All rights reserved. **95** (top) Adapted from illustration by Julie Stacey, *USA Today,* November 24, 1995, p. 3B. Copyright 1995 USA TODAY. Reprinted with permission. **95** (bottom) *New York Times Magazine,* November 5, 1995, pp. 50–57, 64. Copyright © 1995 The New York Times Company. Reprinted by permission. **99** Drawing adapted from Sun Microsystems, Inc., in "A New Model for Personal Computing," *San Jose Mercury News,* August 13, 1995, p. 27A. **116, 117, 287** Adapted from T. O'Leary, B. Williams, & L. O'Leary, *McGraw-Hill Computing Essentials* (New York: McGraw-Hill, 1990), pp. 58, 60, 155, 172. **179** Adapted from Chris Foreman, "Lions and Tigers and . . . Computer Viruses?" Reprinted from *PC Novice,* May 1992, pp. 34–36. PC Novice, 120 W. Harvest Dr., Lincoln, NE 68521. For subscription information, please call 800-472-4100. Please mention code number 4110. **180** Adapted from Janice M. Horowitz, "Crippled by Computers," *Time,* October 12, 1992, pp. 70–72. **203** Reprinted from *PC Novice,* June 1994, pp. 32–33. PC Novice, 120 W. Harvest Dr., Lincoln, NE 68521. For subscription information, please call 800-472-4100. Please mention code number 4110. **239** (top) Network Wizards; (bottom) Adapted from *Der Spiegel,* November, 1996; **241** Adapted from "How Internet Mail Finds Its Way," *PC Magazine,* October 11, 1994, p. 121. **250** Adapted from table "Bandwidth Battle," in Mark Landler, "Jingling the Keys to Cyberspace, Cable Officials Sing a New Tune," *New York Times,* May 9, 1995, pp. C1, C10. **255** Adapted from Jared Schneidman, "How It Works," *Wall Street Journal,* February 11, 1994, p. R5, and *Popular Science.* **289** (top) Adapted from T. O'Leary & B. Williams, *Computers & Information Systems* (Redwood City, CA: Benjamin/Cummings, 1985), p. 171. **297** Adapted from artwork by Kris Strawser/The Chronicle, appearing in article by Don Clark, "New Vision of Communications," *San Francisco Chronicle,* November 23, 1992, p. B1. © SAN FRANCISCO CHRONICLE. **311** *Popular Science,* June 1994, pp. 90–93. Reprinted with permission from Popular Science Magazine, copyright 1994, Times Mirror Magazines Inc. Distributed L.A. Times Syndicate.

Photo Sources

Page 7 IBM; **14** (middle) Apple Computer, Inc.; (bottom) Hewlett-Packard; **17** (top, middle) Hewlett Packard; (bottom left) © Frank Bevans; (bottom right) Quantum; **22** (clockwise from top left) Intel Corp., IBM, Hewlett-Packard, Hewlett-Packard, Intel Corp., IBM, Adastra Systems; **23** Ted Morrison / Still Life Stock; **24** (top) Ted Morrison / Still Life Stock; (bottom) Sanyo-Fisher USA; **26** Paul Chesley Photographers / Aspen; **27** (top) © Monica Limas; (bottom) Frank Pryor / Apple Computer, Inc.; **41** Luciano Galiardi / The Stock Market; **43** Toshiba; **46** San Francisco Chronicle; **49** IBM; **52-53** Microsoft Corp.; **60** Microsoft Corp.; **62** IBM; **68** Hewlett-Packard; **68** (left) Autodesk; (right) IBM; **93** © Frank Bevans; **107** Packard Bell Navigator; **108** © Frank Bevans; **86** © Monica Limas; **88** IBM; **91** (background) Bradbury Building / G. E. Kidder Smith; (insert) Tandy / Radio Shack; **92** (insert) Apple Computer, Inc.; **95** Microsoft Corp.; **98** Microsoft Corp.; **109** IBM Archives; **109** (top, middle) IBM; (bottom) Intel Corp.; **110** (top) Intel Corp.; (bottom insert) Tandy / Radio Shack; **112** SUN; **121** Centon Electronics; **125** Hayes; **141** IBM; **144** (top) Microsoft Corp.; (middle) Kensington; (bottom) Thrustmaster; **144** Microtouch Systems; **145** (top) FTE Data Systems; (bottom) FTG; **145** (bottom left) Hewlett-Packard; (bottom right) Wacon Technology; (margin) Apple Computer, Inc.; **146** (background) Uniform Code Council, Inc.; **147** (top, bottom left) NCR; (bottom right) © Frank Bevans; **147** (bottom) NCR / IBM; **148** Apple Computer, Inc.; **151** John Burdette / U.S. Dept. of the Interior / Geology Survey. Office of Earthquakes, Volcanoes, and Engineering; **164** (top) © Frank Bevans; (middle) IBM; (bottom) Multisync; **166** IBM; **167** (top) Canon Computer Systems, Inc.; (bottom) John Greenleigh / Apple Computer, Inc.; **168** Hewlett-Packard; **170** (left) Calcomp; (right) Hewlett-Packard; **171** Peter Fox / Apple Computer, Inc.; **173** Edward Keating / New York Times; **175** (top) Gryphon Software; (bottom) Paul Higdon / New York Times; **191** IBM; **197** (top) Maxtor; (bottom) John Greenleigh / Apple Computer, Inc.; **198** (left) Mountain Gate; (right) Syquest; (bottom) APS; **199** NCR Corp.; **200** Toshiba; **201** (top) GME Timeline screen; (bottom) © Monica Limas; **202** © Greenlar / The Image Works; **203** Elastic Reality, Inc.; **205** (top) IBM; **205** (bottom) IBM; **206** Exabyte; **207** © Frank Bevans; **209** Pat Rogondino; **232** J. L. Bersuder / © Sipa-Press; **234** (left) Intel Corp.; (right) Sharp Electronics; **236** AT&T; **237** (top) America Online; (middle) CompuServe; (bottom) Prodigy; **242** Network Wizards; **245** Lakewood; **249** (top) Hayes; (bottom) Multitech Systems; **253** (top, bottom right) AT&T; (bottom left) U.S. Sprint; **259** © Hank Morgan / Rainbow; **301** Symantec; **307** © Frank Bevans; **312** IBM; (bottom) © Thomas Dallas.

Index